OXFORD WORLD'S CLASSICS

NATURAL THEOLOGY

WILLIAM PALEY was born in Peterborough in 1743. His father was a clergyman who became a headmaster. In 1759 he went to Christ's College, Cambridge, where he won scholarships and prizes, and graduated as Senior Wrangler, the best student of his year in the prestigious mathematical course. He was ordained deacon in 1766, and elected a fellow of his college. There with his friend John Law he undertook teaching, especially of moral philosophy. In 1775 Law's father, Edmund, Bishop of Carlisle, offered Paley a post as a vicar. In 1776 he left Cambridge, and married Jane Hewitt. In 1785 he published *The Principles of Moral and Political Philosophy*, based on his Cambridge lectures. It sold very well, and made his name as a clear and accessible writer. He was promoted to be Archdeacon of Carlisle. In 1791 his wife died, leaving him with eight children to bring up, and in 1795 he married Catherine Dobinson as his second wife. They moved to Bishop Wearmouth, a well-endowed parish, where he spent the rest of his life.

Paley published *Evidences of Christianity* in 1794 and it rapidly became a classic, dealing with the fulfilment of prophecy, miracles, and the reliability of the Bible. *Natural Theology* appeared in 1802. As a classic statement of the argument for intelligent design, it was a huge success, and a major spur to Charles Darwin's thinking. Paley died in Lincoln in 1805.

MATTHEW D. EDDY is Lecturer in the History and Philosophy of Science and an associate of the Centre for the History of Medicine and Disease at the University of Durham. He has most recently held fellowships at the Dibner Institute (MIT), Harvard University, the Max Planck Institute for the History of Science (Berlin), and with the University of Notre Dame's Erasmus Institute. He has written numerous articles on eighteenth- and nineteenth-century intellectual history. Most recently he has edited (with David M. Knight) *Science and Beliefs: From Natural Philosophy to Natural Science, 1700–1900* (2005) and he is currently writing a book on the interaction between medicine, philosophy, and geology in Enlightenment Edinburgh.

DAVID M. KNIGHT is Emeritus Professor of the History and Philosophy of Science at the University of Durham. He has edited the *British Journal for the History of Science*, and served as President of the British Society for the History of Science. In 1997 and 1998 he was awarded prizes by the Templeton Foundation for his course on Science and Religion in the nineteenth century, and in 2003 he received the American Chemical Society's Edelstein Award for History of Chemistry. He has published numerous works; his single-authored books include *Atoms and Elements* (1967), *The Nature of Science* (1976), *Ordering the World* (1981), *The Age of Science* (1986), *Natural Science Books in English, 1600–1900* (1989), *Science in the Romantic Era* (1998), *and Science and Spirituality* (2004).

OXFORD WORLD'S CLASSICS

*For over 100 years Oxford World's Classics have brought
readers closer to the world's great literature. Now with over 700
titles—from the 4,000-year-old myths of Mesopotamia to the
twentieth century's greatest novels—the series makes available
lesser-known as well as celebrated writing.*

*The pocket-sized hardbacks of the early years contained
introductions by Virginia Woolf, T. S. Eliot, Graham Greene,
and other literary figures which enriched the experience of reading.
Today the series is recognized for its fine scholarship and
reliability in texts that span world literature, drama and poetry,
religion, philosophy and politics. Each edition includes perceptive
commentary and essential background information to meet the
changing needs of readers.*

OXFORD WORLD'S CLASSICS

WILLIAM PALEY

Natural Theology

or

Evidence of the Existence and Attributes of the Deity, collected from the appearances of nature

Edited with an Introduction and Notes by
MATTHEW D. EDDY *and* DAVID KNIGHT

OXFORD
UNIVERSITY PRESS

OXFORD
UNIVERSITY PRESS

Great Clarendon Street, Oxford OX2 6DP

Oxford University Press is a department of the University of Oxford.
It furthers the University's objective of excellence in research, scholarship,
and education by publishing worldwide in

Oxford New York

Auckland Cape Town Dar es Salaam Hong Kong Karachi
Kuala Lumpur Madrid Melbourne Mexico City Nairobi
New Delhi Shanghai Taipei Toronto

With offices in

Argentina Austria Brazil Chile Czech Republic France Greece
Guatemala Hungary Italy Japan Poland Portugal Singapore
South Korea Switzerland Thailand Turkey Ukraine Vietnam

Oxford is a registered trade mark of Oxford University Press
in the UK and in certain other countries

Published in the United States
by Oxford University Press Inc., New York

Editorial material © Matthew D. Eddy and David Knight 2006

British Library Cataloguing in Publication Data

Data available

Library of Congress Cataloging in Publication Data
Paley, William, 1743–1805.
Natural theology : evidence of the existence and attributes of the deity, collected from the
appearances of nature / William Paley ; edited with an introduction and notes by
Matthew D. Eddy and David Knight.
p. cm.
Includes bibliographical references.
1. Natural theology.   I. Eddy, Matthew, 1972–   II. Knight, David M.   III. Title.
BL183.P35 2006   210—dc22   2005026316

Typeset in Ehrhardt
by RefineCatch Limited, Bungay, Suffolk
Printed in Great Britain by
Clays Ltd., St Ives plc
ISBN 0–19–280584–3 978–0–19–280584–3

1

ACKNOWLEDGEMENTS

We offer a special thanks to the many students and colleagues who encouraged us to pursue this project, particularly to those who responded to our initial emails in the early stages of our research and to the following scholars who offered considerable moral support and advice: John H. Brooke, Geoffrey N. Cantor, Beth Rainey, Roger Norris, Jonathan Topham, Aileen Fyfe, William H. Brock, Peter J. Bowler, Andreas-Holger Maehle, E. Jonathan Lowe, Simon P. James, Andreas Pantazatos, Momme von Sydow, Stephen W. Sykes, Paul Murray, Colin G. Crowder, Rob Iliffe, Anthony Grafton, Ann Blair, and Alistair McGrath. Resources and funds for our research were generously supplied by the University of Durham, the Dibner Institute for the History of Science, the Max Planck Institute for the History of Science, the Rotary International Ambassadorial Scholars Programme, Princeton Theological Seminary, the University of Edinburgh's library system (the Special Collections Department and New College Library) and the Harvard University Library system (especially the Hilles Library, which provided a gratis photocopy of the first edition of *Natural Theology*). Finally, our thanks are extended to Dimitrios Grigoropoulos, who graciously translated all Latin and Greek quotations, to Geoffrey Scarre, who lent us his personal copy of John Ray's *The Wisdom of God* (London, 1714), and to Dick Watson, who read the draft of the introduction and made helpful suggestions for its improvement.

CONTENTS

Introduction   ix

Note on the Text   xxx

Select Bibliography   xxxii

A Chronology of William Paley   xxxvi

NATURAL THEOLOGY   I

Appendix: Further Reading   284

Explanatory Notes   294

INTRODUCTION

But can there be any person . . . who can consider the regular
movements of the heavenly bodies, the prescribed courses of the
stars, and see how all is linked and bound into a single system,
and then deny that there is any conscious purpose in this and say
that it is the work of chance?

Marcus Tullius Cicero (106–43 BC), *De Natura Deorum*

[T]he Almighty discovers more of his Wisdom in forming such a
vast multitude of different sorts of Creatures, and all with
admirable and irreprovable Art, than if he had created but a few;
for this declares the greatness and unbounded capacity of his
Understanding.

John Ray, *The Wisdom of God* (1691)

A Brief History of Natural Theology

Natural Theology is the practice of inferring the existence and wisdom of God from the order and beauty of the world. William Paley is
so strongly identified with natural theology that he is sometimes
thought to have invented it when he published *Natural Theology* in
1802. In fact, natural theology has a long history, going back well
before the time of Jesus. Thus, in philosophy of religion courses
today students are taught that there are three different kinds of
arguments that seek to demonstrate the existence of God: cosmological, teleological, and ontological. The first two have been around
since the ancient Greeks, while the last was most clearly formulated
by Anselm of Canterbury in the eleventh century. The *cosmological
argument* holds that the world, and everything in it, depends on
something for its existence. This 'something' must be God. Some
forms of the argument go even further and say that the physical
causes operating in the natural world (*cosmos*) were started by a
divine first cause (God) at some point in the past. The *teleological
argument* holds that the natural world appears to have been designed,
or created, by a designer; some forms of the argument also affirm
that the world was created to serve some sort of divinely inspired
end (*telos*). The *ontological argument* holds that existence is entailed
by the concept of God—a move which inherently assumes that

God exists *a priori* (before experience) and which is dependent upon evidence taken from reason alone (not the physical world). Though dividing up arguments for God's existence into three categories is a helpful heuristic tool, the history of Western thought shows that these arguments did not usually come in neat packages. More often than not, teleological and cosmological premises were combined to form arguments that sought to describe the nature of the divine. A good example of this practice is given in the last dialogue of Plato's *Laws*. There Clinias, one of the characters, exclaims about unbelievers, 'Why, to begin with, think of the earth, and sun, and planets, and everything! And the wonderful and beautiful order of the seasons with its distinctions of years and months!'

Throughout late Antiquity and the Middle Ages, different versions of natural theology were promoted by Christian Churches, but orthodox believers were reminded that such arguments were only supplementary to what was found in the Bible. With the 'scientific revolution' of the seventeenth century, the telescope and microscope opened new and wonderful vistas, and Plato's belief that the wanderings of the planets across the sky would be shown to be orderly was vindicated. Sir Isaac Newton's law of gravity revealed the simplicity and rationality of the solar system, uniting heaven and earth in a new physics. Natural theology was much strengthened. On his Grand Tour, the philosopher and scientist Robert Boyle visited Strasbourg, and likened the universe to the intricate workings of its great cathedral clock. But was the Deity, First Cause, or Supreme Being who had made the immense clockwork universe and presided over it, also the personal God of Abraham, Isaac and Jacob, concerned at the fall of a sparrow, ready to work miracles, and to provide salvation? There were many competing answers offered to this question during the early modern period. One of the most influential thinkers on this topic was Isaac Newton.

Although Newton's natural philosophy would eventually become closely intertwined with natural theology, the actual process of linking his ideas with theological topics was done by others. Robert Boyle, an orthodox and pious Anglican and prominent Fellow of the newly founded Royal Society, bequeathed £50 a year to fund lectures confuting atheism. The first series was delivered in 1692 by Richard Bentley, an ambitious young cleric who would later become Master of Trinity College, Cambridge. Perceiving how Newton's recently published *Principia Mathematica* (1687) strengthened the argument from design, he wrote to Newton for advice on how to exploit this in his lectures. Newton took examples not only from astronomy, but also from anatomy: 'such an usefulness of things or a fitness of means to Ends, as neither proceeds from the necessity of their Beings, nor can happen to them by Chance, doth necessarily infer that there was an Intelligent Being, which was the Author and Con- triver of that Usefulness.'[1] Bentley's project was judged a great success and other works soon followed, each unique in their own way. John Ray's *The Wisdom of God Manifested in the Works of the Creation* (1691) invoked Nature as God's agent, preserving God's wisdom and benevolence while allowing for the explanation of occa- sional apparent mistakes or failures of design in the world. Conversely, Thomas Burnet's *Sacred Theory of the Earth* (1684) suggested that the earth was a ruin, a sphere of punishment, hard labour, pain, disease, and death—a spoilt paradise rather than a magnificent clock. Building on the success of these and other works, the publication of natural theology books continued at a steady pace well into the nineteenth century.

Using evidence harvested from the 'Book of Nature' to supple- ment descriptions of the divine found in the 'Book of Scripture' was a practice that stretched back to the Old and New Testaments through to the Trinitarian debates of the early Christian councils. Early theologians argued that even though the true nature of the divine was beyond human perception or understanding, the personal qualities of God, or the divine attributes, could be inferred from the Bible. Using attributes like wisdom, omniscience, goodness, and immutability as a starting point, Church leaders used the natural world to illuminate these qualities. Personal experience was aug- mented, first by Aristotle's natural philosophy, and then by Lockean empiricism and Newtonian mechanics. In his *Essay* (1689), Locke had suggested that the 'idea' of God was not innate, but learned: 'Since then though the knowledge of a GOD, be the most natural discovery of humane reason, yet *the Idea of him*, is *not innate*.'[2] Though this notion was not fully acceptable to theologians

---

[1] *Isaac Newton's Papers and Letters on Natural Philosophy*, ed. I. Bernard Cohen and Robert E. Schofield (Cambridge: Cambridge University Press, 1958), 393.

[2] John Locke, *An Essay Concerning Human Understanding* (Oxford: Clarendon Press, 1975; reprint of the 4th edn. (1700)), I. IV. X.

(especially since Locke's explication is clouded), it did act to increase the empirical language used to comprehend the attributes of God. Newton made connections between God and physics in the third edition of his *Principia* (1726) and the fourth edition of his *Opticks* (1730). Yet Newton's view of God was unorthodox, and so the 'Newtonization' of the divine attributes was left to apologists like Bentley and Samuel Clarke.

By the 1720s natural theology was considered orthodox in the Church of England and was thus a lantern meant to illuminate, but not replace, the scriptural basis of the divine attributes. Because Hanoverian Britain was permissive of heterodox theological thought, there were poets, philosophers, and pamphleteers (and even priests!) who offered natural religion instead. These should not be confused with Paley, whose interest in the attributes of God is evinced in the very subtitle of his *Natural Theology*: 'Evidences of the Existence and Attributes of the Deity': echoing the subtitle of William Derham's *Physico-Theology* (1713) and other works published throughout the eighteenth century. Paley's orthodoxy explains why he did not directly cite the descriptions of God advanced in works like Burnet's *Sacred Theory*, William Wollaston's *Religion of Nature Delineated* (1725), Alexander Pope's *An Essay on Man* (1733–4), and Joseph Priestley's *Disquisitions Relating to Matter and Spirit* (1777). Though influenced by some of these heterogeneous works, Paley did not seek to give them greater currency, but sifted them for ideas that were compatible with what he already believed. Science in the eighteenth century was saturated with natural theology, which made it seem serious and relevant (rather than a curious hobby) to a wide audience that ranged from the emerging professional classes to the aristocracy.

Natural theology was also something that might have united all Christians in this time of religious controversy and division, when Dissenters from the established Churches were making their presence felt. But for true believers it was suspiciously close to Deism (belief in a remote creator), or to the scepticism of those, like Edward Gibbon the historian, who saw no certainty in religion. For churchmen, such threats to the moral basis of society needed answering; and so especially did the philosopher David Hume, whose irreligion allied with respectability was deeply shocking. His posthumously published *Dialogues Concerning Natural Religion* (1779) were a particular challenge to natural theologians, bringing sceptical doubts to their comparisons of human artefacts and divine creation, and suggesting a brutal world of pain and struggle, ill-adapted to happy life. Across Europe, scepticism and enquiry, disseminated by writers and philosophers such as Diderot, d'Alembert, and Voltaire in revolutionary France, alarmed the governing classes;[3] it was essential to demonstrate that science properly conducted and understood was the handmaid of religion. The stage was set for William Paley.

William Paley

William Paley was born in Peterborough, Cambridgeshire, in July 1743. He was the eldest child of the Revd William Paley and his wife Elizabeth Clapham; his father was a minor canon at Peterborough and from 1745 headmaster of Giggleswick School in Yorkshire, where his son was educated. He was a clumsy but bright boy, developing a lifelong keenness for fishing: Sir Humphry Davy records an anecdote that, when *Natural Theology* was being written, the Bishop of Durham (Shute Barrington, to whom the book was in due course dedicated) asked how it was going, and got the reply: 'My Lord, I shall work steadily at it when the fly-fishing season is over.'[4] He also developed an early interest in the law, attending a murder trial in York shortly before going up to Christ's College, Cambridge, in 1759. There, after a slow start, he worked hard, taking the prestigious mathematical course, winning scholarships and prizes, and emerging as senior wrangler, the best graduate of the year. He then taught at Greenwich for a time, enjoying theatres and attending trials at the Old Bailey, but determined on a career in the Church. He was ordained deacon in 1766, and became a curate in Greenwich; but was soon elected a fellow of his college, and returned to Cambridge, where he was ordained priest.

There he became a close friend of John Law, whose father Edmund became Bishop of Carlisle in 1768. Energetic and able, they divided the instruction in the college between them, raising its reputation. Paley's teaching of moral philosophy from 1768 to 1776 was particularly effective: he stressed the need to make students see the

---

[3] See esp. J. Rose, *The Intellectual Life of the British Working Classes* (New Haven: Yale University Press, 2002), 16–22.

[4] [H. Davy], *Salmonia*, London: Murray (3rd edn., 1832), 7.

problems rather than giving them answers. In 1774 Edmund Law appointed his son to a prebend at Carlisle, and in 1775 presented Paley with a living, a parish in Westmorland. Vicars, but not College fellows, could marry, and in 1776 Paley wed Jane Hewitt, and left Cambridge. In 1785 he published his first book, *The Principles of Moral and Political Philosophy*, based upon his Cambridge lectures. Paley was not an original thinker, and the book's utilitarian philosophy was not new, but he was a wonderfully clear, fresh writer and guide to conduct. The book was a great success; though a new author, he was paid a princely £1,000 for it, and the publisher's investment paid off. By 1793 the fifth edition was pirated in Dublin; and by 1809 it was in its seventeenth edition, with many versions still to come.[5]

Paley was comfortable and well off. He was a cheerful man, who saw Providence in the prevailing happiness of the world, human and animal. From 1789 he had become prominent in the agitation against the slave trade; thus he was neither a closet moralist nor a naive and foolish optimist like Voltaire's Candide. But in 1791 his wife died, leaving him to bring up four sons and four daughters. He had become Chancellor and Archdeacon of Carlisle, holding various other posts in plurality. Content where he was, in 1789 he had turned down the Mastership of Jesus College, but in 1795 was awarded the degree of Doctor of Divinity at Cambridge. In 1795 he married a second wife, Catherine Dobinson of Carlisle, and moved into the magnificent parsonage at Bishop Wearmouth, Sunderland. Paley's attempt to place reason at the centre of Christian ethics in what he took to be a mainstream Anglican tradition incurred suspicion from clergy and faithful: this may be why, seeming too liberal or latitudinarian at a time of evangelical revival, he never got a bishopric or deanery. His preferment culminated in an archdeaconry, and he ended as a vicar and as subdean at Lincoln Cathedral, where he spent three months each year. It was a useful, comfortable, reasonably eminent but not glittering career.

In *Moral and Political Philosophy*, Paley had praised Edmund Law for demonstrating that 'whatever renders religion more rational, renders it more credible', purging it of ignorance and superstition.

[5] William St Clair, *The Reading Nation in the Romantic Period* (Cambridge: Cambridge University Press, 2004), 626; and this as a 'conduct book', pp. 273 ff.

Paley believed that previous writers had divided 'too much of the law of nature from the precepts of revelation';[6] as a good churchman he aimed to keep the two in balance. Because his experience had shown that 'in discoursing to young minds of morality, it required more pains to make them perceive the difficulty, than to understand the solution', he excited curiosity in order to arouse enthusiasm. The morality he sketched might seem simplistic. With statements such as 'So then actions are to be estimated by their tendency. Whatever is expedient is right', his blunt clear style avoided the qualifying or fudging all too often found in such treatises, but alarmed some readers: 'What promotes the public happiness, or happiness upon the whole, is agreeable to the fitness of things, to nature, to reason, and to truth'; and 'such is the divine character, that what promotes the general happiness is required by the will of God.' Morality could not therefore be separated from theology, as Hume had tried to do, without enfeebling it. God, for Paley, showed His benevolence in the ways He made our senses 'instruments of gratification and enjoyment'. He 'might have made, for example, everything we tasted bitter; everything we saw loathsome; everything we touched a sting; every smell a stench; and every sound a discord'—but He wanted us to be happy, and 'contrivance proves design'. We assess human actions in the same way as we look at creation, by their 'tendency' (taking general rules and the long run into account); and that allows us to infer God's love. Paley included powerful images of social injustice, and covered very readably a wide range of topics; the work deserved its success and in many ways it can be seen as a mainstream conduct-book.

The next major task was to authenticate biblical narratives, and in 1790 Paley published his most original book, *Horae Paulinae*, dedicated to John Law who was by then Bishop of Killalla and Achonry in Ireland. This book sought to demonstrate the truth of the story of St Paul in the Bible by close comparisons between his Epistles and the Acts of the Apostles. Here, in contrast to his other works, Paley looked for artlessness, and absence of design and contrivance. Had there been complete harmony between the Epistles and Acts, this would be evidence of 'meditation, artifice, and design'—like too

---

[6] William Paley, *The Principles of Moral and Political Philosophy* (5th edn., Dublin: Byrne, 1793), quoting from pp. iii, xii, 37, 44, 47; references to Hume are on pp. 10, 42.

good a story cooked up by false witnesses in court. Artlessness is the sign of substantial truth; minute, circuitous, and oblique circumstances, which no forger making up a tidy story would have bothered about, are to be ferreted out by detective work: 'If what is here offered shall add one thread to that complication of probabilities by which the Christian history is attested, the reader's attention will be repaid by the supreme importance of the subject, and my design will be fully answered.' Each chapter examines a different Epistle, looking for small links with the others and with Acts, and Paley also sought to establish their independence as texts. He set out to show how unlikely it was that one was derived from another, or that several were forgeries. He wrote in an attractively argumentative style, like a good lawyer. The book is thus one long argument from beginning to end, giving the impression that objections have been foreseen and fairly considered. Paley's sermons were criticized for their lack of peroration and conclusion and so in *Horae Paulinae* he made his case and stopped. The same plain style is to the reader's advantage in his other writings.

Then in 1794 Paley published *Evidences of Christianity*, which rapidly became a classic with a seventh edition appearing in 1800. It was unoriginal, but very clear. It was concerned with the historical evidence, more generally than in *Horae Paulinae*, and thus engaging directly with Gibbon and his ironic account of the early Church in his *Decline and Fall of the Roman Empire* (1776–88). Paley dealt with the fulfilment of prophecy; with the miracles, and especially Jesus' resurrection; with the morality of the Gospels, and the character of Jesus; and then with various popular objections, including the discrepancies between the Gospels. Most agreed that he had clearly, satisfactorily, and judiciously dealt with these questions. These were very traditional and important topics, and the book became a set text in Cambridge. By way of contrast, Paley looked at the success of Islam, 'Mahometanism': like others of his time, he saw it as an imposture, a false religion, not to be compared with Christianity because it was essentially military and political, and its rapid spread had everything to do with conquest, and nothing to do with truth.

Natural Theology, the last of Paley's books, was written during the late 1790s and published in 1802. The atmosphere of the 1790s was fraught. War had been declared with France in 1793 and while bad harvests and high prices added to economic hardship, constant fear of a fifth column of atheistic radicals further destabilized the country. The book was welcome, and it continued to sell throughout the nineteenth century, frequently reprinted by the Religious Tract Society and other publishers. Pantheists and Deists could not ignore Paley's evidence for revealed religion. Now what was required to complete Paley's project was an up-to-date treatise on natural theology, demonstrating how valuable the argument from Design still was, and how science, rightly understood, complemented true religion. After 1800 Paley was often in bad health, probably suffering from kidney stones, but he wrote *Natural Theology* to complete his work of defending and propagating the faith by demonstrating God's work in nature; it was published only three years before his death in 1805. It is an old man's book, referring to works encountered during his Cambridge years; yet also citing up-to-date scientific authors from around 1800. He had in his writings put ethics and then revealed religion on a sound footing, and so one might have expected that he would be writing a theology of nature, demonstrating how the revealed God of the Scriptures, in whom we have faith through experience, made the world. But Paley's book was a genuine natural theology, looking for the God of Nature: 'in which works, such as they are, the public have now before them, the evidences of natural religion, the evidences of revealed religion, and an account of the duties that result from both. It is of small importance, that they have been published in an order, the very reverse of that in which they ought to be read' (p. 4). We may therefore hope with him that readers who begin with *Natural Theology* will be interested enough to read his other works, though there the issues are less current in our new century, where creationism is rife and Design thus controversial.

Natural Theology

The structure of *Natural Theology* is like a sandwich. The first half addresses medicine and natural history and the last half treats of the attributes of God. Wedged in the middle are chapters on the four 'elements', and on astronomy. Like its predecessors, Paley's book is divided into thematic chapters, consisting of strings of examples to convince the reader that the world was designed. In order to turn these examples into convincing 'proofs', Paley uses metaphors,

analogies, and appeals to probability. His use of metaphors and analogies was shaped by his knowledge of classical rhetoric. Though
much neglected by intellectual historians, the influence of com-
positional methods taken from classical rhetoric had a profound
impact upon classification techniques and scientific writing from the
Renaissance until the nineteenth century. During his time at Cambridge, Paley had been an avid reader of Cicero, the great Roman
orator; a pastime, as his son Edmund tells us,[7] that continued all the
way up to the end of his life—and he duly cites Cicero in *Natural
Theology*. Such an awareness of rhetoric allowed him to identify and
redeploy striking metaphors that already had wide circulation in
English anatomy, natural history, and astronomy texts: 'pipes' (for
veins), 'tube' (for a butterfly proboscis), an 'orange' (for an oblate
spheroid), to name just a few. Paley's style also shows him following
the rhetorical practices of his contemporaries in his use of literary
figures of speech.[8]

Yet it was Paley's analogies that most often caught the eye of his
later readers, especially because the acceptance of a resemblance
between two objects is highly dependent upon the intellectual
disposition or training of the observer. The proper use of analogy to
establish a premise within a logical argument was an issue
that remained unresolved from Antiquity to the Enlightenment.
Aristotle's *Posterior Analytics* had stated that it was sometimes
impossible to establish a philosophical principle without the introduction of an analogy. In Paley's day, logic and rhetoric had been
fused together into a polite writing style. As the influential rhetor-
ician George Campbell (1719–96) stated: 'To attain either of these
ends, the speaker must always assume the character of the close
candid reasoner: for though he may be an acute logician who is no
orator, he will never be a consummate orator who is no logician.'[9]

Natural Theology begins with the famous analogy between the
world and a pocket-watch (which, in the form of the chronometer,
was high technology in 1802). Paley states that if one were to
encounter a stone while walking across a heath, one might think it

---

[7] Edmund Paley, *An Account of the Life and Writings of William Paley* (Farnborough: Gregg, 1970), 60.

[8] Jeanne Fahnestock, *Rhetorical Figures in Science* (Oxford: Oxford University Press, 1999).

[9] George Campbell, *The Philosophy of Rhetoric* (Edinburgh, 1776), V. ii. iv.

had always happened to be there. However, if one were to see a watch lying on the ground in the same manner, the intricacy of its parts would surely lead one to conclude that it had been made by an intelligent designer. Following on from this analogy, the rest of the book demonstrates that the world is a great clock, made by a wise and benevolent God. This was an old analogy: behind the 'scientific revolution' and the Enlightenment lurked the idea that we and other creatures, large or microscopic, were little mechanisms living in an immense clockwork universe. This famous analogy, forming the first and very effective sentences of the text, was often singled out by nineteenth- and twentieth-century authors, most famously perhaps in the title to Richard Dawkins's best-selling evolutionary work *The Blind Watchmaker* (1986). However, *Natural Theology* is packed with one analogy after another, thus creating 'the argument cumulative', as Paley explains in Chapter VI. So, as nineteenth-century commentators and editors realized, it is not crucial that biological and geological knowledge was rapidly increasing around the time of Paley's death. When he wrote the book he was not, and was not trying to be, at the frontier of scientific knowledge. In selecting examples that best fitted the analogies used in the book, Paley relied upon familiar and tested science, taken from well-established sources.

Although he does discuss chemistry and astronomy, the bulk of Paley's analogies come from anatomy: particular structures adapted to the curious ways of life of various creatures; and prospective contrivances, where organs were provided which were no use to the infant animal but which would appear at the appropriate moment and be valuable when it grew up, like our second teeth. Paley was impressed by relations, the correlation of the various parts of organisms; and he looked at instincts, further evidence for him of God's foresight and benevolence in the provision of what was necessary. But God was not only wise, but also good: so the contrivances He had supplied were beneficial, and He had also 'superadded *pleasure* to animal sensations' (p. 237). Paley, by analogy, followed many Enlightenment natural historians and imputed the human emotions (especially happiness) to animals, thereby allowing him to assert that 'It is a happy world after all' (p. 238).

From Francis Bacon's time up to the nineteenth century, the notion of 'evidence' in British philosophical argument experienced a slow redefinition. By the late seventeenth century most followed

Bacon's notion that 'evidence' must be based on personal experience and empirical observation, challenging many canonical Greek and Roman natural philosophers. With the writings of John Locke and others, evidence in the early eighteenth century was directly linked to the five senses. During the eighteenth century another significant form of evidence emerged: *probability*. Sometimes this could be quantitative. Arguments relevant to mortality rates, population growth, and agricultural production began to be based more heavily on statistics, which affected the related topics of moral philosophy and political economy. Sometimes, however, probability had to be informal and qualitative. Throughout *Natural Theology*, Paley uses informal probability to support his argument. This was a tactic he imported from his *Principles of Moral and Political Philosophy* and, to an extent, from Thomas Malthus's *An Essay on the Principle of Population* (1798).

Throughout his writings, Paley accepts that God's existence cannot be rigorously proved like a theorem in geometry. Bishop Joseph Butler, in his famous *Analogy of Religion* (1736), had written that 'probable evidence is essentially distinguished from demonstrative by this, that it admits of degrees; and all variety of them, from the highest moral certainty, to the very lowest presumption'. Referring to John Locke, he continued:

Probable evidence, in its very nature, affords but an imperfect kind of information; and is to be considered as relative only to beings of limited capacities. For nothing which is the possible object of knowledge, whether past, present, or future, can be probable to an infinite Intelligence; since it cannot but be discerned absolutely as it is in itself, certainly true, or certainly false. But to Us, probability is the very guide to life.[10]

For Paley as for Butler, science (natural history and natural philosophy) was therefore important in telling us what God had, in all probability, actually done.

These uncertainties meant that natural theology entailed a less-than-rigorous argument. Deductive reasoning, for which Euclid's geometry was the great example, depends upon the acceptance of axioms. If readers accept that if a is greater than b, and b is greater than c, then a is greater than c; and (more contentiously) that parallel

---

[10] Joseph Butler, *The Analogy of Religion, Natural and Revealed, to the Constitution and Course of Nature* (new edn., London: Rivington, 1791), 1, 3.

lines never meet; then they cannot doubt surprising conclusions about squares on hypotenuses, or centres of gravity of curiously shaped solids. Each theorem depends on those which have gone before. This kind of reasoning is much beloved of logicians. Indeed, in the 'ontological argument' that God must necessarily exist, because to deny Him is absurd, Anselm and Descartes tried to apply it to religion. Kant would eventually show that this sort of reasoning did not stand up: that nobody can be brought to faith via a fine-drawn argument about 'necessary' existence. Yet Kant was not translated into English until the early nineteenth century and it is unlikely that Paley had ever read him. In Paley's time, mathematicians were beginning to quantify exactly some aspects of probability, like dice or roulette, but in the affairs of humankind, the scope for tight deductive arguments is (despite Sherlock Holmes) rather small. Butler was right that probability is the guide to life.

Paley's argument could not therefore be of the knock-down kind, but had to be cumulative. His *Natural Theology* is not a chain of reasoning like Euclid's, where one weak link would spoil the whole argument. It is instead a rope, where the various fibres are in themselves weak, but twisted together will support a great weight. If the rope is worn and a few fibres have frayed or broken, it does not matter too much. Practical reasoning, where we decide what to do, depends on this kind of thinking. We weigh up the data, and the likely consequences of doing this or that, in ordinary life or indeed in science. Paley's book was one long argument, but unlike Euclid's it took the form of a series of converging inferences, where if the reader were to feel that one or two were weak or unsatisfactory, the conclusion might still stand.

Paley had a legal cast of mind. When he was a young man, and later in Durham, he attended the law courts for entertainment. As he knew well, legal arguments are often probabilistic. The jury has to make up its mind, from the evidence given and the lawyers' arguments, whether it is beyond reasonable doubt that the accused committed the crime. Paley sought to prove the existence of a Designer beyond reasonable doubt—knowing that deductive logical proof was not possible, but that while sceptical logic-choppers could never be silenced, they could be made to look absurd.

Paley's probabilistic approach was closely linked to his perception of moral philosophy. During the Enlightenment, this subject was

xxii *Introduction*

closely tied to theological doctrine. In Paley's case the divine attribute of goodness gave him the ideal space to combine his utilitarian notion of happiness with the rising tide of probabilistic thinking in science and medicine;[11] especially when he suggested that life, on average, was filled more with pleasure than pain (a move that allowed him to use 'benevolent' as an adjective for the divine). Yet within theology's wide-ranging field of relevance were cracks that would widen into disciplinary fissures and eventually lead to the secularization of notions like providence, suffering, free will, and benevolence. Although such a result was not what Paley, an Anglican priest, would have intended, the seeds for this transformation can be seen in the very pages of the books that he wrote; particularly in his *Moral and Political Philosophy*, where he argued that the government was obliged to protect the collective needs of the population via its regulation of property, contracts, and lending. But Paley's ideas were sometimes thought to be too liberal and the following pigeon analogy ruffled the feathers of George III to the extent that (it was rumoured) he prevented Paley from becoming a bishop:

If you should see a flock of pigeons in a field of corn: and if (instead of each picking where and what it liked, taking just as much as it wanted, and no more) you should see ninety-nine of them gather all they got, into a heap; reserving nothing for themselves, but the chaff and the refuse; keeping this heap for one, and that the weakest, perhaps worst, pigeon of the flock; sitting round, and looking on, all the winter, whilst this one was devouring, throwing about, and wasting it; and if a pigeon more hardy or hungry than the rest, touched a grain of the hoard, all the others flying upon it, and tearing it to pieces; if you should see this, you would see nothing more than what is every day practised and established among men.[12]

Within Britain there had been calls for secularizing the government ever since John Locke's *Letter Concerning Toleration* (1689). In Paley and his contemporaries, we see the start of a slow reconfiguring of the political stage in which the expediency of discourse of the divine

[11] See Lorraine Daston, *Classical Probability in the Enlightenment* (Princeton: Princeton University Press, 1988), and Ulrich Tröhler, *'To Improve the Evidence of Medicine': The 18th Century British Origins of a Critical Approach* (Edinburgh: Royal College of Physicians of Edinburgh, 2000).
[12] Paley, *Principles of Moral and Political Philosophy*, book III, part I, chapter I.

was pushed further back into the scenery, to be replaced by the invisible hand of the market, or the promotion of common sense.

Audience and Reception

As a successful author, Paley knew his audience, and the high price of *Natural Theology* meant that it would most probably line the library shelves of the nobility and gentry. But, more specifically, Paley crafted his argument to appeal to fellow clergymen and literate parishioners who were already familiar with his other books and sermons. Additionally, the book's references to provincial natural history and local mechanical wonders, like the bridge over the Wear at Sunderland, made it relevant to the gentlemen living in the seats of Britain's numerous philosophical societies, and to the other members of the reading public whom he encountered when travelling to and from Carlisle, Bishop Wearmouth, Durham, and Lincoln. To advance his argument he used examples from every branch of science known in his day. One of the reasons why Paley was able to cover so many different topics rests in the tools that he used to compose the book. Stylistically, as reviewers noted, it was famously clear, and Paley cited authors like Jean-Jacques Rousseau, Oliver Goldsmith, and Jacques Bernardin de Saint-Pierre whose literary merits were well known to British readers.

To support many of his philosophical or theological points, Paley drew from a vast array of authors who wrote about the human body and the natural world. Since many of these writers do not fit comfortably into the simplified caricature of Enlightenment 'science' used to set the stage for the chemical and Darwinian 'revolutions', the scientific relevance of *Natural Theology* has often been ignored. Contrary to modern popular histories of biology, Charles Darwin was not the first to address the nature of morphological change in organisms, nor was the idea of spontaneous generation a new one.[13] In Paley's day a wide range of biological theories addressed everything from the spontaneous generation of life to the behaviour of humans and insects. Notions of biological change and causality had been part of Western thinking since the ancient Greeks, so much so

---

[13] James E. Strick, *Sparks of Life: Darwinism and the Victorian Debates over Spontaneous Generation* (Chicago: University of Chicago Press, 2002), 1–34.

that they were deeply engrained into the Enlightenment's philosophical consciousness. The authors and theories mentioned (or avoided) by Paley in *Natural Theology* show that he was familiar with many of these ideas and how they would be received by his polite audience.

Within the larger context of Enlightenment medicine and natural history, it seemed obvious, as Paley points out, that careful experiments were bound to disprove spontaneous generation, but that 'degeneration', as of a webbed foot into a claw, made much more sense than would sudden innovations. Such assertions appealed to his audience's sense of natural order, a notion that ran particularly strong in Britain at the time, especially for the many followers of Newtonian natural theology. At the end of the eighteenth century this sense of order was also closely related to aesthetic notions of nature frequently expressed in novels, poems, and hymns. By drawing attention to the beauty and symmetry of nature, Paley's argument not only appealed to the Church of England's long tradition of natural theology, but also to the deep interest in contemplation and piety shown by Evangelicals and Dissenters. Indeed, several authors whom he cites were familiar to Evangelicals. Charles Bonnet, for example, was a Swiss author who incorporated natural theology into natural history, and whose works were such an inspiration to John Wesley that he published a translated version of Bonnet's *Contemplation* in his *A Compendium of Natural Philosophy, Being a Survey of the Wisdom of God in the Creation* (1763).

Natural Theology was an enormous success. Robert Faulder published the book in London during the early months of 1802. He printed 1,000 copies, sold at 12*s*. Paley would have signed the copyright over to Faulder in exchange for being paid a set price at the start.[14] A 1,000-copy run at the beginning of the nineteenth century was common for the first edition of a book that had been written by an established author like Paley. The first run of Ann Radcliffe's third novel, *The Italian* (1797), stood at a remarkable 2,000 copies; Lord Byron was even more successful. But unlike novels or poetry that came and went with the weather, *Natural Theology* stood the test of time. Its first run sold out almost immediately and the second,

[14] Publication figures and print runs for *Natural Theology*'s first decade are given in St Clair, *Reading Nation*, 626–7.

third, and fourth editions (each 1,000 copies) were released during the same year. Fifth and sixth editions appeared by 1808 and by 1809 the run of the twelfth edition stood at 2,000 copies.

After Paley's death at least twelve more editions appeared between 1816 and 1822.[15] Increased availability drove up competition, which in turn drove down the price. By 1818 the book could be bought for 7s.—a bargain for a gentleman, but still a princely sum for Britain's numerous day labourers, many of whom were only paid around 10s. per week. Although the book's format continued to change, the content of the text remained relatively untouched until the mid-1820s, when political events inspired several updated editions.

During the 1810s and 1820s *Natural Theology* became a political lightning rod. At this time, the medical community in London was experiencing growing pains. Most of the hospitals and lucrative medical practices were directly linked to London's established hierarchy of aristocrats, gentry, or graduates from Oxford or Cambridge. Such a situation made it very hard for ambitious middle-class physicians, surgeons, and apothecaries (many of whom were trained either in Edinburgh or in Continental Europe) to gain a foothold. Under the guidance of leaders like Thomas Wakley, they founded their own periodicals (like the *Lancet*) and attacked the London establishment. Many of them, well versed in the new anatomical methods being taught in France, used this knowledge as a weapon against the conservative curriculum of Oxford and Cambridge. As the political jousting became more pointed, Wakley's radicals hit upon an idea: they would use Paley as a representation, a symbol, of all that was wrong with the universities. They pounced on *Natural Theology*'s ageing scientific examples, especially those from physiology. In their rhetoric, a 'Paleyite' was an ignorant, outdated sycophant dependent upon the patronage of the establishment.[16]

As many biographies of Darwin have noted, the book was read in Cambridge and Oxford colleges during the 1820s, and this meant that the 'Paleyite' missiles launched from London stung the ego of many a university don. Even so, *Natural Theology* thrived in

[15] Aileen Fyfe, 'Publishing and the Classics: Paley's Natural Theology and the Nineteenth-Century Scientific Canon', *Studies in the History and Philosophy of Science*, 33 (2002), 433–55.

[16] Adrian Desmond, *The Politics of Evolution: Morphology, Medicine and Reform in Radical London* (Chicago: University of Chicago Press, 1989), chapters 2 and 3.

England's ancient seats of learning. Some tutors and professors even saw it as a platform from which they could argue for new scientific subjects in the curriculum: Oxford's John Kidd (1775–1851) introduced it as an anatomy text, publishing *An Introductory Lecture to a Course in Comparative Anatomy, Illustrative of Paley's Natural Theology* in 1826; and John Duncan (1769–1844) used it as a guide when he restructured some of the arrangements of the Ashmolean Museum at Oxford. Overall, despite Wakley and other London radicals, the general consensus in the universities was that Paley's argument was right, but that his examples needed to be updated. This resulted in the book's first significantly revised edition in 1826. James Paxton (1786–1860), an Oxford physician, added thirty-seven plates that illustrated various mechanical, anatomical, and botanical examples in the book. He added an appendix and expanded the work into two volumes priced at 24*s*. As well as adding material, he also removed parts of the text, particularly Paley's negative comments about Buffon and several of the citations from pre-1800 authors.

At the beginning of the nineteenth century there was a second scientific revolution in progress, spreading from secularized France and associated with specialization and the opening up of professional careers, and for one author to attempt to cover the whole of science began to seem absurd. To meet these changes, the Earl of Bridgewater, who died in 1829, bequeathed £8,000 to the Royal Society to publish treatises demonstrating the goodness and wisdom of God. During the following decade eight treatises were duly commissioned from authors active in various fields in the scientific community. They were a tremendous success, to the astonishment of the world of publishing: natural theology was not only a pervasive ethos, but also an excellent way of popularizing scientific, theological, and philosophical ideas.[17] It made authors focus on the big picture, rather than (in writing for their peers) on details of observation, experiment, or mathematical equations.

The publication of the Bridgewater treatises during the early 1830s allayed the need for Paley's text to be revised beyond what had occurred in Paxton's edition. However, it was not long before Henry Brougham (1778–1868) and Charles Bell (1774–1842), two Scots

---

[17] Jonathan R. Topham, 'Beyond the "Common Context": The Production and Reading of the Bridgewater Treatises', *Isis*, 89 (1998), 233–62.

prominent in the London establishment, began to talk of seriously updating the text. Both were unabashedly committed to natural theology, both in principle and as a way of attacking radicals. Brougham, a politician, had published *A Discourse of Natural Theology, Showing the Nature of the Evidence and the Advantages of the Study* (1835) and Bell, a surgeon, a Bridgewater treatise, *The Hand: Its Mechanism and Vital Endowments as Evincing Design* (1833, and later editions in the 1830s). Brougham and Bell set themselves the task of revising Paley's text in order to reflect the most up-to-date science available. Their edition (1836–9) was, like Paxton's, jammed with new footnotes, editorial commentaries, illustrations, and additional essays that expanded the book into two volumes (with a 21*s.* price tag).

For some years the Bridgewater treatises were expensive; and they mostly lacked the attractive style of Paley. His book went on selling, especially in cheap editions. Robert Chambers, who wrote (anonymously) the sensational evolutionary book *Vestiges of the Natural History of Creation* (1844), was with his brother William an important Edinburgh publisher. In 1849 they brought out an edition of Paley's *Natural Theology*, small enough to fit into a (large) pocket, and updated by a surgeon, Thomas Smibert. There were other cheap editions in various forms, available for example in the libraries of Mechanics' Institutes.[18] But the most famous reader was the young Charles Darwin at Cambridge (occupying Paley's room in Christ's College in the 1820s). He found it one of the few stimulating books he had to read there and wrestled with finding an alternative to Paley's vision of Design. He found it in the hidden hand of natural selection: the survival of the fittest in the struggle for existence. The *Origin of Species* follows a similar plan to *Natural Theology*, cumulative, based upon probabilities, and facing difficulties squarely. But Darwin and his contemporaries were children not only of the Enlightenment but also of the Romantic Movement,[19] when the clockwork universe was rejected in favour of a 'dynamic' world-view, based upon forces and equilibria, where change was to be expected,

---

[18] Rose, *Intellectual Life of the British Working Classes*, 64, 226.
[19] Robert J. Richards, *The Romantic Conception of Life: Science and Philosophy in the Age of Goethe* (Chicago: Chicago University Press, 2002), 6–14; Dov Ospovat, *The Devlelopment of Darwin's Theory: Natural History, Natural Theology and Natural Selection, 1838–1859* (Cambridge: Cambridge University Press, 1995), 60–73.

and imagination had a role in science. Nevertheless, while there is no doubt that Paley would have believed that Darwinian evolution was a specious hypothesis, there is a sense in which Darwin was his disciple. It is in this sense that Darwinians will read the text with interest and pleasure. The conclusion of the *Origin* is distinctly Paleyan:

As natural selection works solely by and for the good of each being, all corporeal and mental endowments will tend to progress towards perfection. It is interesting to contemplate an entangled bank, clothed with many plants of many kinds, with birds singing on the bushes, with various insects flitting about, and with worms crawling through the damp earth, and to reflect that these elaborately constructed forms, so different from each other, and dependent on each other in so complex a manner, have all been produced by laws acting around us. . . . Thus, from the war of nature, from famine and death, the most exalted object which we are capable of conceiving, namely, the production of higher animals directly follows. There is grandeur in this view of life, with its several powers, having been originally breathed into a few forms or into one; and that, whilst this planet has gone cycling on according to the fixed law of gravity, from so simple a beginning endless forms most beautiful and most wonderful have been, and are being, evolved.[20]

Darwin was a great original thinker, whereas Paley was not; but Paley was the better writer, as Darwin recognized, and his arguments cannot be despised. Most men of science in the earlier nineteenth century were generally supportive, even if, emphasizing law and regularity, they drifted towards Unitarianism, pantheism, or deism.

But not everybody admired Paley, or believed that his work supported true religion: he had many critics, from both within and without the religious world. Thus, the utilitarian ethics that he had taken up from Wollaston the Deist and Priestley the Unitarian, and which was then prominent in the sceptics Jeremy Bentham and James Mill, did not go down well with many churchmen. It meant that Paley was never as popular in Oxford, where humanities and High Church theology were central, as he was in more mathematical Cambridge, for expediency and calculation of consequences seemed to many thoroughly unethical, and in contrast with biblical teaching and examples of doing the right thing whatever came of it. On the

---

[20] Charles Darwin, *On the Origin of Species* (London: Murray, 1859), 489–90.

other side, Paley's theologizing made Benthamites like James Mill uneasy: he saw so much evil in the world that scepticism and the rejection of all dogmatic religious belief (including atheism) was necessary.[21]

The whole project of supporting religion legalistically with evidences revolted those like Samuel Taylor Coleridge, for whom feeling the need for God and making the leap of faith were crucial. He knew 'of no religion not revealed', and saw a place only for theology of nature: 'Assume the existence of God,—and then the harmony and fitness of the physical creation may be shown to correspond with and support such an assumption;—but to set about *proving* the existence of God by such means is a mere circle, a delusion.'[22] Coleridge was strongly opposed to 'bibliolatry', prosy literalism in reading the Bible, and he was an advocate of a broad church: but his uneasy view of natural theology was shared by evangelicals like Thomas Gisborne, who in his *Natural Theology* (1818) refused to accept Paley's optimistic vision. Like Burnet, he saw the world in a grimmer light as a place of punishment for mankind's sins, a fallen world where pain and death intruded on God's original plan. He saw refusal to believe as a crime. Gisborne's queasy feelings were echoed in publications further down the social scale, although the Religious Tract Society continued to publish versions of Paley.

Because it was controversial and not just reassuring, *Natural Theology* was one of the most published books of the nineteenth century. Revised editions appeared and the process of adding and removing information from *Natural Theology* was repeated time and again by editors seeking to sell the book to different segments of an increasingly literate and affluent population, in both Britain and America. Today, however, the book is of most interest precisely because of its historical moment, as an important contributor to the debate on science and religion that continues unabated to this day.

[21] John Stuart Mill, *Autobiography*, ed. Harold Laski (London: Oxford University Press, 1924), 32–3.
[22] Samuel Taylor Coleridge, *Confessions of an Enquiring Spirit* (1853; ed. H. St J. Hart, London: Black, 1956), 79; Kathleen Coburn, *Inquiring Spirit* (London: Routledge, 1951), 381.

NOTE ON THE TEXT

Over the course of the nineteenth century *Natural Theology* went through numerous editions. Each printing, especially after the 1820s, brought changes and this means that the text, format, and footnotes of mid- to late-nineteenth-century editions are not the same as those which appeared in the first edition. The present edition reproduces the text of the first, 1802 edition. By going back to the original work, we hope that the following text will serve as a standard by which all later versions can be compared. As is the case for many books published during the Georgian period, only a few first editions of *Natural Theology* are now extant. Our research in American and British libraries only turned up a dozen or so copies. Though we looked at several of these, we eventually settled on a copy housed in the Hilles Library at Harvard University. The original purchaser of this text is unknown, but the front of the book contains two signatures: 'William J. Rutledge' and Wm J. Rutledge Armagh'; so it is highly likely that the book went from London to Armagh, Ireland, before it made its way to North America. It was deposited in the library of Harvard's Radcliffe College sometime in the late nineteenth century. Luckily for us, Harvard's staff and students treated it well until it was transferred into the special collections cabinets of the Hilles Library. We are grateful to the Hilles staff for allowing us to photocopy the text and to compare it to other editions housed in their collection. For our research back at the University of Durham, the Cathedral Library allowed us to use an excellent second edition.

Most of the spelling, hyphenation, and punctuation of the 1802 edition has been retained (the 'long s' has been converted to modern 's', however). Double quotation marks have been changed to single, and full points are omitted after abbreviations and headings, in line with series style. A few obvious spelling errors have been silently corrected.

Paley's footnotes are retained, cued by superior figure. Where possible, the bibliographical details of his sources are expanded and this information is given in the Explanatory Notes. These are cued by asterisk.

For more advanced researchers, suggestions for further reading are given in an appendix.

SELECT BIBLIOGRAPHY

Primary Sources

Addison, J., *The Spectator*, No. 120, Wednesday, 18 July 1710.

—— *The Evidences of the Christian Religion* (London, 1730).

Balguy, T., *Divine Benevolence Asserted; and Vindicated from the Objections of Ancient and Modern Sceptics* (London, 1781).

Bentley, R., *A Confutation of Atheism from the Origin and Frame of the World* (London, 1693).

Bonnet, C., *The Contemplation of Nature* (London, 1766).

Butler, J., *The Analogy of Religion, Natural and Revealed, to the Constitution and Course of Nature* (1736; new edn. London, 1791).

Cheselden, W., *The Anatomy of the Humane Body: Illustrated with Twenty-Three Copper-plates of the Most Considerable Parts; All Done after the Life* (London, 1713).

Cicero, M. T., *De Natura Deorum*, trans. H. Rackham (London, 1932).

Clark, S., *Demonstration of the Being and Attributes of God* (London, 1705).

Derham, W., *The Artificial Clock-Maker: A Treatise of Watch and Clock-work* (London, 1696).

—— *Physico-Theology: or, A Demonstration of the Being and Attributes of God from his Works of Creation* (London, 1713).

—— *Astro-Theology: or A Demonstration of the Being and Attributes of God from a Survey of the Heavens* (London, 1714).

—— *Christo-Theology: or, a Demonstration of the Divine Authority of the Christian Religion* (London, 1729).

Hartley, D., *Observations on Man, his Frame, his Duty, and his Expectations* (London, 1749).

Harvey, W., *Anatomical Exercitations, Concerning the Generation of Living Creatures* (London, 1653).

Keill, J., *Anatomy of the Human Body Abridg'd* (London, 1708).

Nieuwentyjt, B., *A Religious Philosopher, or the Right Use of Contemplating the Works of the Creator* (London, 1718).

Priestley, J., *Disquisitions Relating to Matter and Spirit: To which is Added, the History of the Philosophical Doctrine Concerning the Origin of the Soul, and the Nature of Matter* ... (London, 1777).

Ray, J., *The Wisdom of God Manifested in the Works of the Creation* (London, 1691).

—— *Three Physico-Theological Discourses, Concerning I. The Primitive Chaos, and Creation of the World. II. The General Deluge ... III. The Dissolution of the World, and Future Conflagration* (London, 1693).

Wesley, J. (ed.), *A Compendium of Natural Philosophy Being a Survey of the Wisdom of God in the Creation*, vols. I and II (Bristol, 1763).

Background and Critical Studies

Barr, J., *Biblical Faith and Natural Theology: The Gifford Lectures for 1991* (Oxford, 1993).

Brooke, J., *Science and Religion: Some Historical Perspectives* (Cambridge, 1991).

—— and Cantor, G., *Reconstructing Nature: The Engagement of Science and Religion* (Oxford, 1998).

Burbridge, D., 'William Paley Confronts Erasmus Darwin: Natural Theology and Evolutionism in the Eighteenth Century', *Science & Christian Belief*, 10 (1998), 49–71.

Clarke, M. L., *Paley: Evidences for the Man* (Toronto, 1974).

Daston, L., *Classical Probability in the Enlightenment* (Princeton, 1988).

Desmond, A., *The Politics of Evolution: Morphology, Medicine, and Reform in Radical London* (Chicago, 1992).

Dodds, G. L., *Paley, Wearside and Natural Theology* (Sunderland, 2003).

Eddy, M. D., 'The Science and Rhetoric of Paley's Natural Theology', *Literature and Theology*, 18 (2004), 1–22.

Fox, J. J., 'Divine Attributes', in C. G. Herbermann, E. A. Pace, C. B. Pallen, T. J. Shahan, and J. J. Wynne (eds.), *The Catholic Encyclopedia: An International Work of Reference on the Constitution, Doctrine, Discipline, and History of the Catholic Church*, vol. ii (New York, 1907).

Fyfe, A., 'Publishing and the Classics: Paley's *Natural Theology* and the Nineteenth-Century Scientific Canon', *Studies in the History and Philosophy of Science*, 33 (2002), 433–55.

Gascoigne, J., 'Rise and Fall of British Newtonian Natural Theology', *Science in Context*, 2 (1988), 219–56.

Gillespie, N. C., 'Divine Design and the Industrial Revolution: William Paley's Abortive Reform of *Natural Theology*', *Isis*, 81 (1990), 214–29.

Gilson, E., *From Aristotle to Darwin and Back Again: A Journey in Final Causality, Species, and Evolution*, trans. John Lyon (Notre Dame, Ind., 1984).

Heavner, E., 'Malthus and the Secularization of Political Ideology', *History of Political Thought*, 17 (1996), 408–30.

Hoffman, J., and Rosenkrantz, G. S., *The Divine Attributes* (Oxford, 2002).

Knight, D. M., *Science and Spirituality: The Volatile Connection* (London, 2004).

Lemahieu, D. L., *The Mind of William Paley: A Philosopher and His Age* (London, 1976).

Lindberg, D. C., and Numbers, R. L. (eds.), *When Science and Christianity Meet* (Chicago, 2003).

Louth, A., *The Origins of the Christian Mystical Tradition: From Plato to Denys* (Oxford, 1981).

McAdoo, H. R., *The Spirit of Anglicanism: A Survey of Anglican Theological Method in the Seventeenth Century* (London, 1965).

McGrath, A. E., *A Scientific Theology*, vol. i: *Nature* (Edinburgh, 2001).

Meadley, G. W., *Memoirs of William Paley, to which is Added an Appendix* (London, 1809).

Ospovat, D., *The Development of Darwin's Theory: Natural History, Natural Theology and Natural Selection, 1838–1859* (Cambridge, 1995).

Paley, E., *An Account of the Life and Writings of William Paley* (Farnborough, 1970; originally the first volume of *The Works of William Paley*, London, 1825).

Pelikan, J., *Christianity and Classical Culture: The Metamorphosis of Natural Theology in the Christian Encounter with Hellenism* (New Haven, 1993).

Philipp, W., 'Physicotheology in the Age of Enlightenment: Appearance and History', *Studies on Voltaire and the Eighteenth Century*, 57 (1967), 1233–67.

Porter, R., 'Creation and Credence', in B. Barnes and S. Shapin (eds.), *Natural Order: Historical Studies of Scientific Culture* (Beverly Hills, Calif., 1979).

Raven, C., *Natural Religion and Christian Theology* (Cambridge, 1953).

Richards, R. J., *The Romantic Conception of Life: Science and Philosophy in the Age of Goethe* (Chicago, 2002).

Rose, J., *The Intellectual Life of the British Working Classes* (New Haven, 2002).

Rousseau, G. S., and Porter, R. (eds.), *The Ferment of Knowledge: Studies in the Historiography of Eighteenth-Century Science* (Cambridge, 1980).

St Clair, W., *The Reading Nation in the Romantic Period* (Cambridge, 2004).

Topham, J. R., 'Beyond the "Common Context": The Production and Reading of the Bridgewater Treatises', *Isis*, 89 (1998), 233–62.

Viner, J., *The Role of Providence in the Social Order* (Philadelphia, 1972).

Webb, C. C. J., *Studies in Natural Theology* (Oxford, 1915).

Further Reading in Oxford World's Classics

Darwin, Charles, *The Origin of Species*, ed. Gillian Beer.

Gosse, Edmund, *Father and Son*, ed. Michael Newton.

Hume, David, *Dialogues Concerning Natural Religion / The Natural History of Religion*, ed. J. C. A. Gaskin.

Literature and Science in the Nineteenth Century, ed. Laura Otis.

Malthus, Thomas, *An Essay on the Principle of Population*, ed. Geoffrey Gilbert.

Rousseau, Jean-Jacques, *Confessions*, trans. Angela Scholar, ed. Patrick Coleman.

A CHRONOLOGY OF WILLIAM PALEY

1743 July: Paley born to William Paley (1711–99) and Elizabeth Clapham (c.1713–96) of Giggleswick in the West Riding of Yorkshire. Baptized in Peterborough Cathedral.

1758 Admitted to Christ's College, Cambridge. Studies mathematics with William Howarth at Topcliffe.

1759 Matriculates at Christ's. Studies algebra, geometry, natural philosophy, logic, metaphysics, and moral philosophy.

1757 Joins the Hyson Club and begins lifelong friendship with John Law.

1760 George III crowned King of Great Britain and Ireland.

1761 Awarded the Bunting scholarship, the highest prize for mathematics at Cambridge.

1763 Graduates from Cambridge as senior wrangler; becomes a schoolmaster's assistant and then also a Latin tutor at Bracken's academy in Greenwich. John Wesley, *A Compendium of Natural Philosophy Being a Survey of the Wisdom of God in the Creation*.

1765 Wins prize for best Latin prose essay written by a Cambridge graduate.

1766 Returns to Christ's for his MA, becomes a fellow, ordained priest, and starts to teach. English edition of Charles Bonnet, *The Contemplation of Nature*.

1767 Becomes praelector, junior dean, and catechist at Christ's; ordained priest in the Church of England by the Bishop of London.

1768 Appointed steward, Mildmay preacher, Hebrew lecturer, and senior Greek lecturer; tutors in ethics, metaphysics, and the Greek New Testament.

1769 Appointed as chaplain to Edmund Law, Bishop of Carlisle.

1770 Appointed as taxor of Cambridge University.

1771 Appointed to be a preacher at the Royal Chapel, Whitehall.

1774 *A Defence of the 'Considerations on the Propriety of Requiring a Subscription to Articles of Faith'*, a pamphlet defending Edmund Law's views of religious toleration.

1775 Appointed Senior Dean of Christ's; given his first benefice in Appleby, Westmorland (near Great Musgrave), by Bishop Law.

1776 Marries Jane Hewitt (*c.*1751–91). *Observations upon the Character and Example of Christ*. Appointed by Bishop Law to a more substantial benefice in St Lawrence's, Appleby. William Withering, *The Botanical Arrangement of All the Vegetables Naturally Growing in Great Britain*.

1779 David Hume, *Dialogues Concerning Natural Religion*, published posthumously.

1780 Appointed to a prebendal stall in Carlisle Cathedral.

1781 Thomas Balguy, *Divine Benevolence Asserted*.

1782 Succeeds his friend John Law as Archdeacon of Carlisle and acquires the vicarage of Great Salkeld, Cumberland.

1785 *The Principles of Moral and Political Philosophy*. Succeeds Richard Burn as chancellor of the diocese of Carlisle.

1787 *Principles* becomes mandatory for Cambridge examinations.

1789 Declines the mastership of Jesus College, Cambridge. Storming of the Bastille starts the French Revolution. Jeremy Bentham, *An Introduction to Principles of Morals and Legislation*.

1790 *Horae Paulinae*; *Reasons for Contentment, Addressed to the Labouring Part of the British Public* (a tract). Edmund Burke, *Reflections on the French Revolution*.

1791 Jane Paley dies. Erasmus Darwin, *The Botanic Garden*.

1792 Condemns slavery as the 'diabolical traffic' at a public meeting in Carlisle.

1793 War with France begins and continues until 1815.

1794 *A View of the Evidences of Christianity*. Made prebendary of St Pancras at St Paul's Cathedral, London.

1795 Appointed subdean of Lincoln Cathedral. Awarded a Doctorate of Divinity by Cambridge University. Bishop Barrington of Durham gives him the scenic rectory of Bishop Wearmouth. Marries Catherine Dobinson (*c.*1748–1819) of Carlisle.

1798 George Canning attacks Erasmus Darwin's proto-evolutionary thought in the *Anti-Jacobin Review*. Malthus, *An Essay on the Principle of Population*.

1801 Suffers incapacitating attack from a kidney stone.

1802 *Natural Theology*.

1805 Dies at Bishop Wearmouth on 25 May.

1806 *Sermons* published posthumously.

NATURAL THEOLOGY

MY LORD,

THE following Work was undertaken at your Lordship's recommendation; and, amongst other motives, for the purpose of making the most acceptable return I could make for a great and important benefit conferred upon me.

It may be unnecessary, yet not, perhaps, quite impertinent, to state to your Lordship and to the reader, the several inducements that have led me once more to the press. The favor of my first and ever honored patron had put me in possession of so liberal a provision in the church, as abundantly to satisfy my wants, and much to exceed my pretensions. Your Lordship's munificence, in conjunction with that of some other excellent Prelates, who regarded my services with the partiality with which your Lordship was pleased to consider them, hath since placed me in ecclesiastical situations, more than adequate to every object of reasonable ambition. In the mean time, a weak, and, of late, a painful state of health, deprived me of the power of discharging the duties of my station, in a manner at all suitable, either to my sense of those duties, or to my most anxious wishes concerning them. My inability for the public functions of my profession, amongst other consequences, left me much at leisure. That leisure was not to be lost. It was only in my study that I could repair my deficiencies in the church. It was only through the press that I could speak. These circumstances, in particular, entitled your Lordship to call upon me for the only species of exertion of which I was capable, and disposed me without hesitation to obey the call in the best manner that I could. In the choice of a subject I had no place left for doubt: in saying which, I do not so much refer, either to the supreme importance of the subject, or to any scepticism concerning it with which the present times are charged, as I do, to its connection with the subjects treated of in my former publications. The following discussion alone was wanted to make up my works into a system:*

in which works, such as they are, the public have now before them, the evidences of natural religion, the evidences of revealed religion, and an account of the duties that result from both. It is of small importance, that they have been written in an order, the very reverse of that in which they ought to be read. I commend therefore the present volume to your Lordship's protection, not only as, in all probability, my last labor, but as the completion of a consistent and comprehensive design.

Hitherto, My Lord, I have been speaking of myself and not of my Patron. Your Lordship wants not the testimony of a dedication; nor any testimony from me: I consult therefore the impulse of my own mind alone when I declare, that in no respect has my intercourse with your Lordship been more gratifying to me, than in the opportunities, which it has afforded me, of observing your earnest, active, and unwearied solicitude, for the advancement of substantial Christianity; a solicitude, nevertheless, accompanied with that candor of mind, which suffers no subordinate differences of opinion, when there is a coincidence in the main intention and object, to produce any alienation of esteem, or diminution of favor. It is fortunate for a country, and honorable to its government, when qualities and dispositions like these are placed in high and influencing stations. Such is the sincere judgment which I have formed of your Lordship's character, and of its public value: my personal obligations I can never forget. Under a due sense of both these considerations, I beg leave to subscribe myself, with great respect and gratitude,

My Lord,
Your Lordship's faithful
And most devoted servant,

Bishop Wearmouth,*   WILLIAM PALEY
July 1802

TABLE OF CONTENTS

I.	State of the Argument	7
II.	State of the Argument Continued	11
III.	Application of the Argument	16
IV.	Of the Succession of Plants and Animals	32
V.	Application of the Argument Continued	35
VI.	The Argument Cumulative	45
VII.	Of the *Mechanical* and *Immechanical* Functions of Animals and Vegetables	47
VIII.	Of *Mechanical* Arrangement in the Human Frame— Of the Bones	54
IX.	Of the Muscles	69
X.	Of the Vessels of Animal Bodies	82
XI.	Of the Animal Structure Regarded as a Mass	101
XII.	Comparative Anatomy	114
XIII.	Peculiar Organizations	129
XIV.	Prospective Contrivances	135
XV.	Relations	140
XVI.	Compensation	147
XVII.	The Relation of Animated Bodies to Inanimate Nature	155
XVIII.	Instincts	160
XIX.	Of Insects	170
XX.	Of Plants	183
XXI.	The Elements	194
XXII.	Astronomy	199
XXIII.	Of the Personality of the Deity	213
XXIV.	Of the Natural Attributes of the Deity	230
XXV.	The Unity of the Deity	234
XXVI.	The Goodness of the Deity	237
XXVII.	Conclusion	277

CHAPTER I

STATE OF THE ARGUMENT

IN crossing a heath, suppose I pitched my foot against a *stone*, and were asked how the stone came to be there, I might possibly answer, that, for any thing I knew to the contrary, it had lain there for ever: nor would it perhaps be very easy to shew the absurdity of this answer. But suppose I had found a *watch** upon the ground, and it should be enquired how the watch happened to be in that place, I should hardly think of the answer which I had before given, that, for any thing I knew, the watch might have always been there. Yet why should not this answer serve for the watch, as well as for the stone? Why is it not as admissible in the second case, as in the first? For this reason, and for no other, viz. that, when we come to inspect the watch, we perceive (what we could not discover in the stone) that its several parts are framed and put together for a purpose, e.g. that they are so formed and adjusted as to produce motion, and that motion so regulated as to point out the hour of the day; that, if the several parts had been differently shaped from what they are, of a different size from what they are, or placed after any other manner, or in any other order, than that in which they are placed, either no motion at all would have been carried on in the machine, or none which would have answered the use, that is now served by it. To reckon up a few of the plainest of these parts, and of their offices, all tending to one result:—We see a cylindrical box containing a coiled elastic spring, which, by its endeavour to relax itself, turns round the box. We next observe a flexible chain (artificially wrought for the sake of flexure) communicating the action of the spring from the box to the fusee. We then find a series of wheels, the teeth of which catch in, and apply to, each other, conducting the motion from the fusee to the balance, and from the balance to the pointer; and at the same time, by the size and shape of those wheels, so regulating that motion, as to terminate in causing an index, by an equable and measured progression, to pass over a given space in a given time. We take notice that the wheels are made of brass, in order to keep them from rust; the springs of steel, no other metal being so elastic; that over the face of the watch there is placed a glass, a material employed in no other part of the work, but, in the room of which, if there had been any other than a transparent substance, the hour could not be seen without opening the case. This mechanism* being observed (it requires indeed an examination of the instrument, and perhaps some previous knowledge of the subject, to perceive and understand it; but being once, as we have said, observed and understood), the inference, we think, is inevitable; that the watch must have had a maker; that there must have existed, at some time and at some place or other, an artificer or artificers who formed it for the purpose which we find it actually to answer; who comprehended its construction, and designed its use.

I. Nor would it, I apprehend, weaken the conclusion, that we had never seen a watch made; that we had never known an artist capable of making one; that we were altogether incapable of executing such a piece of workmanship ourselves, or of understanding in what manner it was performed: all this being no more than what is true of some exquisite remains of ancient art, of some lost arts, and, to the generality of mankind, of the more curious productions of modern manufacture. Does one man in a million know how oval frames are turned? Ignorance of this kind exalts our opinion of the unseen and unknown artist's skill, if he be unseen and unknown, but raises no doubt in our minds of the existence and agency of such an artist, at some former time, and in some place or other. Nor can I perceive that it varies at all the inference, whether the question arise concerning a human agent, or concerning an agent of a different species, or an agent possessing, in some respects, a different nature.

II. Neither, secondly, would it invalidate our conclusion, that the watch sometimes went wrong, or that it seldom went exactly right. The purpose of the machinery, the design, and the designer, might be evident, and in the case supposed would be evident, in whatever way we accounted for the irregularity of the movement, or whether we could account for it or not. It is not necessary that a machine be perfect, in order to shew with what design it was made: still less necessary, where the only question is, whether it were made with any design at all.

III. Nor, thirdly, would it bring any uncertainty into the argument, if there were a few parts of the watch, concerning which we could not discover, or had not yet discovered, in what manner they conduced to the general effect; or even some parts, concerning which we could not ascertain, whether they conduced to that effect in any manner whatever. For, as to the first branch of the case; if, by the loss, or disorder, or decay of the parts in question, the movement of the watch were found in fact to be stopped, or disturbed, or retarded, no doubt would remain in our minds as to the utility or intention of these parts, although we should be unable to investigate the manner according to which, or the connection by which, the ultimate effect depended upon their action or assistance: and the more complex is the machine, the more likely is this obscurity to arise. Then, as to the second thing supposed, namely, that there were parts, which might be spared without prejudice to the movement of the watch, and that we had proved this by experiment,—these superfluous parts, even if we were completely assured that they were such, would not vacate the reasoning which we had instituted concerning other parts. The indication of contrivance remained, with respect to them, nearly as it was before.

IV. Nor, fourthly, would any man in his senses think the existence of the watch, with its various machinery, accounted for, by being told that it was one out of possible combinations of material forms; that whatever he had found in the place where he found the watch, must have contained some internal configuration or other; and that this configuration might be the structure now exhibited, viz. of the works of a watch, as well as a different structure.

V. Nor, fifthly, would it yield his enquiry more satisfaction to be answered, that there existed in things a principle of order,* which had disposed the parts of the watch into their present form and situation. He never knew a watch made by the principle of order; nor can he even form to himself an idea of what is meant by a principle of order, distinct from the intelligence of the watch-maker.

VI. Sixthly, he would be surprised to hear, that the mechanism of the watch was no proof of contrivance, only a motive to induce the mind to think so:

VII. And not less surprised to be informed, that the watch in his hand was nothing more than the result of the laws of *metallic* nature. It is a perversion of language to assign any law, as the efficient, operative, cause of any thing. A law presupposes an agent; for it is only the mode, according to which an agent proceeds: it implies a power; for it is the order, according to which that power acts. Without this agent, without this power, which are both distinct from itself, the *law* does nothing; is nothing. The expression, 'the law of metallic nature,' may sound strange and harsh to a philosophic ear, but it seems quite as justifiable as some others which are more familiar to him, such as 'the law of vegetable nature'—'the law of animal nature,' or indeed as 'the law of nature' in general, when assigned as the cause of phænomena, in exclusion of agency and power; or when it is substituted into the place of these.

VIII. Neither, lastly, would our observer be driven out of his conclusion, or from his confidence in its truth, by being told that he knew nothing at all about the matter. He knows enough for his argument. He knows the utility of the end: he knows the subserviency and adaptation of the means to the end. These points being known, his ignorance of other points, his doubts concerning other points, affect not the certainty of his reasoning. The consciousness of knowing little, need not beget a distrust of that which he does know.

CHAPTER II

STATE OF THE ARGUMENT CONTINUED

SUPPOSE, in the next place, that the person, who found the watch, should, after some time, discover, that, in addition to all the properties which he had hitherto observed in it, it possessed the unexpected property of producing, in the course of its movement, another watch like itself; (the thing is conceivable;) that it contained within it a mechanism, a system of parts, a mould for instance, or a complex adjustment of laths, files, and other tools, evidently and separately calculated for this purpose; let us enquire, what effect ought such a discovery to have upon his former conclusion?

I. The first effect would be to increase his admiration of the contrivance, and his conviction of the consummate skill of the contriver. Whether he regarded the object of the contrivance, the distinct apparatus, the intricate, yet in many parts intelligible, mechanism by which it was carried on, he would perceive, in this new observation, nothing but an additional reason for doing what he had already done; for referring the construction of the watch to design, and to supreme art. If that construction *without* this property, or, which is the same thing, before this property had been noticed, proved intention and art to have been employed about it; still more strong would the proof appear, when he came to the knowledge of this further property, the crown and perfection of all the rest.

II. He would reflect, that though the watch before him were, *in some sense*, the maker of the watch, which was fabricated in the course of its movements, yet it was in a very different sense from that, in which a carpenter, for instance, is the maker of a chair; the author of its contrivance, the cause of the relation of its parts to their use. With respect to these, the first watch was no cause at all to the second: in no such sense as this was it the author of the constitution and order, either of the parts which the new watch contained, or of the parts by the aid and instrumentality of which it was produced. We might possibly say, but with great latitude of expression, that a stream of water ground corn:* but no latitude of expression would allow us to say, no stretch of conjecture could lead us to think, that the stream of water built the mill, though it were too ancient for us to know who the builder was. What the stream of water does in the affair is neither more nor less than this: by the application of an unintelligent impulse to a mechanism previously arranged, arranged independently of it, and arranged by intelligence, an effect is produced, viz. the corn is ground. But the effect results from the arrangement. The force of the stream cannot be said to be the cause or author of the effect, still less of the arrangement. Understanding and plan in the formation of the mill were not the less necessary, for any share which the water has in grinding the corn: yet is this share the same, as that which the watch would have contributed to the production of the new watch, upon the supposition assumed in the last section. Therefore,

III. Though it be now no longer probable, that the individual watch which our observer had found, was made immediately by the hand of an artificer, yet doth not this alteration in any wise affect the inference, that an artificer had been originally employed and concerned in the production. The argument from design remains as it was. Marks of design and contrivance are no more accounted for now, than they were before. In the same thing, we may ask for the cause of different properties. We may ask for the cause of the colour of a body, of its hardness, of its heat; and these causes may be all different. We are now asking for the cause of that subserviency to an use, that relation to an end, which we have remarked in the watch before us. No answer is given to this question by telling us that a preceding watch produced it. There cannot be design without a designer;* contrivance without a contriver; order without choice; arrangement, without any thing capable of arranging; subserviency and relation to a purpose, without that which could intend a purpose; means suitable to an end, and executing their office in accomplishing that end, without the end ever having been contemplated, or the means accommodated to it. Arrangement, disposition of parts, subserviency of means to an end, relation of instruments to an use, imply the presence of intelligence and mind. No one, therefore, can rationally believe, that the insensible, inanimate watch, from which the watch before us issued, was the proper cause of the mechanism we so much admire in it; could be truly said to have constructed the instrument, disposed its parts, assigned their office, determined their order, action, and mutual dependency, combined their several motions into one result, and that also a result connected with the utilities of other beings. All these properties, therefore, are as much unaccounted for, as they were before.

IV. Nor is any thing gained by running the difficulty further back, i. e. by supposing the watch before us to have been produced from another watch, that from a former, and so on indefinitely. Our going back ever so far brings us no nearer to the least degree of satisfaction upon the subject. Contrivance is still unaccounted for. We still want a contriver. A designing mind is neither supplied by this supposition, nor dispensed with. If the difficulty were diminished the further we went back, by going back indefinitely we might exhaust it. And this is the only case to which this sort of reasoning applies. Where there is a tendency, or, as we increase the number of terms, a continual approach towards a limit, *there*, by supposing the number of terms to be what is called infinite, we may conceive the limit to be attained: but where there is no such tendency or approach, nothing is effected by lengthening the series. There is no difference as to the point in question, (whatever there may be as to many points) between one series and another; between a series which is finite, and a series which is infinite. A chain, composed of an infinite number of links, can no more support itself, than a chain composed of a finite number of links. And of this we are assured, (though we never *can* have tried the experiment), because, by increasing the number of links, from ten for instance to a hundred, from a hundred to a thousand, etc. we make not the smallest approach, we observe not the smallest tendency, towards self-support. There is no difference in this respect (yet there may be a great difference in several respects) between a chain of a greater or less length, between one chain and another, between one that is finite and one that is indefinite. This very much resembles the case before us. The machine, which we are inspecting, demonstrates, by its construction, contrivance and design. Contrivance must have had a contriver; design, a designer; whether the machine immediately proceeded from another machine, or not. That circumstance alters not the case. That other machine may, in like manner, have proceeded from a former machine: nor does that alter the case: contrivance must have had a contriver. That former one from one preceding it: no alteration still: a contriver is still necessary. No tendency is perceived, no approach towards a diminution of this necessity. It is the same with any and every succession of these machines; a succession of ten, of a hundred, of a thousand; with one series as with another; a series which is finite, as with a series which is infinite. In whatever other respects they may differ, in this they do not. In all equally, contrivance and design are unaccounted for.

The question is not simply, How came the first watch into existence? which question, it may be pretended, is done away by supposing the series of watches thus produced from one another to have been infinite, and consequently to have had no such *first*, for which it was necessary to provide a cause. This, perhaps, would have been nearly the state of the question, if nothing had been before us but an unorganised, unmechanised, substance, without mark or indication of contrivance. It might be difficult to shew that such substance could not have existed from eternity, either in succession (if it were possible, which I think it is not, for unorganised bodies to spring from one another), or by individual perpetuity. But that is not the question now. To suppose it to be so, is to suppose that it made no difference whether we had found a watch or a stone. As it is, the metaphysics* of that question have no place; for, in the watch which we are examining, are seen contrivance, design; an end, a purpose; means for the end, adaptation to the purpose. And the question, which irresistibly presses upon our thoughts, is, whence this contrivance and design. The thing required is the intending mind, the adapting hand, the intelligence by which that hand was directed. This question, this demand, is not shaken off, by increasing a number of succession of substances, destitute of these properties; nor the more, by increasing that number to infinity. If it be said, that, upon the supposition of one watch being produced from another in the course of that other's movements, and by means of the mechanism within it, we have a cause for the watch in my hand, viz. the watch from which it proceeded, I deny, that for the design, the contrivance, the suitableness of means to an end, the adaptation of instruments to an use (all which we discover in the watch), we have any cause whatever. It is in vain, therefore, to assign a series of such causes; or to alledge that a series may be carried back to infinity; for I do not admit that we have yet any cause at all of the phænomena, still less any series of causes either finite or infinite. Here is contrivance, but no contriver: proofs of design, but no designer.

V. Our observer would further also reflect, that the maker of the watch before him, was, in truth and reality, the maker of every watch produced from it; there being no difference (except that the latter manifests a more exquisite skill) between the making of another watch with his own hands by the mediation of files, laths, chisels, etc. and the disposing, fixing, and inserting, of these instruments, or of others equivalent to them, in the body of the watch already made, in such a manner, as to form a new watch in the course of the movements which he had given to the old one. It is only working by one set of tools, instead of another.

The conclusion which the *first* examination of the watch, of its works, construction, and movement suggested, was, that it must have had, for the cause and author of that construction, an artificer, who understood its mechanism, and designed its use. This conclusion is invincible. A *second* examination presents us with a new discovery. The watch is found, in the course of its movement, to produce another watch, similar to itself: and not only so, but we perceive in it a system of organization, separately calculated for that purpose. What effect would this discovery have, or ought it to have, upon our former inference? What, as hath already been said, but to increase, beyond measure, our admiration of the skill, which had been employed in the formation of such a machine? Or shall it, instead of this, all at once turn us round to an opposite conclusion, viz. that no art or skill whatever has been concerned in the business, although all other evidences of art and skill remain as they were, and this last and supreme piece of art be now added to the rest? Can this be maintained without absurdity? Yet this is atheism.*

CHAPTER III

APPLICATION OF THE ARGUMENT

THIS is atheism: for every indication of contrivance, every manifestation of design, which existed in the watch, exists in the works of nature; with the difference, on the side of nature, of being greater and more, and that in a degree which exceeds all computation. I mean that the contrivances of nature surpass the contrivances of art, in the complexity, subtlety, and curiosity of the mechanism; and still more, if possible, do they go beyond them in number and variety: yet, in a multitude of cases, are not less evidently mechanical, not less evidently contrivances, not less evidently accommodated to their end, or suited to their office, than are the most perfect productions of human ingenuity.

I know no better method of introducing so large a subject, than that of comparing a single thing with a single thing; an eye, for example, with a telescope.* As far as the examination of the instrument goes, there is precisely the same proof that the eye was made for vision, as there is that the telescope was made for assisting it. They are made upon the same principles; both being adjusted to the laws by which the transmission and refraction of rays of light are regulated. I speak not of the origin of the laws themselves; but, such laws being fixed,* the construction, in both cases, is adapted to them. For instance; these laws require, in order to produce the same effect, that the rays of light, in passing from water into the eye, should be refracted by a more convex surface, than when it passes out of air into the eye. Accordingly we find, that the eye of a fish, in that part of it called the crystalline lense, is much rounder than the eye of terrestrial animals. What plainer manifestation of design can there be than this difference? What could a mathematical instrument-maker have done more, to shew his knowledge of his principle, his application of that knowledge, his suiting of his means to his end; I will not say to display the compass or excellency of his skill and art, for in these all comparison is indecorous, but to testify counsel, choice, consideration, purpose?

To some it may appear a difference sufficient to destroy all similitude between the eye and the telescope, that the one is a perceiving organ, the other an unperceiving instrument. The fact is, that they are both instruments. And, as to the mechanism, at least as to mechanism being employed, and even as to the kind of it, this circumstance varies not the analogy at all. For observe, what the constitution of the eye is. It is necessary, in order to produce distinct vision, that an image or picture of the object be formed at the bottom of the eye. Whence this necessity arises, or how the picture is connected with the sensation, or contributes to it, it may be difficult, nay we will confess, if you please, impossible for us to search out. But the present question is not concerned in the enquiry. It may be true, that, in this, and in other instances, we trace mechanical contrivance a certain way; and that then we come to something which is not mechanical, or which is inscrutable. But this affects not the certainty of our investigation, as far as we have gone. The difference between an animal and an automatic statue,* consists in this,—that, in the animal, we trace the mechanism to a certain point, and then we, are stopped; either the mechanism becoming too subtile for our discernment, or something else beside the known laws of mechanism taking place; whereas, in the automaton, for the comparatively few motions of which it is capable, we trace the mechanism throughout. But, up to the limit, the reasoning is as clear and certain in the one case as the other. In the example before us, it is a matter of certainty, because it is a matter which experience and observation demonstrate,* that the formation of an image at the bottom of the eye is necessary to perfect vision. The image itself can be shewn. Whatever affects the distinctness of the image, affects the distinctness of the vision. The formation then of such an image being necessary (no matter how), to the sense of sight, and to the exercise of that sense, the apparatus by which it is formed is constructed and put together, not only with infinitely more art, but upon the self-same principles of art, as in the telescope or the camera obscura.* The perception arising from the image may be laid out of the question: for the production of the image, these are instruments of the same kind. The end is the same; the means are the same. The purpose in both is alike; the contrivance for accomplishing that purpose is in both alike. The lenses of the telescope, and the humours of the eye bear a complete resemblance to one another, in their figure, their position, and in their power over the rays of light, viz. in bringing each pencil to a point at the right distance from the lense; namely, in the eye, at the exact place where the membrane is spread to receive it. How is it possible, under circumstances of such close affinity, and under the operation of equal evidence, to exclude contrivance from the one, yet to acknowledge the proof of contrivance having been employed, as the plainest and clearest of all propositions in the other?

The resemblance between the two cases is still more accurate, and obtains in more points than we have yet represented, or than we are, on the first view of the subject, aware of. In dioptric telescopes there is an imperfection of this nature. Pencils of light, in passing through glass lenses, are separated into different colours, thereby tingeing the object, especially the edges of it, as if it were viewed through a prism. To correct this inconvenience had been long a desideratum in the art. At last it came into the mind of a sagacious optician, to enquire how this matter was managed in the eye; in which there was exactly the same difficulty to contend with, as in the telescope. His observation taught him, that, in the eye, the evil was cured by combining together lenses composed of different substances, i. e. of substances which possessed different refracting powers. Our artist borrowed from thence his hint; and produced a correction of the defect by imitating, in glasses made from different materials, the effects of the different humours through which the rays of light pass before they reach the bottom of the eye. Could this be in the eye without purpose, which suggested to the optician the only effectual means of attaining that purpose?

But further; there are other points, not so much perhaps of strict resemblance between the two, as of superiority of the eye over the telescope; yet, of a superiority, which being founded in the laws that regulate both, may furnish topics of fair and just comparison. Two things were wanted to the eye, which were not wanted, at least in the same degree, to the telescope; and these were, the adaptation of the organ,* first, to different degrees of light; and, secondly, to the vast diversity of distance at which objects are viewed by the naked eye, viz. from a few inches to as many miles. These difficulties present not themselves to the maker of the telescope. He wants all the light he can get; and he never directs his instrument to objects near at hand. In the eye, both these cases were to be provided for; and for the purpose of providing for them a subtile and appropriate mechanism is introduced.

I. In order to exclude excess of light, when it is excessive, and to render objects visible under obscurer degrees of it, when no more can be had; the hole or aperture in the eye, through which the light enters, is so formed, as to contract or dilate itself for the purpose of admitting a greater or less number of rays at the same time. The chamber of the eye is a camera obscura, which, when the light is too small, can enlarge its opening; when too strong, can again contract it; and that without any other assistance than that of its own exquisite machinery. It is further also, in the human subject, to be observed, that this hole in the eye, which we call the pupil, under all its different dimensions, retains its exact circular shape. This is a structure extremely artificial. Let an artist only try to execute the same. He will find that his threads and strings must be disposed with great consideration and contrivance, to make a circle, which shall continually change its diameter, yet preserve its form. This is done in the eye by an application of fibres, i. e. of strings, similar, in their position and action, to what an artist would and must employ, if he had the same piece of workmanship to perform.

II. The second difficulty which has been stated, was the suiting of the same organ to the perception of objects that lie near at hand, within a few inches, we will suppose, of the eye, and of objects which were placed at a considerable distance from it, that, for example, of as many furlongs* (I speak in both cases of the distance at which distinct vision can be exercised). Now, this, according to the principles of optics, that is, according to the laws by which the transmission of light is regulated, (and these laws are fixed,) could not be done, without the organ itself undergoing an alteration, and receiving an adjustment, that might correspond with the exigency of the case, that is to say, with the different inclination to one another under which the rays of light reached it. Rays issuing from points placed at a small distance from the eye, and which consequently must enter the eye in a spreading or diverging order, cannot, by the same optical instrument in the same state, be brought to a point, i. e. be made to form an image, in the same place with rays proceeding from objects situated at a much greater distance, and which rays arrive at the eye in directions nearly, and physically speaking, parallel. It requires a rounder lense to do it. The point of concourse behind the lense must fall critically upon the retina, or the vision is confused; yet, this point, by the immutable properties of light, is carried further back, when the rays proceed from a near object, than when they are sent from one that is remote. A person, who was using an optical instrument, would manage this matter by changing, as the occasion required, his lense or his telescope; or by adjusting the distance of his glasses with his hand or his screw: but how is it to be managed in the eye? What the alteration was, or in what part of the eye it took place, or by what means it was effected (for, if the known laws which govern the refraction of light be maintained, some alteration in the state of the organ there must be), had long formed a subject of enquiry and conjecture.* The change, though sufficient for the purpose, is so minute as to elude ordinary observation. Some very late discoveries, deduced from a laborious and most accurate inspection of the structure and operation of the organ, seem at length to have ascertained the mechanical alteration which the parts of the eye undergo. It is found, that by the action of certain muscles, called the straight muscles, and which action is the most advantageous that could be imagined for the purpose,—it is found, I say, that, whenever the eye is directed to a near object, three changes are produced in it at the same time, all severally contributing to the adjustment required. The cornea, or outermost coat of the eye, is rendered more round and prominent; the crystalline lense underneath is pushed forwards; and the axis of vision, as the depth of the eye is called, is elongated. These changes in the eye vary its power over the rays of light in such a manner and degree as to produce exactly the effect which is wanted, viz. the formation of an image *upon the retina*, whether the rays come to the eye in a state of divergency, which is the case when the object is near to the eye, or come parallel to one another, which is the case when the object is placed at a distance. Can any thing be more decisive of contrivance than this is? The most secret laws of optics must have been known to the author of a structure endowed with such a capacity of change. It is, as though an optician, when he had a nearer object to view, should *rectify* his instrument by putting in another glass, at the same time drawing out also his tube to a different length.

Observe a new-born child first lifting up its eyelids. What does the opening of the curtain discover? The anterior part of two pellucid globes, which, when they come to be examined, are found to be constructed upon strict optical principles; the self-same principles upon which we ourselves construct optical instruments. We find them perfect for the purpose of forming an image by refraction; composed of parts executing different offices; one part having fulfilled its office upon the pencil of light, delivering it over to the action of another part; that to a third, and so onward: the progressive action depending for its success upon the nicest, and minutest adjustment of the parts concerned; yet, these parts so in fact adjusted, as to produce, not by a simple action or effect, but by a combination of actions and effects, the result which is ultimately wanted. And forasmuch as this organ would have to operate under different circumstances, with strong degrees of light, and with weak degrees, upon near objects, and upon remote ones, and these differences demanded, according to the laws by which the transmission of light is regulated, a corresponding diversity of structure; that the aperture, for example, through which the light passes, should be larger or less; the lenses rounder or flatter, or that their distance from the tablet, upon which the picture is delineated, should be shortened or lengthened: this, I say, being the case and the difficulty, to which the eye was to be adapted, we find its several parts capable of being occasionally changed, and a most artificial apparatus provided to produce that change. This is far beyond the common regulator of a watch, which requires the touch of a foreign hand to set it; but is not altogether unlike Harrison's contrivance* for making a watch regulate itself, by inserting within it a machinery, which, by the artful use of the different expansion of metals, preserves the equability of the motion under all the various temperatures of heat and cold in which the instrument may happen to be placed. The ingenuity of this last contrivance has been justly praised. Shall, therefore, a structure which differs from it, chiefly by surpassing it, be accounted no contrivance at all? or, if it be a contrivance, that it is without a contriver?

But this, though much, is not the whole: by different species of animals the faculty we are describing is possessed, in degrees suited to the different range of vision which their mode of life, and of procuring their food, requires. Birds, for instance, in general, procure their food by means of their beak; and the distance between the eye and the point of the beak being small, it becomes necessary that they should have the power of seeing very near objects distinctly. On the other hand, from being often elevated much above the ground, living in air, and moving through it with great velocity, they require, for their safety, as well as for assisting them in descrying their prey, a power of seeing at great distance; a power, of which, in birds of rapine, surprising examples are given. The fact accordingly is, that two peculiarities are found in the eyes of birds, both tending to facilitate the change upon which the adjustment of the eye to different distances depends. The one is a bony, yet, in most species, a flexible rim or hoop, surrounding the broadest part of the eye; which, confining the action of the muscles to that part, increases the effect of their lateral pressure upon the orb, by which pressure its axis is elongated for the purpose of looking at very near objects. The other is, an additional muscle called the marsupium, to draw, upon occasion, the crystal-line lense *back*, and so fit the same eye for the viewing of very distant objects. By these means the eyes of birds can pass from one extreme to another of their scale of adjustment, with more ease and readiness than the eyes of other animals.

The eyes of fishes also, compared with those of terrestrial animals, exhibit certain distinctions of structure, adapted to their state and element. We have already observed upon the figure of the crystalline compensating by its roundness the density of the medium through which their light passes. To which we have to add, that the eyes of fish, in their natural and indolent state, appear to be adjusted to near objects, in this respect differing from the human eye, as well as those of quadrupeds and birds. The ordinary shape of the fish's eye being in a much higher degree convex than that of land animals, a corresponding difference attends its muscular conformation, viz. that it is throughout calculated for *flattening* the eye.

The iris also in the eyes of fish does not admit of contraction. This is a great difference, of which the probable reason is, that the diminished light in water is never too strong for the retina.

In the eel, which has to work its head through sand and gravel, the roughest and harshest substances, there is placed before the eye, and at some distance from it, a transparent, horny, convex case or covering, which, without obstructing the sight, defends the organ. To such an animal, could any thing be more wanted, or more useful?

Thus, in comparing together the eyes of different kinds of animals, we see, in their resemblances and distinction, one general plan laid down, and that plan varied with the varying exigences to which it is to be applied.

There is one property, however, common, I believe, to all eyes, at least to all which have been examined,[1] namely, that the optic nerve enters the bottom of the eye, not in the centre or middle, but a little on one side; not in the point where the axis of the eye meets the retina, but between that point and the nose.—The difference which this makes is, that no part of an object is unperceived by both eyes at the same time.

In considering vision as achieved by the means of an image formed at the bottom of the eye, we can never reflect without wonder upon the smallness, yet correctness, of the picture, the subtility of the touch, the fineness of the lines. A landscape of five or six square leagues* is brought into a space of half an inch diameter; yet the multitude of objects which it contains are all preserved; are all discriminated in their magnitudes, positions, figures, colours. The prospect from Hampstead-Hill is compressed into the compass of a sixpence, yet circumstantially represented. A stage coach travelling at its ordinary speed for half an hour, passes, in the eye, only over one-twelfth of an inch, yet is this change of place in the image distinctly perceived throughout its whole progress; for it is only by means of that perception that the motion of the coach itself is made sensible to the eye. If any thing can abate our admiration of the smallness of the visual tablet compared with the extent of vision, it is a reflection, which the view of nature leads us, every hour, to make, viz. that, in the hands of the Creator, great and little are nothing.

Sturmius* held, that the examination of the eye was a cure for atheism. Beside that conformity to optical principles which its internal constitution displays, and which alone amounts to a manifestation of intelligence having been exerted in its structure; beside this, which forms, no doubt, the leading character of the organ, there is to be seen, in every thing belonging to it and about it, an extraordinary degree of care, an anxiety for its preservation, due, if we may so speak, to its value and its tenderness. It is lodged in a strong, deep, bony socket, composed by the junction of seven different bones,[2] hollowed out at their edges. In some few species, as that of the coatimondi,[3]* the orbit is not bony throughout; but whenever this is the case, the upper, which is the deficient part, is supplied by a

---

[1] The eye of the seal or sea calf, I understand, is an exception. Mem. Acad. Paris, 1701,* p. 123.

[2] Heister,* sect. 89.

[3] Mem. R. Ac. Paris, p. 117.

cartilaginous ligament; a substitution which shews the same care. Within this socket it is imbedded in fat, of all animal substances the best adapted both to its repose and motion. It is sheltered by the eyebrows, an arch of hair, which, like a thatched penthouse, prevents the sweat and moisture of the forehead from running down into it.

But it is still better protected by its *lid*. Of the superficial parts of the animal frame, I know none which, in its office and structure, is more deserving of attention than the eyelid. It defends the eye; it wipes it; it closes it in sleep. Are there, in any work of art whatever, purposes more evident than those which this organ fulfills; or an apparatus for executing those purposes more intelligible, more appropriate, or more mechanical? If it be overlooked by the observer of nature, it can only be because it is obvious and familiar. This is a tendency to be guarded against. We pass by the plainest instances, whilst we are exploring those which are rare and curious: by which conduct of the understanding, we sometimes neglect the strongest observations, being taken up with others, which, though more recondite and scientific,* are, as solid arguments, entitled to much less consideration.

In order to keep the eye moist and clean, which qualities are necessary to its brightness and its use, a wash is constantly supplied by a secretion for the purpose; and the superfluous brine is conveyed to the nose through a perforation in the bone as large as a goose quill. When once the fluid has entered the nose, it spreads itself upon the inside of the nostril, and is evaporated by the current of warm air, which, in the course of respiration, is continually passing over it. Can any pipe or outlet for carrying off the waste liquor from a dye-house or a distillery, be more mechanical than this is? It is easily perceived that the eye must want moisture; but could the want of the eye generate the gland which produces the tear, or bore the hole by which it is discharged—a hole through a bone?

It is observable that this provision is not found in fish, the element in which they live supplying a constant lotion to the eye.

It were, however, injustice to dismiss the eye as a piece of mechanism, without noticing that most exquisite of all contrivances, the nictitating membrane,* which is found in the eyes of birds and of many quadrupeds. Its use is to sweep the eye, which it does in an instant; to spread over it the lachrymal humor;* to defend it also from sudden injuries; yet not totally, when drawn upon the pupil, to shut out the light. The commodiousness with which it lies folded up in the upper corner of the eye, ready for use and action, and the quickness with which it executes its purpose, are properties known and obvious to every observer; but, what is equally admirable, though not quite so obvious, is the combination of two different kinds of substance, muscular and elastic, and of two different kinds of action, by which the motion of this membrane is performed. It is not, as in ordinary cases, by the action of two antagonist muscles, one pulling forward and the other backward, that a reciprocal change is effected; but it is thus: The membrane itself is an elastic substance, capable of being drawn out by force like a piece of elastic gum, and by its own elasticity returning, when the force is removed, to its former position. Such being its nature, in order to fit it up for its office it is connected by a tendon or thread with a muscle in the back part of the eye; this tendon or thread, though strong, is so fine, as not to obstruct the sight, even when it passes across it; and the muscle itself being placed in the *back* part of the eye, derives from its situation the advantage, not only of being secure, but of being out of the way; which it would hardly have been in any position that could be assigned to it in the anterior part of the orb, where its function lies. When the muscle behind the eye contracts, the membrane, by means of the communicating thread, is instantly drawn over the forepart of it. When the muscular contraction (which is a positive, and, most probably, a voluntary effort,) ceases to be exerted, the elasticity alone of the membrane brings it back again to its position.[1] Does not this, if any thing can do it, bespeak an artist, master of his work, acquainted with his materials? 'Of a thousand other things,' say the French Academicians, 'we perceive not the contrivance, because we understand them only by the effects, of which we know not the causes; but we here treat of a machine, all the parts whereof are visible; and which need only be looked upon to discover the reasons of its motion and action.'[2]

In the configuration of the muscle, which, though placed behind the eye, draws the nictitating membrane over the eye, there is, what the authors, just now quoted, deservedly call a marvellous

---

[1] Phil. Trans. 1796.*
[2] Memoirs for a Natural History of Animals by the Royal Academy of Sciences at Paris, done into English by Order of the Royal Society, 1701, p. 249.

mechanism. I suppose this structure to be found in other animals; but, in the Memoirs from which this account is taken, it is anatomic-ally demonstrated only in the cassowary.* The muscle is *passed through a loop formed by another muscle*; and is there inflected, as if it were round a pulley. This is a peculiarity; and observe the advantage of it. A single muscle with a straight tendon, which is the common muscular form, would have been sufficient, if it had had power to draw far enough. But the contraction, necessary to draw the mem-brane over the whole eye, required a longer muscle than could lie straight at the bottom of the eye. Therefore, in order to have a greater length in a less compass, the cord of the main muscle makes an angle. This, so far, answered the end; but; still further, it makes an angle, not round a fixed pivot, but round a loop formed by another muscle; which second muscle, whenever it contracts, of course twitches the first muscle at the point of inflection, and thereby assists the action designed by both.

One question may possibly have dwelt in the reader's mind during the perusal of these observations, namely, Why should not the Deity have given to the animal the faculty of vision *at once*? Why this circuitous perception; the ministry of so many means? an element provided for the purpose; reflected from opaque; substances, refracted through transparent ones; and both according to precise laws: then, a complex organ, an intricate and artificial apparatus, in order, by the operation of this element, and in conformity with the restrictions of these laws, to produce an image upon a membrane communicating with the brain? Wherefore all this? Why make the difficulty in order only to surmount it? If to perceive objects by some other mode than that of touch, or objects which lay out of the reach of the sense, were the thing purposed, could not a simple volition of the Creator have communicated the capacity? Why resort to contriv-ance, where power is omnipotent? Contrivance, by its very definition and nature, is the refuge of imperfection. To have recourse to expedients, implies difficulty, impediment, restraint, defect of power. This question belongs to the other senses, as well as to sight; to the general functions of animal life, as nutrition, secretion, respir-ation; to the œconomy of vegetables; and indeed to almost all the operations of nature. The question therefore is of very wide extent; and, amongst other answers which may be given to it, beside reasons of which probably we are ignorant, one answer is this. It is only by the display of contrivance, that the existence, the agency, the wisdom of the Deity, *could* be testified to his rational creatures. This is the scale by which we ascend to all the knowledge of our Creator which we possess, so far as it depends upon the phænomena, or the works of nature. Take away this, and you take away from us every subject of observation, and ground of reasoning; I mean as our rational faculties are formed at present. Whatever is done, God could have done, without the intervention of instruments or means: but it is in the construction of instruments, in the choice and adaptation of means, that a creative intelligence is seen. It is this which constitutes the order and beauty of the universe. God, therefore, has been pleased to prescribe limits to his own power, and to work his ends within those limits.* The general laws of matter have perhaps the nature of these limits; its inertia, its reaction; the laws which govern the communication of motion, the refraction and reflection of light, the constitution of fluids non-elastic and elastic, the transmission of sound through the latter; the laws of magnetism, of electricity; and probably others yet undiscovered. These are general laws; and when a particular purpose is to be effected, it is not by making a new law, nor by the suspension of the old ones, nor by making them wind and bend and yield to the occasion (for nature with great steadiness adheres to, and supports them), but it is, as we have seen in the eye, by the interposition of an apparatus corresponding with these laws, and suited to the exigency which results from them, that the purpose is at length attained. As we have said, therefore, God prescribes limits to his power, that he may let in the exercise, and thereby exhibit demonstrations, of his wisdom. For then, i. e. such laws and limitations being laid down, it is as though one Being should have fixed certain rules; and, if we may so speak, provided certain materials; and, afterwards, have committed to another Being, out of these materials, and in subordination to these rules, the task of drawing forth a creation: a supposition which evidently leaves room, and induces indeed a necessity, for contrivance. Nay, there may be many such agents, and many ranks of these. We do not advance this as a doctrine either of philosophy or of religion; but we say that the subject may safely be represented under this view, because the Deity, acting himself by general laws, will have the same consequences upon our reasoning, as if he had prescribed these laws to another. It has been said, that the problem of creation was, 'attraction and matter being given, to make a world out of them:' and, as above explained, this statement perhaps does not convey a false idea.

We have made choice of the eye as an instance upon which to rest the argument of this chapter. Some single example was to be proposed; and the eye offered itself under the advantage of admitting of a strict comparison with optical instruments. The ear,* it is probable, is no less artificially and mechanically adapted to its office, than the eye. But we know less about it: we do not so well understand the action, the use, or the mutual dependency of its internal parts. Its general form, however, both external and internal, is sufficient to shew that it is an instrument adapted to the reception of sound; that is to say, already knowing that sound consists in pulses of the air, we perceive, in the structure of the ear, a suitableness to receive impressions from this species of action, and to propagate these impressions to the brain. For of what does this structure consist? An external ear (the concha), calculated, like an ear-trumpet, to catch and collect the pulses of which we have spoken;—in large quadrupeds, turning to the sound, and possessing a configuration, as well as motion, evidently fitted for the office: of a tube which leads into the head, lying at the root of this outward ear, the folds and sinuses thereof tending and conducting the air towards it: of a thin membrane, like the pelt of a drum, stretched across this passage upon a bony rim: of a chain of moveable, and infinitely curious, bones, forming a communication, and the only communication that can be observed, between the membrane last mentioned and the interior channels and recesses of the skull: of cavities, similar in shape and form to wind instruments of music, being spiral or portions of circles: of the eustachian tube, like the hole in a drum, to let the air pass freely into and out of the barrel of the ear, as the covering membrane vibrates, or as the temperature may be altered: the whole labyrinth hewn out of a rock: that is, wrought into the substance of the hardest bone of the body. This assemblage of connected parts constitutes together an apparatus, plainly enough relative to the transmission of sound, or of the impulses received from sound, and only to be lamented in not being better understood.

The communication within, formed by the small bones of the ear, is, to look upon, more like what we are accustomed to call machinery, than any thing I am acquainted with in animal bodies. It seems evidently designed to continue towards the sensorium the tremulous motions which are excited in the 'membrane of the tympanum,' or what is better known by the name of the 'drum of the ear.' The compages of bones consists of four, which are so disposed, and so hinge upon one another, as that, if the membrane, the drum of the ear, vibrate, all the four are put in motion together; and, by the result of their action, work the base of that which is the last in the series, upon an aperture which it closes, and upon which it plays, and which aperture opens into the tortuous canals that lead to the brain. This last bone of the four is called the stapes. The office of the drum of the ear is to spread out an extended surface, capable of receiving the impressions of sound, and of being put by them into a state of vibration. The office of the stapes is to repeat these vibrations. It is a repeating frigate, stationed more within the line. From which account of its action may be understood, how the sensation of sound will be excited, by any thing which communicates a vibratory motion to the stapes, though not, as in all ordinary cases, through the inter-vention of the membrana tympani.* This is done by solid bodies applied to the bones of the skull, as by a metal bar held at one end between the teeth, and touching at the other end a tremulous body. It likewise appears to be done, in a considerable degree, by the air itself, even when this membrane, the drum of the ear, is greatly damaged. Either in the natural or præternatural state of the organ, the use of the chain of bones is to propagate the impulse in a direction towards the brain,* and to propagate it with the advantage of a lever; which advantage consists in increasing the force and strength of the vibra-tion, and at the same time diminishing the space through which it oscillates: both of which changes may augment or facilitate the still deeper action of the auditory nerves.

The benefit of the eustachian tube to the organ, may be made out upon known pneumatic principles.* Behind the drum of the ear is a second cavity or barrel, called the tympanum. The eustachian tube is a slender pipe, but sufficient for the passage of air, leading from this cavity into the back part of the mouth. Now it would not have done to have had a vacuum in this cavity; for, in that case, the pressure of the atmosphere from without would have burst the membrane which covered it. Nor would it have done to have filled the cavity with lymph or any other secretion; which would necessarily have obstructed, both the vibration of the membrane, and the play of the small bones. Nor, lastly would it have done to have occupied the space with confined air, because the expansion of that air by heat, or its contraction by cold would have distended or relaxed the covering membrane, in a degree inconsistent with the purpose which it was assigned to execute. The only remaining expedient, and that for which the eustachian tube serves, is to open to this cavity a communication with the external air. In one word; it exactly answers the purpose of the hole in a drum.

The membrana tympani itself, likewise, deserves all the examination which can be made of it. It is not found in the ears of fish; which furnishes an additional proof of what indeed is indicated by every thing about it, that it is appropriated to the action of air, or of an elastic medium. It bears an obvious resemblance to the pelt or head of a drum, from which it takes its name. It resembles also a drum head in this principal property, that its use depends upon its tension. Tension is the state essential to it. Now we know that, in a drum, the pelt is carried over a hoop, and braced, as occasion requires, by the means of strings attached to its circumstance. In the membrane of the ear, the same purpose is provided for, more simply, but not less mechanically, nor less successfully, by a different expedient, viz. by the end of a bone (the handle of the malleus) pressing upon its centre. It is only in very large animals that the texture of this membrane can be discerned. In the Philosophical Transactions for the year 1800, (vol. i.) Mr Everard Home* has given some curious observations upon the ear, and the drum of the ear, of an *elephant*. He discovered in it, what he calls a radiated muscle, that is, straight muscular fibres, passing along the membrane from the circumference to the centre; from the bony rim which surrounds it, towards the handle of the malleus to which the central part is attached. This muscle he supposes to be designed to bring the membrane into unison with different sounds: but then he also discovered, that this muscle itself cannot act, unless the membrane be drawn to a stretch, and kept in a due state of tightness, by what may be called a foreign force, viz. the action of the muscles of the malleus. Our author, supposing his explanation of the use of the parts to be just, is well founded in the reflection which he makes upon it; 'that this mode of adapting the ear to different sounds, is one of the most beautiful applications of muscles in the body; *the mechanism is so simple, and the variety of effects so great.*'

In another volume of the Transactions above referred to, and of the same year, two most curious cases are related, of persons who retained the sense of hearing, not in a perfect, but in a very considerable degree, notwithstanding the almost total loss of the membrane we have been describing. In one of these cases, the use here assigned to that membrane, of modifying the impressions of sound by change of tension, was attempted to be supplied by straining the muscles of the outward ear. 'The external ear,' we are told, 'had acquired a distinct motion upward and backward, which was observable whenever the patient listened to any thing which he did not distinctly hear; when he was addressed in a whisper, the ear was seen immediately to move; when the tone of voice was louder, it then remained altogether motionless.'

It appears probable, from both these cases, that a collateral, if not principal, use of the membrane, is to cover and protect the barrel of the ear which lies behind it. Both the patients suffered from cold; one, 'a great increase of deafness from catching cold;' the other, 'very considerable pain from exposure to a stream of cold air.' Bad effects therefore followed from this cavity being left open to the external air; yet, had the author of nature shut it up by any other cover, than what was capable, by its texture, of receiving vibrations from sound, and, by its connection with the interior parts, of transmitting those vibrations to the brain, the use of the organ, so far as we can judge, must have been entirely obstructed.

CHAPTER IV

OF THE SUCCESSION OF PLANTS AND ANIMALS

THE generation of the animal no more accounts for the contrivance of the eye or ear, than, upon the supposition stated in a preceding chapter, the production of a watch by the motion and mechanism of a former watch, would account for the skill and intention evidenced in the watch so produced; than it would account for the disposition of the wheels, the catching of their teeth, the relation of the several parts of the works to one another and to their common end, for the suitableness of their forms and places to their offices, for their connection, their operation, and the useful result of that operation. I do insist most strenuously upon the correctness of this comparison; that it holds as to every mode of specific propagation; and that whatever was true of the watch, under the hypothesis above mentioned, is true of plants and animals.

I. To begin with the fructification of plants. Can it be doubted but that the seed contains a particular organization? Whether a latent plantule with the means of temporary nutrition, or whatever else it be, it incloses an organization suited to the germination of a new plant. Has the plant which produced the feed any thing more to do with that organization, than the watch would have had to do with the structure of the watch which was produced in the course of its mechanical movement? I mean, Has it any thing at all to do with the *contrivance*? The maker and contriver of one watch, when he inserted within it a mechanism suited to the production of another watch, was, in truth, the maker and contriver of that other watch. All the properties of the new watch were to be referred to his agency: the design manifested in it, to his intention: the art, to him as the artist: the collocation of each part, to his placing: the action, effect, and use, to his counsel, intelligence, and workmanship. In producing it by the intervention of a former watch, he was only working by one set of tools instead of another. So it is with the plant, and the feed produced by it. Can any distinction be assigned between the two cases; between the producing watch, and the producing plant?* both passive, unconscious substances; both, by the organization which was given to them, producing their like, without understanding or design; both, that is, instruments.

II. From plants we may proceed to oviparous animals; from seeds to eggs.* Now I say, that the bird has the same concern in the formation of the egg which she lays, as the plant has in that of the seed which it drops; and no other, nor greater. The internal constitution of the egg is as much a secret to the hen, as if the hen were inanimate. Her will cannot alter it, or change a single feather of the chick. She can neither foresee nor determine of which sex her brood shall be, or how many of either: yet the thing produced shall be, from the first, very different in its make, according to the sex which it bears. So far therefore from adapting the means, she is not beforehand apprized of the effect. If there be concealed within that smooth shell a provision and a preparation for the production and nourishment of a new animal, they are not of her providing or preparing: if there be contrivance, it is none of hers. Although, therefore, there be the difference of life and perceptivity between the animal and the plant, it is a difference which enters not into the account. It is a foreign circumstance. It is a difference of properties not employed. The animal function and the vegetable function are alike destitute of any design which can operate upon the form of the thing produced. The plant has no design in producing the seed, no comprehension of the nature or use of what it produces: the bird with respect to its egg, is not above the plant with respect to its seed. Neither the one nor the other bears that sort of relation to what proceeds from them, which a joiner does to the chair which he makes. Now a cause, which bears this relation to the effect, is what we want, in order to account for the suitableness of means to an end, the fitness and fitting of one thing to another, and this cause the parent plant or animal does not supply.

It is further observable concerning the propagation of plants and animals, that the apparatus employed exhibits no resemblance to the thing produced; in this respect holding an analogy with instruments and tools of art. The filaments, antheræ, and stigmata of flowers, bear no more resemblance to the young plant, or even to the seed, which is formed by their intervention, than a chisel or a plane does to a table or a chair. What then are the filaments, antheræ, and stigmata of plants, but instruments, strictly so called?

III. We may advance from animals which bring forth eggs, to

animals which bring forth their young alive;* and, of this latter class, from the lowest to the highest; from irrational to rational life, from brutes to the human species; without perceiving, as we proceed, any alteration whatever in the terms of the comparison. The rational animal does not produce its offspring with more certainty or success than the irrational animal; a man than a quadruped, a quadruped than a bird; nor (for we may follow the gradation through its whole scale) a bird than a plant; nor a plant than a watch, a piece of dead mechanism, would do, upon the supposition which has already so often been repeated. Rationality therefore has nothing to do in the business. If an account must be given of the contrivance which we observe; if it be demanded, whence arose either the contrivance by which the young animal is produced, or the contrivance manifested in the young animal itself, it is not from the reason of the parent that any such account can be drawn. He is the cause of his offspring in the same sense as that in which a gardener is the cause of the tulip which grows upon his parterre, and in no other. We admire the flower; we examine the plant; we perceive the conduciveness of many of its parts to their end and office; we observe a provision for its nourishment, growth, protection, and fecundity: but we never think of the gardener in all this. We attribute nothing of this to his agency; yet it may still be true, that, without the gardener, we should not have had the tulip. Just so is it with the succession of animals even of the highest order. For the contrivance discovered in the structure of the thing produced, we want a contriver. The parent is not that contriver. His consciousness decides that question. He is in total ignorance why that which is produced took its present form rather than any other. It is for him only to be astonished by the effect. We can no more look therefore to the intelligence of the parent animal for what we are in search of, a cause of relation and of subserviency of parts to their use, which relation and subserviency we see in the procreated body, than we can refer the internal conformation of an acorn to the intelligence of the oak from which it dropped, or the structure of the watch to the intelligence of the watch which produced it; there being no difference, as far as argument is concerned, between an intelligence which is not exerted, and an intelligence which does not exist.

CHAPTER V

APPLICATION OF THE ARGUMENT CONTINUED

EVERY observation which was made, in our first chapter, concerning the watch, may be repeated with strict propriety concerning the eye; concerning animals; concerning plants; concerning, indeed, all the organized parts of the works of nature.* As,

I. When we are enquiring simply after the *existence* of an intelligent Creator, imperfection, inaccuracy, liability to disorder, occasional irregularities,* may subsist, in a considerable degree, without inducing any doubt into the question: just as a watch may frequently go wrong, seldom perhaps exactly right, may be faulty in some parts, defective in some, without the smallest ground of suspicion from thence arising, that it was not a watch; not made; or not made for the purpose ascribed to it. When faults are pointed out, and when a question is started concerning the skill of the artist, or the dexterity with which the work is executed, then indeed, in order to defend these qualities from accusation, we must be able, either to expose some intractableness and imperfection in the materials, or point out some invincible difficulty in the execution, into which imperfection and difficulty the matter of complaint may be resolved; or, if we cannot do this, we must adduce such specimens of consummate art and contrivance proceeding from the same hand, as may convince the enquirer, of the existence, in the case before him, of impediments like those which we have mentioned, although, what from the nature of the case is very likely to happen, they be unknown and unperceived by him. This we must do in order to vindicate the artist's skill, or, at least, the perfection of it; as we must also judge of his intention, and of the provisions employed in fulfilling that intention, not from an instance in which they fail, but from the great plurality of instances in which they succeed. But, after all, these are different questions from the question of the artist's existence; or, which is the same, whether the thing before us be a work of art or not: and the questions ought always to be kept separate in the mind. So likewise it is in the works of nature. Irregularities and imperfections are of little or no weight in the consideration, when that consideration relates simply to the existence of a Creator. When the argument respects his attributes,* they are of weight; but are then to be taken in conjunction (the attention is not to rest upon them, but they are to be taken in conjunction) with the unexceptionable evidences which we possess, of skill, power, and benevolence, displayed in other instances; which evidences may, in strength, number, and variety be such, and may so overpower apparent blemishes, as to induce us, upon the most reasonable ground, to believe, that these last ought to be referred to some cause, though we be ignorant of it, other than defect of knowledge or of benevolence in the author.

II. There may be also parts of plants and animals, as there were supposed to be of the watch, of which, in some instances, the operation, in others, the use is unknown. These form different cases; for the operation may be unknown, yet the use be certain. Thus it is with the lungs of animals. It does not, I think, appear, that we are acquainted with the action of the air upon the blood, or in what manner that action is communicated by the lungs; yet we find that a very short suspension of their office destroys the life of the animal. In this case, therefore, we may be said to know the use, nay we experience the necessity, of the organ, though we be ignorant of its operation. Nearly the same thing may be observed of what is called the lymphatic system.* We suffer grievous inconveniences from its disorder, without being informed of the office which it sustains in the œconomy of our bodies. There may possibly also be some few examples of the second class, in which not only the operation is unknown, but in which experiments may seem to prove that the part is not necessary; or may leave a doubt, how far it is even useful to the plant or animal in which it is found. This is said to be the case with the spleen; which has been extracted from dogs, without any sensible injury to their vital functions. Instances of the former kind, namely, in which we cannot explain the operation, may be numerous; for they will be so in proportion to our ignorance. They will be more or fewer to different persons, and in different stages of science. Every improvement of knowledge diminishes their number. There is hardly, perhaps, a year passes, that does not, in the works of nature, bring some operation, or some mode of operation, to light, which was before undiscovered, probably unsuspected. Instances of the second kind, namely, where the part appears to be totally useless, I believe to be extremely rare: compared with the number of those, of which the use is evident, they are beneath any assignable proportion;

and, perhaps, have never been submitted to a trial and examination sufficiently accurate, long enough continued, or often enough repeated. No accounts which I have seen are satisfactory. The mutilated animal may live and grow fat, as was the case of the dog deprived of its spleen, yet may be defective in some other of its functions; which, whether they can all, or in what degree of vigour and perfection, be performed, or how long preserved, without the extirpated organ, does not seem to be ascertained by experiment. But to this case, even were it fully made out, may be applied the consideration which we suggested concerning the watch, viz. that these superfluous parts do not negative the reasoning which we instituted concerning those parts which are useful, and of which we know the use. The indication of contrivance, with respect to them, remains as it was before.

III. One atheistic way of replying to our observations upon the works of nature, and to the proofs of a Deity which we think that we perceive in them, is to tell us, that all which we see must necessarily have had some form, and that it might as well be its present form as any other. Let us now apply this answer to the eye, as we did before to the watch. Something or other must have occupied that place in the animal's head; must have filled up, we will say, that socket: we will say also, that it must have been of that sort of substance which we call animal substance, as flesh, bone, membrane, cartilage, etc. but that it should have been an *eye*, knowing as we do what an eye comprehends, viz. that it should have consisted, first, of a series of transparent lenses (very different, by the bye, even in their substance, from the opaque materials of which the rest of the body is, in general at least, composed; and with which the whole of its surface, this single portion of it excepted, is covered): secondly, of a black cloth or canvass (the only membrane of the body which is black) spread out behind these lenses, so as to receive the image formed by pencils of light transmitted through them; and placed at the precise geometrical distance at which, and at which alone, a distinct image could be formed, namely, at the concourse of the refracted rays: thirdly, of a large nerve communicating between this membrane and the brain; without which the action of light upon the membrane, however modified by the organ, would be lost to the purposes of sensation. That this fortunate conformation of parts should have been the lot, not of one individual out of many thousand individuals, like the great prize in a lottery, or like some singularity in nature, but the happy chance of a whole species; nor of one species out of many thousand species, with which we are acquainted, but of by far the greatest number of all that exist; and that under varieties, not casual or capricious, but bearing marks of being suited to their respective exigences; that all this should have taken place, merely because something must have occupied those points in every animal's fore-head; or, that all this should be thought to be accounted for, by the short answer, 'that whatever was there must have had some form or other,' is too absurd to be made more so by any argumentation. We are not contented with this answer, we find no satisfaction in it, by way of accounting for appearances of organization far short of those of the eye, such as we observe in fossil shells, petrified bones, or other substances which bear the vestiges of animal or vegetable rec-rements, but which, either in respect of utility, or of the situation in which they are discovered, may seem accidental enough. It is no way of accounting even for these things, to say that the stone, for instance, which is shewn to us, (supposing the question to be con-cerning a petrification,) must have contained some internal con-formation or other. Nor does it mend the answer to add, with respect to the singularity of the conformation, that, after the event, it is no longer to be computed what the chances were against it. This is always to be computed, when the question is whether an useful or imitative conformation be the produce of chance or not. I desire no greater certainty in reasoning, than that by which chance is excluded from the present disposition of the natural world.* Universal experi-ence is against it. What does chance ever do for us? In the human body, for instance, chance, i. e. the operation of causes without design, may produce a wen, a wart, a mole, a pimple, but never an eye. Amongst inanimate substances, a clod, a pebble, a liquid drop, might be; but never was a watch, a telescope, an organized body of any kind, answering a valuable purpose by a complicated mechan-ism, the effect of chance. In no assignable instance hath such a thing existed without intention somewhere.

IV. There is another answer which has the same effect as the resolving of things into chance; which answer would persuade us to believe, that the eye, the animal to which it belongs, every other animal, every plant, indeed every organized body which we see, are only so many out of the possible varieties and combinations of being, which the lapse of infinite ages has brought into existence;* that the present world is the relict of that variety; millions of other bodily forms and other species having perished, being by the defect of their constitution incapable of preservation, or of continuance by generation. Now there is no foundation whatever for this conjecture in anything which we observe in the works of nature: no such experiments are going on at present; no such energy operates as that which is here supposed, and which should be constantly pushing into existence new varieties of beings, nor are there any appearances to support an opinion, that every possible combination of vegetable or animal structure has formerly been tried. Multitudes of conformations, both of vegetables and animals, may be conceived capable of existence and succession, which yet do not exist. Perhaps almost as many forms of plants might have been found in the fields, as figures of plants can be delineated upon paper. A countless variety of animals might have existed which do not exist. Upon the supposition here stated, we should see unicorns and mermaids, sylphs and centaurs; the fancies of painters and the fables of poets realized by examples. Or, if it be alledged that these may transgress the limits of possible life and propagation, we might, at least, have nations of human beings without nails upon their fingers, with more or fewer fingers and toes than ten, some with one eye, others with one ear, with one nostril, or without the sense of smelling at all. All these, and a thousand other imaginable varieties, might live and propagate. We may modify any one species many different ways, all consistent with life, and with the actions necessary to preservation, although affording different degrees of conveniency and enjoyment to the animal. And if we carry these modifications through the different species which are known to subsist, their number would be incalculable. No reason can be given why, if these deperdits ever existed, they have now disappeared. Yet, if all possible existences have been tried, they must have formed part of the catalogue.

But, moreover, the division of organized substances into animals and vegetables, and the distribution and sub-distribution of each into genera and species, which distribution is not an arbitrary act of the mind, but is founded in the order which prevails in external nature, appear to me to contradict the supposition of the present world being the remains of an indefinite variety of existences; of a variety which rejects all plan. The hypothesis teaches, that every possible variety of being hath, at one time or other, found its way into existence (by what cause or in what manner is not said), and that those which were badly formed, perished: but how or why those which survived should be cast, as we see that plants and animals are cast, into regular classes, the hypothesis does not explain; or rather the hypothesis is inconsistent with this phænomenon.

The hypothesis, indeed, is hardly deserving of the consideration which we have given to it. What should we think of a man, who, because we had never ourselves seen watches, telescopes, stocking-mills,* steam-engines, etc. made; knew not how they were made; or could prove by testimony when they were made, or by whom;— would have us believe that these machines, instead of deriving their curious structures from the thought and design of their inventors and contrivers, in truth derive them from no other origin than this; that, a mass of metals and other materials having run when melted into all possible figures, and combined themselves in all possible forms and shapes and proportions, these things which we see, are what were left from the accident, as best worth preserving; and, as such, are become the remaining stock of a magazine, which, at one time or other, has, by this means, contained every mechanism, useful and useless, convenient and inconvenient, into which such like materials could be thrown? I cannot distinguish the hypothesis as applied to the works of nature, from this solution, which no one would accept, as applied to a collection of machines.

V. To the marks of contrivance discoverable in animal bodies, and to the argument deduced from them, in proof of design, and of a designing Creator, this turn is sometimes attempted to be given, viz. that the parts were not intended for the use, but that the use arose out of the parts. This distinction is intelligible. A cabinet-maker rubs his mahogany with fish-skin;* yet it would be too much to assert that the skin of the dog fish was made rough and granulated on purpose for the polishing of wood, and the use of cabinet-makers. Therefore the distinction is intelligible. But I think that there is very little place for it in the works of nature. When roundly and generally affirmed of them, as it hath sometimes been, it amounts to such another stretch of assertion, as it would be to say, that all the implements of the cabinet-maker's workshop, as well as his fish-skin, were substances accidentally configurated, which he had picked up, and converted to his use; that his adzes, saws, planes, and gimlets, were not made, as we suppose, to hew, cut, smooth, shape out, or bore wood with; but that, these things being made, no matter with what design, or whether with any, the cabinet-maker perceived that they were applicable to his purpose, and turned them to account.

But, again; so far as this solution is attempted to be applied to those parts of animals the action of which does not depend upon the will of the animal, it is fraught with still more evident absurdity. Is it possible to believe that the eye was formed without any regard to vision; that it was the animal itself which found out, that, though formed with no such intention, it would serve to see with; and that the use of the eye, as an organ of sight, resulted from this discovery, and the animal's application of it? The same question may be asked of the ear; the same of all the senses. None of the senses fundamentally depend upon the election of the animal: consequently neither upon his sagacity, nor his experience. It is the impression which objects make upon them that constitutes their use. Under that impression he is passive. He may bring objects to the sense, or within its reach; he may select these objects; but over the impression itself he has no power, or very little; and that properly is the sense.

Secondly, there are many parts of animal bodies which seem to depend upon the will of the animal in a greater degree than the senses do, and yet with respect to which this solution is equally unsatisfactory. If we apply the solution to the human body, for instance, it forms itself into questions upon which no reasonable mind can doubt; such as, whether the teeth were made expressly for the mastication of food, the feet for walking, the hands for holding; or whether, these things being as they are, being in fact in the animal's possession, his own ingenuity taught him that they were convertible to these purposes, though no such purposes were contemplated in their formation.

All that there is of the appearance of reason in this way of considering the subject is, that, in some cases, the organization seems to determine the habits of the animal, and its choice, to a particular mode of life; which, in a certain sense, may be called 'the use arising out of the part.' Now to all the instances, in which there is any place for this suggestion, it may be replied, that the organization determines the animal to habits beneficial and salutary to itself; and that this effect would not be seen so regularly to follow, if the several organizations did not bear a concerted and contrived relation to the substances by which the animal was surrounded. They would, otherwise, be capacities without objects; powers without employment. The web foot determines, you say, the duck to swim: but what would that avail, if there were no water to swim in? The strong, hooked bill, and sharp talons, of one species of bird, determine it to prey upon animals; the soft straight bill, and weak claws, of another species, determine it to pick up seeds: but neither determination could take effect in providing for the sustenance of the birds, if animal bodies and vegetable seeds did not lie within their reach. The peculiar conformation of the bill, and tongue, and claws of the woodpecker, determines that bird to search for his food amongst the insects lodged behind the bark, or in the wood, of decayed trees; but what would this profit him if there were no trees, no decayed trees, no insects lodged under their bark, or in their trunk? The proboscis with which the bee is furnished, determines him to seek for honey; but what would that signify, if flowers supplied none? Faculties thrown down upon animals at random, and without reference to the objects amidst which they are placed, would not produce to them the services and benefits which we see: and if there be that reference, then there is intention.

Lastly, the solution fails entirely when applied to plants. The parts of plants answer their uses, without any concurrence from the will or choice of the plant.

VI. Others have chosen to refer every thing to a *principle of order* in nature. A principle of order is the word: but what is meant by a principle of order, as different from an intelligent Creator, has not been explained either by definition or example: and, without such explanation, it should seem to be a mere substitution of words for reasons, names for causes. Order itself is only the adaptation of means to an end: a principle of order therefore can only signify the mind and intention which so adapts them. Or, were it capable of being explained in any other sense, is there any experience, any analogy, to sustain it? Was a watch ever produced by a principle of order? and why might not a watch be so produced, as well as an eye?

Furthermore, a principle of order, acting blindly and without choice, is negatived by the observation, that order is not universal; which it would be, if it issued from a constant and necessary principle; nor indiscriminate, which it would be, if it issued from an unintelligent principle. Where order is wanted, there we find it;

where order is not wanted, i. e. where, if it prevailed, it would be useless, there we do not find it. In the structure of the eye (for we adhere to our example), in the figure and position of its several parts, the most exact order is maintained. In the forms of rocks and mountains, in the lines which bound the coasts of continents and islands, in the shape of bays and promontories, no order whatever is perceived, because it would have been superfluous. No useful purpose would have arisen from moulding rocks and mountains into regular solids, bounding the channel of the ocean by geometrical curves, or from the map of the world resembling a table of diagrams in Euclid's Elements or Simpson's Conic Sections.*

VII. Lastly, the confidence which we place in our observations upon the works of nature, in the marks which we discover of contrivance, choice, and design, and in our reasoning upon the proofs afforded us, ought not to be shaken, as it is sometimes attempted to be done, by bringing forward to our view our own ignorance, or rather the general imperfection of our knowledge, of nature. Nor, in many cases, ought this consideration to affect us, even when it respects some parts of the subject immediately under our notice. True fortitude of understanding consists in not suffering what we know to be disturbed by what we do not know. If we perceive an useful end, and means adapted to that end, we perceive enough for our conclusion. If these things be clear, no matter what is obscure. The argument is finished. For instance; if the utility of vision to the animal which enjoys it, and the adaptation of the *eye* to this office be evident and certain (and I can mention nothing which is more so), ought it to prejudice the inference which we draw from these premises, that we cannot explain the use of the spleen? Nay more; if there be parts of the eye, viz. the cornea, the crystalline, the retina, in their substance, figure and position, manifestly suited to the formation of an image by the refraction of rays of light, at least as manifestly as the glasses and tubes of a dioptric telescope are suited to that purpose, it concerns not the proof which these afford of design and of a designer, that there may perhaps be other parts, certain muscles, for instance, or nerves, in the same eye, of the agency or effect of which we can give no account; any more than we should be inclined to doubt, or ought to doubt, about the construction of a telescope, viz. for what purpose it was constructed, or whether it were constructed at all, because there belonged to it certain screws and pins, the use or action of which we did not comprehend. I take it to be a general way of infusing doubts and scruples into the mind, to recall to it its own ignorance, its own imbecility; to tell us that upon these subjects we know little; that little imperfectly; or rather, that we know nothing properly about the matter. These suggestions so fall in with our consciousnesses, as sometimes to produce a general distrust of our faculties and our conclusions. But this is an unfounded jealousy. The uncertainty of one thing, does not necessarily affect the certainty of another thing. Our ignorance of many points need not suspend our assurance of a few. Before we yield, in any particular instance, to the scepticism which this sort of insinuation would induce, we ought accurately to ascertain, whether our ignorance or doubt concern those precise points upon which our conclusion rests. Other points are nothing. Our ignorance of other points may be of no consequence to these; though they be points, in various respects, of great importance. A just reasoner removes from his consideration, not only what he knows, but what he does not know, touching matters not strictly connected with his argument, i. e. not forming the very steps of his deduction: beyond these, his knowledge and his ignorance are alike irrelative.

CHAPTER VI

THE ARGUMENT CUMULATIVE

WERE there no example in the world of contrivance except that of the *eye*, it would be alone sufficient to support the conclusion which we draw from it, as to the necessity of an intelligent Creator. It could never be got rid of: because it could not be accounted for by any other supposition, which did not contradict all the principles we possess of knowledge; the principles according to which, things do, as often as they can be brought to the test of experience, turn out to be true or false. Its coats and humours, constructed, as the lenses of a telescope are constructed, for the refraction of rays of light to a point, which forms the proper action of the organ; the provision in its muscular tendons for turning its pupil to the object, similar to that which is given to the telescope by screws, and upon which power of direction in the eye, the exercise of its office as an optical instrument depends; the further provision for its defence, for its constant lubricity and moisture, which we see in its socket and its lids, in its gland for the secretion of the matter of tears, its outlet or communication with the nose for carrying off the liquid after the eye is washed with it; these provisions compose altogether an apparatus, a system of parts, a preparation of means, so manifest in their design, so exquisite in their contrivance, so successful in their issue, so precious and so infinitely beneficial in their use, as, in my opinion, to bear down all doubt that can be raised upon the subject. And what I wish, under the title of the present chapter, to observe, is, that, if other parts of nature were inaccessible to our enquiries, or even if other parts of nature presented nothing to our examination but disorder and confusion, the validity of this example would remain the same. If there were but one watch in the world, it would not be less certain that it had a maker. If we had never in our lives seen any but one single kind of hydraulic machine;* yet, if of that one kind we understood the mechanism and use, we should be as perfectly assured that it proceeded from the hand, and thought, and skill of a workman, as if we visited a museum of the arts, and saw collected there twenty different kinds of machines for drawing water, or a thousand different kinds for other purposes. Of this point each machine is a proof, independently of all the rest. So it is with the evidences of a divine agency. The proof is not a conclusion, which lies at the end of a chain of reasoning, of which chain each instance of contrivance is only a link, and of which, if one link fail, the whole falls; but it is an argument separately supplied by every separate example. An error in stating an example affects only that example. The argument is cumulative in the fullest sense of that term.* The eye proves it without the ear; the ear without the eye. The proof in each example is complete; for when the design of the part, and the conduciveness of its structure to that design, is shewn, the mind may set itself at rest: no future consideration can detract any thing from the force of the example.

CHAPTER VII

OF THE MECHANICAL AND IMMECHANICAL PARTS AND
FUNCTIONS OF ANIMALS AND VEGETABLES

IT is not that *every* part of an animal or vegetable has not proceeded from a contriving mind; or that every part is not constructed with a view to its proper end and purpose, according to the laws belonging to, and governing, the substance or the action made use of in that part; or that each part is not so constructed, as to effectuate its purpose whilst it operates according to these laws: but it is, because these laws themselves are not in all cases equally understood; or, what amounts to nearly the same thing, are not equally exemplified in more simple processes, and more simple machines; that we lay down the distinction, here proposed, between the mechanical parts, and other parts, of animals and vegetables.*

For instance; the principle of muscular motion, viz. upon what cause the swelling of the belly of the muscle, and consequent contraction of its tendons, either by an act of the will or by involuntary irritation, depends, is wholly unknown to us. The substance employed, whether it be fluid, gaseous, elastic, electrical,* or none of these, or nothing resembling these, is also unknown to us: of course the laws belonging to that substance, and which regulate its action, are unknown to us. We see nothing similar to this contraction in any machine which we can make, or any process which we can execute. So far (it is confessed) we are in ignorance: but no further. This power and principle, from whatever cause it proceeds, being assumed, the collocation of the fibres to receive the principle, the disposition of the muscles for the use and application of the power, is mechanical; and is as intelligible as the adjustment of the wires and strings by which a puppet is moved. We see therefore, as far as respects the subject before us, what is not mechanical in the animal frame, and what is. The nervous influence (for we are often obliged to give names to things which we know little about)—I say the nervous influence, by which the belly or middle of the muscle is swelled, is not mechanical. The utility of the effect we perceive; the means, or the preparation of means, by which it is produced, we do not. But obscurity as to the origin of muscular motion brings no doubtfulness into our observations upon the sequel of the process. Which observations relate, 1st, to the constitution of the muscle; in consequence of which constitution, the swelling of the belly or middle part is necessarily and mechanically followed by a contraction of the tendons: 2dly, to the number and variety of the muscles, and the corresponding number and variety of useful powers which they supply to the animal; which is astonishingly great: 3dly, to the judicious (if we may be permitted to use that term, in speaking of the author, or of the works, of nature), to the wise and well contrived disposition of each muscle for its specific purpose; for moving the joint this way, and that way, and the other way; for pulling and drawing the part, to which it is attached, in a determinate and particular direction; which is a mechanical operation, exemplified in a multitude of instances. To mention only one; The tendon of the trochlear muscle of the eye, to the end that it may draw in the line required, is passed through a cartilaginous ring, at which it is reverted, exactly in the same manner as a rope in a ship is carried over a block or round a stay, in order to make it pull in the direction which is wanted. All this, as we have said, is mechanical; and is as accessible to inspection, as capable of being ascertained, as the mechanism of the automaton in the Strand.* Suppose the automaton to be put in motion by a magnet (which is probable), it will supply us with a comparison very apt for our present purpose. Of the magnetic effluvium* we know perhaps as little as we do of the nervous fluid. But magnetic attraction being assumed (it signifies nothing from what cause it proceeds), we can trace, or there can be pointed out to us, with perfect clearness and certainty, the mechanism, viz. the steel bars, the wheels, the joints, the wires, by which the motion so much admired is communicated to the fingers of the image: and to make any obscurity, or difficulty, or contraversy in the doctrine of magnetism, an objection to our knowledge or our certainty concerning the contrivance, or the marks of contrivance, displayed in the automaton, would be exactly the same thing, as it is to make our ignorance (which we acknowledge) of the cause of nervous agency,* or even of the substance and structure of the nerves themselves, a ground of question or suspicion as to the reasoning which we institute concerning the mechanical part of our frame. That an animal is a machine, is a proposition neither correctly true, nor wholly false. The distinction which we have been discussing will serve to shew how far the comparison, which this expression implies, holds; and wherein it fails. And, whether the distinction be thought of importance or not, it is certainly of importance to remember, that there is neither truth nor justice in endeavouring to bring a cloud over our understandings, or a distrust into our reasonings upon this subject, by suggesting that we know nothing of voluntary motion, of irritability, of the principle of life, of sensation, of animal heat,* upon all which the animal functions depend; for our ignorance of these parts of the animal frame concerns not at all our knowledge of the mechanical parts of the same frame. I contend, therefore, that there is mechanism in animals; that this mechanism is as properly such, as it is in machines made by art; that this mechanism is intelligible and certain; that it is not the less so, because it often begins or terminates with something which is not mechanical; that whenever it is intelligible and certain, it demonstrates intention and contrivance, as well in the works of nature as in those of art; and that it is the best demonstration which either can afford.

But, whilst I contend for these propositions, I do not exclude myself from asserting that there may be, and that there are, other cases, in which, although we cannot exhibit mechanism, or prove indeed that mechanism is employed, we want not sufficient evidence to conduct us to the same conclusion.

There is what may be called the *chymical* part of our frame; of which, by reason of the imperfection of our chymistry, we can attain to no distinct knowledge: I mean, not to a knowledge, either in degree or kind, similar to that which we possess of the mechanical part of our frame. It does not therefore afford the same species of argument as that which mechanism affords; and yet it may afford an argument in a high degree satisfactory. The *gastric juice*, or the liquor which digests the food in the stomachs of animals, is of this class. Of all menstrua* it is the most active, the most universal. In the human stomach, for instance, consider what a variety of strange substances, and how widely different from one another, it, in a few hours, reduces to one uniform pulp, milk, or mucilage. It seizes upon every thing, it dissolves the texture of almost every thing, that comes in its way. The flesh of perhaps all animals; the seeds and fruits of the greatest number of plants; the roots and stalks and leaves of many, hard and tough as they are, yield to its powerful pervasion. The change wrought by it is different from any chymical solution which we can produce, or with which we are acquainted, in this respect as well as many others, that, in our chymistry, particular menstrua act only upon particular substances. Consider moreover that this fluid, stronger in its operation than a caustic alkali or mineral acid, than red precipitate or aqua fortis itself,* is nevertheless as mild, and bland, and inoffensive to the touch or taste, as saliva or gum water, which it much resembles. Consider, I say, these several properties of the digestive organ, and of the juice with which it is supplied, or rather with which it is made to supply itself, and you will confess it to be entitled to a name, which it has sometimes received, that of 'the chymical wonder of animal nature.'

Still we are ignorant of the composition of this fluid, and of the mode of its action; by which is meant that we are not capable, as we are in the mechanical part of our frame, of collating it with the operations of art. And this I call the imperfection of our chymistry; for, should the time ever arrive, which is not perhaps to be despaired of, when we can compound ingredients, so as to form a solvent, which will act in the manner in which the gastric juice acts, we may be able to ascertain the chymical principles upon which its efficacy depends, as well as from what part, and by what concoction, in the human body, these principles are generated and derived.

In the mean time, ought that, which is in truth the defect of our chymistry, to hinder us from acquiescing in the inference, which a production of nature, by its place, its properties, its action, its surprising efficacy, its invaluable use, authorises us to draw in respect of a creative design?

Another most subtle and curious function of animal bodies is secretion. This function is semi-chymical and semi-mechanical; exceedingly important and diversified in its effects, but obscure in its process and in its apparatus. The importance of the secretory organs is but too well attested by the diseases, which an excessive, a deficient, or a vitiated secretion is almost sure of producing. A single secretion being wrong, is enough to make life miserable, or sometimes to destroy it. Nor is the variety less than the importance. From one and the same blood (I speak of the human body) about twenty different fluids are separated; in their sensible properties, in taste, smell, colour, and consistency, the most unlike one another that is possible; thick, thin, salt, bitter, sweet: and, if from our own we pass to other species of animals, we find amongst their secretions not only the most various, but the most opposite properties; the most nutritious aliment, the deadliest poison; the sweetest perfumes, the most fetid odours. Of these the greater part, as the gastric juice, the saliva, the bile, the slippery mucilage which lubricates the joints, the tears which moisten the eye, the wax which defends the ear, are, after they are secreted, made use of in the animal œconomy; are evidently subservient, and are actually contributing to the utilities of the animal itself. Other fluids seem to be separated only to be rejected. That this also is necessary (though why it was originally necessary, we cannot tell) is shewn by the consequence of the separation being long suspended; which consequence is disease and death. Akin to secretion, if not the same thing, is assimilation, by which one and the same blood is converted into bone, muscular flesh, nerves, membranes, tendons; things as different as the wood and iron, canvass and cordage, of which a ship with its furniture is composed. We have no operation of art wherewith exactly to compare all this, for no other reason perhaps than that all operations of art are exceeded by it. No chymical election, no chymical analysis* or resolution of a substance into its constituent parts, no mechanical sifting or division, that we are acquainted with, in perfection or variety come up to animal secretion. Nevertheless the apparatus and process are obscure; not to say, absolutely concealed from our enquiries. In a few, and only a few instances, we can discern a little of the constitution of a gland. In the kidneys of large animals we can trace the emulgent artery* dividing itself into an infinite number of branches; their extremities every where communicating with little round bodies, in the substance of which bodies the secret of the machinery seems to reside, for there the change is made. We can discern pipes laid from these round bodies towards the pelvis, which is a bason within the solid of the kidney. We can discern these pipes joining and collecting together into larger pipes; and when so collected, ending in innumerable papillæ,* through which the secreted fluid is continually oozing into its receptacle. This is all we know of the mechanism of a gland, even in the case in which it seems most capable of being investigated. Yet to pronounce that we know nothing of animal secretion, or nothing satisfactorily, and with that concise remark to dismiss the article from our argument, would be to dispose of the subject very hastily and very irrationally. For the purpose which we want, that of evincing intention, we know a great deal. And what we know is this. We see the blood carried by a pipe, conduit, or duct, *to* the gland.* We see an organized apparatus, be its construction or action what it will, which we call that gland. We see the blood, or part of the blood, after it has passed through and undergone the action of the gland, coming *from* it by an emulgent vein or artery, i. e. by another pipe or conduit. And we see also at the same time a new and specific fluid issuing from the same gland by its excretory duct, i. e. by a third pipe or conduit; which new fluid is in some cases discharged out of the body, in more cases retained within it, and there executing some important and intelligible office. Now supposing, or admitting, that we know nothing of the proper internal constitution of a gland, or of the mode of its acting upon the blood; then our situation is precisely like that of an unmechanical looker-on, who stands by a stocking-loom, a corn-mill, a carding-machine, or a threshing-machine* at work, the fabric and mechanism of which, as well as all that passes within, is hidden from his sight by the outside case; or, if seen, would be too complicated for his uninformed, uninstructed understanding to comprehend. And what is that situation? This spectator, ignorant as he is, sees at one end a material enter the machine, as unground grain the mill, raw cotton the carding-machine, sheaves of unthreshed corn the threshing-machine; and, when he casts his eye to the other end of the apparatus, he sees the material issuing from it in a new state; and, what is more, in a state manifestly adapted to future uses; the grain in meal fit for the making of bread, the wool in rovings* ready for spinning into threads, the sheaf in corn dressed for the mill. Is it necessary that this man, in order to be convinced, that design, that intention, that contrivance has been employed about the machine, should be allowed to pull it in pieces; should be enabled to examine the parts separately; explore their action upon one another, or their operation, whether simultaneous or successive, upon the material which is presented to them? He may long to do this to gratify his curiosity; he may desire to do it to improve his theoretic knowledge; or he may have a more substantial reason for requesting it, if he happen, instead of a common visitor, to be a mill-wright* by profession, or a person sometimes called in to repair such-like machines when out of order; but, for the purpose of ascertaining the existence of counsel and design in the formation of the machine, he wants no such intromission or privity. What he sees is sufficient. The effect upon the material, the change produced in it, the utility of that change for future applications, abundantly testify, be the concealed part of the machine or of its construction what it will, the hand and agency of a contriver. If any confirmation were wanting to the evidence which the animal secretions afford of design, it may be derived, as hath been already hinted, from their variety, and from their appropriation to their place and life. They all come from the same blood; they are all drawn off by glands; yet the produce is very different, and the difference exactly adapted to the work which is to be done, or the end to be answered. No account can be given of this without resorting to appointment. Why, for instance, is the saliva, which is diffused over the seat of taste, insipid, whilst so many others of the secretions, the urine, the tears, and the sweat, are salt? Why does the gland within the ear separate a viscid substance, which defends that passage; the gland in the upper angle of the eye, a thin brine, which washes the ball? Why is the synovia of the joints mucilaginous; the bile bitter, stimulating, and soapy? Why does the juice, which flows into the stomach, contain powers, which make that bowel, the great laboratory, as it is by its situation the recipient, of the materials of future nutrition? These are all fair questions; and no answer can be given to them, but what calls in intelligence and intention.

My object in the present chapter has been to teach three things: first, that it is a mistake to suppose, that, in reasoning from the appearances of nature, the imperfection of our knowledge proportionably affects the certainty of our conclusion; for in many cases it does not affect it at all: secondly, that the different parts of the animal frame may be classed and distributed, according to the degree of exactness with which we can compare them with works of art: thirdly, that the mechanical parts of our frame, or, those in which this comparison is most complete, although constituting, probably, the coarsest portions of nature's workmanship, are the properest to be alledged as proofs and specimens of design.

CHAPTER VIII

OF MECHANICAL ARRANGEMENT IN THE
HUMAN FRAME

WE proceed therefore to propose certain examples taken out of this class; making choice of such, as, amongst those which have come to our knowledge, appear to be the most striking, and the best understood; but obliged, perhaps, to postpone both these recommendations to a third, that of the example being capable of explanation without plates or figures, or technical language.*

Of the Bones

I. I challenge any man to produce, in the joints and pivots of the most complicated, or the most flexible, machine, that was ever contrived, a construction more artificial, or more evidently artificial, than that which is seen in the vertebræ of the *human neck*. Two things were to be done. The head was to have the power of bending forward and backward, as in the act of nodding, stooping, looking upward or downward; and, at the same time, of turning itself round upon the body to a certain extent, the quadrant* we will say, or rather, perhaps, a hundred and twenty degrees of a circle. For these two purposes, two distinct contrivances are employed. First, The head rests immediately upon the uppermost of the vertebræ, and is united to it by a *hinge* joint; upon which joint the head plays freely forward and backward, as far either way as is necessary, or as the ligaments allow: which was the first thing required. But then the rotatory motion is unprovided for. Therefore, secondly, to make the head capable of this, a further mechanism is introduced; not between the head and the uppermost bone of the neck, where the hinge is, but between that bone, and the bone next underneath it. It is a mechanism resembling a *tenon and mortice*.* This second, or uppermost bone but one, has what anatomists call a process, viz. a projection, somewhat similar, in size and shape, to a tooth; which tooth, entering a corresponding hole or socket in the bone above it, forms a pivot or axle, upon which that upper bone, together with the head which it supports, turns freely in a circle; and as far in the circle, as the attached muscles permit the head to turn. Thus are both motions perfect, without interfering with each other. When we nod the head, we use the hinge joint, which lies between the head and the first bone of the neck. When we turn the head round, we use the tenon and mortice, which runs between the first bone of the neck and the second. We see the same contrivance, and the same principle, employed in the frame or mounting of a telescope. It is occasionally requisite, that the object end of the instrument be moved up and down, as well as horizontally, or equatorially. For the vertical motion there is a hinge upon which the telescope plays: for the horizontal or equatorial motion, an axis upon which the telescope and the hinge turn round together. And this is exactly the mechanism which is applied to the motion of the head: nor will any one here doubt of the existence of counsel and design, except it be by that debility of mind, which can trust to its own reasonings in nothing.

We may add, that it was, on another account also, expedient, that the motion of the head backward and forward should be performed upon the upper surface of the first vertebra: for, if the first vertebra itself had bent forward, it would have brought the spinal marrow, at the very beginning of its course, upon the point of the tooth.

II. Another mechanical contrivance, not unlike the last in its object, but different and original in its means, is seen in what anatomists call the *fore-arm*; that is, in the arm between the elbow and the wrist. Here, for the perfect use of the limb, two motions are wanted; a motion at the elbow backward and forward, which is called a reciprocal motion; and a rotatory motion, by which the palm of the hand, as occasion requires, may be turned upward. How is this managed? The fore-arm, it is well known, consists of two bones, lying along-side each other, but touching only towards the ends. One, and only one, of these bones, is joined to the cubit, or upper part of the arm, at the elbow; the other alone, to the hand at the wrist. The first, by means, at the elbow, of a hinge joint (which allows only of motion in the same plane), swings backward and forward, carrying along with it the other bone, and the whole fore-arm. In the mean time, as often as there is occasion to turn the palm upward, that other bone, to which the hand is attached, rolls upon the first, by the help of a groove or hollow near each end of one bone, to which is fitted a corresponding prominence in the other. If both bones had been joined to the cubit or upper arm at the elbow, or both to the hand at the wrist, the thing could not have been done. The first was to be at liberty at one end, and the second at the other: by which means the two actions may be performed together. The great bone, which carries the fore-arm, may be swinging upon its hinge at the elbow, at the very time, that the lesser bone, which carries the hand, may be turning round it in the grooves. The management also of these grooves, or rather of the tubercles and grooves, is very observable. The two bones are called the radius and the ulna. Above, i. e. towards the elbow, a tubercle of the radius plays into a socket of the ulna; whilst below, i. e. towards the wrist, the radius finds the socket, and the ulna the tubercle. A single bone in the fore-arm, with a ball and socket joint at the elbow, which admits of motion in all directions, might, in some degree, have answered the purpose, of both moving the arm, and turning the hand. But how much better it is accomplished by the present mechanism, any person may convince himself, who puts the ease and quickness, with which he can shake his hand at the wrist circularly (moving likewise, if he please, his arm at the elbow at the same time), in competition with the comparatively slow and laborious motion, with which his arm can be made to turn round at the shoulder, by the aid of a ball and socket joint.

III. The *spine* or back bone is a chain of joints of very wonderful construction. Various, difficult, and almost inconsistent offices were to be executed by the same instrument. It was to be firm, yet flexible; now I know no chain made by art, which is both these; for by firmness I mean, not only strength, but stability); *firm*, to support the erect position of the body; *flexible*, to allow of the bending of the trunk in all degrees of curvature. It was further also, which is another, and quite a distinct purpose from the rest, to become a pipe or conduit for the safe conveyance from the brain of the most important fluid of the animal frame, that, namely, upon which all voluntary motion depends, the spinal marrow; a substance, not only of the first necessity to action, if not to life, but of a nature so delicate and tender, so susceptible and so impatient of injury, as that any unusual pressure upon it, or any considerable obstruction of its course, is followed by paralysis or death. Now the spine was not only to furnish the main trunk for the passage of the medullary substance* from the brain, but to give out, in the course of its progress, small pipes therefrom, which, being afterwards indefinitely subdivided, might, under the name of nerves, distribute this exquisite supply to every part of the body. The same spine was also to serve another use not less wanted than the preceding, viz. to afford a fulcrum, stay, or basis (or more properly speaking a series of these) for the insertion of the muscles which are spread over the trunk of the body; in which trunk there are not, as in the limbs, cylindrical bones, to which they can be fastened, and, likewise, which is a similar use, to furnish a support for the ends of the ribs to rest upon.

Bespeak of a workman a piece of mechanism which shall comprise all these purposes, and let him set about to contrive it; let him try his skill upon it; let him feel the difficulty of accomplishing the task, before he be told how the same thing is effected in the animal frame. Nothing will enable him to judge so well of the wisdom which has been employed: nothing will dispose him to think of it so truly. First, for the firmness, yet flexibility, of the spine, it is composed of a great number of bones (in the human subject of twenty-four) joined to one another, and compacted together, by broad bases. The breadth of the bases upon which the parts severally rest, and the closeness of the junction, give to the chain its firmness and stability: the number of parts, and consequent frequency of joints, its flexibility. Which flexibility, we may also observe, varies in different parts of the chain: is least in the back, where strength more than flexure is wanted; greater in the loins, which it was necessary should be more supple than the back; and greatest of all in the neck, for the free motion of the head. Then, secondly, in order to afford a passage for the descent of the medullary substance, each of these bones is bored through in the middle in such a manner, as that, when put together, the hole in one bone falls into a line, and corresponds, with the holes in the two bones contiguous to it. By which means, the perforated pieces, when joined, form an entire, close, uninterrupted channel: at least whilst the spine is upright and at rest. But, as a settled posture is inconsistent with its use, a great difficulty still remained, which was to prevent the vertebræ shifting upon one another, so as to break the line of the canal as often as the body moves or twists; or the joints gaping externally, whenever the body is bent forward, and the spine, thereupon, made to take the form of a bow. These dangers, which are mechanical, are mechanically provided against. The vertebræ, by means of their processes and projections, and of the articulations which some of these form with one another at their extremities, are so locked in and confined, as to maintain, in what are called the bodies or broad surfaces of the bones, the relative position nearly unaltered; and to throw the change and the pressure, produced by flexion, almost entirely upon the intervening cartilages, the springiness and yielding nature of whose substance admits of all the motion which is necessary to be performed upon them, without any chasm being produced by a separation of the parts. I say of all the motion which is necessary; for, although we bend our backs to every degree almost of inclination, the motion of each vertebra is very small; such is the advantage which we receive from the chain being composed of so many links, the spine of so many bones. Had it consisted of three or four bones only, in bending the body the spinal marrow must have been bruised at every angle. The reader need not be told that these intervening cartilages are gristles; and he may see them in perfection in a loin of veal. Their form also favors the same intention. They are thicker before than behind, so that, when we stoop forward, the compressible substance of the cartilage, yielding in its thicker and anterior part to the force which squeezes it, brings the surfaces of the adjoining vertebræ nearer to the being parallel with one another than they were before, instead of increasing the inclination of their planes, which must have occasioned a fissure or opening between them. Thirdly, For the medullary canal giving out in its course, and in a convenient order, a supply of nerves to different parts of the body, notches are made in the upper and lower edge of every vertebra; two on each edge; equidistant on each side from the middle line of the back. When the vertebræ are put together, these notches, exactly fitting, form small holes; through which the nerves, at each articulation, issue out in pairs, in order to send their branches to every part of the body, and with an equal bounty to both sides of the body. The fourth purpose assigned to the same instrument, is the insertion of the bases of the muscles, and the support of the ends of the ribs: and for this fourth purpose, especially the former part of it, a figure, specifically suited to the design, and unnecessary for the other purposes, is given to the constituent bones. Whilst they are plain, and round, and smooth towards the front, where any roughness or projection might have wounded the adjacent viscera, they run out, behind, and on each side, into long processes, to which processes the muscles necessary to the motions of the trunk are fixed; and fixed with such art, that, whilst the vertebræ supply a basis for the muscles, the muscles help to keep these bones in their position, or by their tendons to tie them together.

That most important, however, and general property, viz. the strength of the compages,* and the security against luxation,* was to be still more specially consulted; for where so many joints were concerned, and where, in every one, derangement would have been fatal, it became a subject of studious precaution. For this purpose, the vertebræ are articulated, that is, the moveable joints between them are formed, by means of those projections of their substance, which we have mentioned under the name of processes; and these so lock in with, and overwrap, one another, as to secure the body of the vertebra, not only from accidentally slipping, but even from being pushed, out of its place, by any violence short of that which would break the bone. I have often remarked and admired this structure in the chine of a hare.* In this, as in many instances, a plain observer of the animal œconomy may spare himself the disgust of being present at human dissections, and yet learn enough for his information and satisfaction, by even examining the bones of the animals which come upon his table. Let him take, for example, into his hands, a piece of the clean-picked bone of a hare's back; consisting, we will suppose, of three vertebræ. He will find the middle bone of the three, so implicated, by means of its projections or processes, with the bone on each side of it, that no pressure which he can use, will force it out of its place between them. It will give way neither forward, nor backward, nor on either side. In whichever direction he pushes, he perceives, in the form, or junction, or overlapping of the bones, an impediment opposed to his attempt; a check and guard against dis-location. In one part of the spine, he will find a still further fortifying expedient, in the mode according to which the ribs are annexed to the spine. Each rib rests upon two vertebræ. That is the thing to be remarked, and any one may remark it in carving a neck of mutton. The manner of it is this: the end of the rib is divided by a middle ridge into two surfaces, which surfaces are joined to the bodies of two contiguous vertebræ, the ridge applying itself to the intervening cartilage. Now this is the very contrivance which is employed in the famous iron bridge at my door at Bishop-Wearmouth; and for the same purpose of stability; viz. the cheeks of the bars, which pass between the arches, ride across the joints, by which the pieces composing each arch are united. Each cross bar rests upon two of these pieces at their place of junction; and by that position resists, at least in one direction, any tendency in either piece to slip out of its place.

Thus perfectly, by one means or the other, is the danger of slipping laterally, or of being drawn aside out of the *line* of the back provided against: and, to withstand the bones being pulled asunder longitudinally, or in the direction of the line, a strong membrane runs from one end of the chain to the other, sufficient to resist any force which is ever likely to act in the direction of the back, or parallel to it, and consequently to secure the whole combination in their places. The general result is, that not only the motions of the human body necessary for the ordinary offices of life are performed with safety, but that it is an accident hardly ever heard of, that even the gesticulations of a harlequin distort his spine.

Upon the whole, and as a guide to those who may be inclined to carry the consideration of this subject further, there are three views under which the spine ought to be regarded, and in all which it cannot fail to excite our admiration. These views relate to its articulations, its ligaments, and its perforation; and to the corresponding advantages which the body derives from it, for action, for strength, and for that, which is essential to every part, a secure communication with the brain.

The structure of the spine is not in general different in different animals. In the serpent tribe,* however, it is considerably varied; but with a strict reference to the conveniency of the animal. For, whereas in quadrupeds the number of vertebræ is from thirty to forty, in the serpent it is nearly one hundred and fifty: whereas in men and quadrupeds the surfaces of the bones are flat, and these flat surfaces laid one against the other, and bound tight by sinews; in the serpent, the bones play one within another like a ball and socket,[1] so that they have a free motion upon one another in every direction: that is to say, in men and quadrupeds firmness is more consulted; in serpents, pliancy. Yet even pliancy is not obtained at the expense of safety. The back bone of a serpent, for coherence and flexibility, is one of the most curious pieces of animal mechanism, with which we are acquainted. The chain of a watch, (I mean the chain which passes between the spring-barrel and the fusee*) which aims at the same properties, is but a bungling piece of workmanship in comparison with that of which we speak.

IV. The reciprocal enlargement and contraction of the *chest* to

---

[1] Der. Phys. Theol.* p. 396.

allow for the play of the lungs, depends upon a simple yet beautiful mechanical contrivance, referable to the structure of the bones which inclose it. The ribs are articulated to the back bone, or rather to its side projections, *obliquely*; that is, in their natural position they bend or slope from the place of articulation downwards. But the basis upon which they rest at this end being fixed, the consequence of the obliquity, or the inclination downwards, is, that, when they come to move, whatever pulls the ribs upwards, necessarily, at the same time, draws them out; and that, whilst the ribs are brought to a right angle with the spine behind, the sternum, or part of the chest to which they are attached in front, is thrust forward. The simple action, therefore, of the elevating muscles does the business; whereas, if the ribs had been articulated with the bodies of the vertebræ at right angles, the cavity of the thorax could never have been further enlarged by a change of their position. If each rib had been a rigid bone, articulated at both ends to fixed bases, the whole chest had been immovable. Keill* has observed, that the breast-bone, in an easy inspiration, is thrust out one tenth of an inch; and he calculates that this, added to what is gained to the space within the chest by the flattening or descent of the diaphragm, leaves room for forty-two cubic inches of air to enter at every drawing in of the breath. When there is a necessity for a deeper and more laborious inspiration, the enlargement of the capacity of the chest may be so increased by effort, as that the lungs may be distended with seventy or a hundred such cubic inches.[1] The thorax, says Schelhammer,* forms a kind of bellows, such as never have been, nor probably will be, made by any artificer.

V. The *patella*, or knee-pan,* is a curious little bone; in its form and office unlike any other bone of the body. It is circular; the size of a crown piece; pretty thick; a little convex on both sides, and covered with a smooth cartilage. It lies upon the front of the knee; and the powerful tendons, by which the leg is brought forward, pass through it (or rather it makes a part of their continuation) from their origin in the thigh to their insertion in the tibia. It protects both the tendon and the joint from any injury which either might suffer, by the rubbing of one against the other, or by the pressure of unequal surfaces. It also gives to the tendons a very considerable mechanical

[1] Anat. p. 229.

advantage by altering the line of their direction, and by advancing it
further out from the centre of motion; and this upon the principles
of the resolution of force, upon which principles all machinery is
founded. These are its uses. But what is most observable in it is, that
it appears to be supplemental, as it were, to the frame; added, as it
should almost seem, afterward; not quite necessary, but very convenient. It is separate from the other bones; that is, it is not connected with any other bones by the common mode of union. It is
soft, or hardly formed, in infancy; and produced by an ossification,*
of the inception or progress of which, no account can be given from
the structure or exercise of the part.

VI. The *shoulder-blade* is, in some material respects, a very singular bone: it appearing to be made so expressly for its own purpose,
and so independently of every other reason. In such quadrupeds as
have no collar-bones, which are by far the greater number, the
shoulder-blade has no bony communication with the trunk, either by
a joint, or process, or in any other way. It does not grow to, or out of,
any other bone of the trunk. It does not apply to any other bone of
the trunk (I know not whether this be true of any second bone in the
body, except perhaps the os hyoides.* In strictness, it forms no part of
the skeleton. It is bedded in the flesh; attached only to the muscles. It
is no other than a foundation bone for the arm, laid in, separate, as it
were, and distinct, from the general ossification. The lower limbs
connect themselves at the hip with bones which form part of the
skeleton; but, this connection, in the upper limbs, being wanting, a
basis, where upon the arm might be articulated, was to be supplied
by a detached ossification for the purpose.

I. THE ABOVE are a few examples of bones made remarkable by
their configuration: but to almost all the bones belong *joints*; and in
these, still more clearly than in the form or shape of the bones
themselves, are seen both contrivance and contriving wisdom. Every
joint is a curiosity, and is also strictly mechanical. There is the hinge
joint, and the mortice and tenon joint; each as manifestly such, and
as accurately defined, as any which can be produced out of a cabinet-
maker's shop. And one or the other prevails, as either is adapted to
the motion which is wanted: e.g. a mortice and tenon, or ball and
socket joint, is not required at the knee, the leg standing in need only
of a motion backward and forward in the same plane, for which a
hinge joint is sufficient: a mortice and tenon, or ball and socket joint,

is wanted at the hip, that not only the progressive step may be provided for, but the interval between the limbs may be enlarged or contracted at pleasure. Now observe what would have been the inconveniency, i. e. both the superfluity and the defect of articulation, if the case had been inverted; if the ball and socket joint had been at the knee, and the hinge joint at the hip. The thighs must have been kept constantly together, and the legs have been loose and straddling. There would have been no use that we know of, in being able to turn the calves of the legs before; and there would have been great confinement by restraining the motion of the thighs to one plane. The disadvantage would not have been less, if the joints at the hip and the knee had been both of the same sort; both balls and sockets, or both hinges: yet why, independently of utility, and of a Creator who consulted that utility, should the same bone (the thigh-bone) be rounded at one end, and channelled at the other?

The *hinge joint* is not formed by a bolt passing through the two parts of the hinge, and thus keeping them in their places; but by a different expedient. A strong, tough, parchment-like membrane, rising from the receiving bones, and inserted all round the received bones a little below their heads, incloses the joint on every side. This membrane ties, confines, and holds the ends of the bones together; keeping the corresponding parts of the joint, i. e. the relative convexities and concavities, in close application to each other.

For the *ball and socket joint*, beside the membrane already described, there is in some important joints, as an additional security, a short, strong, yet flexible ligament, inserted, by one end into the head of the ball, by the other into the bottom of the cup; which ligament keeps the two parts of the joint so firmly in their place, that none of the motions which the limb naturally performs, none of the jerks and twists to which it is ordinarily liable, nothing less indeed than the utmost and the most unnatural violence, can pull them asunder. It is hardly indeed imaginable, how great a force is necessary, even to stretch, still more to break, this ligament; yet so flexible is it, as to oppose no impediment to the suppleness of the joint. By its situation also, it is inaccessible to injury from sharp edges. As it cannot be ruptured, such is its strength; so it cannot be cut, except by an accident which would sever the limb. If I had been permitted to frame a proof of contrivance, such as might satisfy the most distrustful enquirer, I know not whether I could have chosen an example of mechanism more unequivocal, or more free from objection, than this ligament. Nothing can be more mechanical; nothing, however subservient to the safety, less capable of being generated by the action of the joint. I would particularly solicit the reader's attention to this provision, as it is found in the head of the *thigh bone*; to its strength, its structure, and its use. It is an instance upon which I lay my hand. One single fact, weighed by a mind in earnest, leaves oftentimes the deepest impression. For the purpose of addressing different understandings and different apprehensions, for the purpose of sentiment, for the purpose of exciting admiration of the Creator's works, we diversify our views, we multiply examples; but, for the purpose of strict argument, one clear instance is sufficient: and not only sufficient, but capable perhaps of generating a firmer assurance than what can arise from a divided attention.

The *ginglymus*,* or hinge joint, does not, it is manifest, admit of a ligament of the same kind with that of the ball and socket joint, but it is always fortified by the species of ligament of which it does admit. The strong, firm, investing membrane above described, accompanies it in every part: and, in particular joints, this membrane, which is properly a ligament, is considerably stronger on the sides than either before or behind, in order that the convexities may play true in their concavities, and not be subject to slip sideways, which is the chief danger; for the muscular tendons generally restrain the parts from going further than they ought to go in the plane of their motion. In the *knee*, which is a joint of this form, and of great importance, there are superadded to the common provisions for the stability of the joint, two strong ligaments which cross each other; and cross each other in such a manner, as to secure the joint from being displaced in any assignable direction. 'I think,' says Cheselden,* 'that the knee cannot be completely dislocated without breaking the *cross* ligaments.'[1] We can hardly help comparing this with the binding up of a fracture, where the fillet is almost always strapped across, for the sake of giving firmness and strength to the bandage.

Another no less important joint, and that also of the ginglymus sort, is the *ankle*; yet, though important, (in order, perhaps, to preserve the symmetry and lightness of the limb) *small*, and, on that account, more liable to injury. Now this joint is strengthened, i. e. is

---

[1] Ches. Anat. ed. 7th, p. 45.

defended from dislocation, by two remarkable processes or pro-longations of the bones of the leg, which processes form the pro-tuberances that we call the inner and outer ankle. It is part of each bone going down lower than the other part, and thereby overlapping the joint: so that, if the joint be in danger of slipping outward, it is curbed by the inner projection, i. e. that of the tibia; if inward, by the outer production, i. e. that of the fibula. Between both, it is locked in its position. I know no account that can be given of this structure except its utility. Why should the tibia terminate, at its lower extrem-ity with a double end, and the fibula the same, but to barricade the joint on both sides by a continuation of part of the thickness of the bone over it?

The joint at the *shoulder* compared with the joint at the *hip*, though both ball and socket joints, discover a difference in their form and proportions, well suited to the different offices which the limbs have to execute. The cup or socket at the shoulder is much shallower and flatter than it is at the hip, and is also in part formed of cartilage set round the rim of the cup. The socket, into which the head of the thigh-bone is inserted, is deeper, and made of more solid materials. This agrees with the duties assigned to each part. The arm is an instrument of motion, principally, if not solely. Accordingly the shallowness of the socket at the shoulder, and the yieldingness of the cartilaginous substance with which its edge is set round, and which in fact composes a considerable part of its concavity, are excellently adapted for the allowance of a freer motion and a wider range; both which the arm wants. Whereas the lower limb, forming a part of the column of the body; having to support the body, as well as to be the means of its locomotion; firmness was to be consulted as well as action. With a capacity for motion, in all directions indeed, as at the shoulder, but not in any direction to the same extent as in the arm, was to be united stability, or resistance to dislocation. Hence the deeper excavation of the socket; and the presence of a less proportion of cartilage upon the edge.

The suppleness and pliability of the joints we every moment experience; and the *firmness* of animal articulation, the property we have hitherto been considering, may be judged of from this single observation, that, at any given moment of time, there are millions of animal joints in complete repair and use, for one that is dislocated;

and this notwithstanding the contortions and wrenches to which the limbs of animals are continually subject.

II. The *joints*, or rather the ends of the bones which form them, display also, in their configuration, another use. The nerves, blood-vessels, and tendons, which are necessary to the life, or for the motion, of the limbs, must, it is evident, in their way from the trunk of the body to the place of their destination, travel over the moveable joints; and it is no less evident, that, in this part of their course, they will have, from sudden motions and from abrupt changes of curvature, to encounter the danger of compression, attrition, or laceration. To guard fibres so tender against consequences so injurious, their path is in those parts protected with peculiar care: and that by a provision in the figure of the bones themselves. The nerves which supply the *fore-arm*, especially the inferior cubital nerves, are at the elbow conducted, by a kind of covered way, between the condyls, or rather under the inner extuberances of the bone, which composes the upper part of the arm.[1] At the *knee* the extremity of the thighbone is divided by a sinus or cliff into two heads or protuberances; and these heads on the back part stand out beyond the cylinder of the bone. Through the hollow, which lies between the hind parts of these two heads, that is to say, under the ham, between the hamstrings, and within the concave recess of the bone formed by the extuberances on each side; in a word, along a defile, between rocks, pass the great vessels and nerves which go to the leg.[2] Who led these vessels by a road so defended and secured? In the joint at the *shoulder*, in the edge of the cup which receives the head of the bone, is a *notch* which is joined or covered at the top with a ligament. Through this hole, thus guarded, the blood-vessels steal to their destination in the arm, instead of mounting over the edge of the concavity.[3]

III. In all joints, the ends of the bones, which work against each other, are tipped with *gristle*.* In the ball and socket joint, the cup is lined, and the ball capped with it. The smooth surface, the elastic and unfriable nature of cartilage, render it of all substances the properest for the place and purpose. I should therefore have pointed this out amongst the foremost of the provisions which have been made in the joints for the facilitating of their action, had it not been alledged,

---

[1] Ches. An. p. 255, ed. 7th.
[2] Ib. p. 35.
[3] Ib. p. 30.

that cartilage in truth is only nascent or imperfect bone; and that the bone in these places is kept soft and imperfect, in consequence of a more complete and rigid ossification being prevented from taking place by the continual motion and rubbing of the surfaces. Which being so, what we represent as a designed advantage, is an unavoidable effect. I am far from being convinced that this is a true account of the fact; or that, if it were so, it answers the argument. To me, the surmounting of the ends of the bones with gristle, looks more like a plating with a different metal, than like the same metal kept in a different state by the action to which it is exposed. At all events we have a great particular benefit, though arising from a general constitution: but this last not being quite what my argument requires, lest I should seem by applying the instance, to overrate its value, I have thought it fair to state the question which attends it.

IV. In some joints, very particularly in the knees, there are loose cartilages or gristles between the bones, and within the joint, so that the ends of the bones, instead of working upon one another, work upon the intermediate cartilages. Cheselden has observed,[1] That the contrivance of a loose ring is practised by mechanics, where the friction of the joints of any of their machines is great; as between the parts of crook hinges of large gates, or under the head of the male screw of large vices. The cartilages of which we speak have very much of the form of these rings. The comparison moreover shews the reason why we find them in the knees rather than in other joints. It is an expedient, we have seen, which a mechanic resorts to, only when some strong and heavy work is to be done. So here the thigh bone has to achieve its motion at the knee, with the whole weight of the body pressing upon it, and often, as in rising from our seat, with the whole weight of the body to lift. It should seem also from Cheselden's account, that the slipping and sliding of the loose cartilages, though it be probably a small and obscure change, humoured the motion of the end of the thigh bone, under the particular configuration which was necessary to be given to it for the commodious action of the tendons; and which configuration requires what he calls a variable socket, that is, a concavity, the lines of which assume a different curvature in different inclinations of the bones.

V. We have now done with the configuration; but there is also in

[1] Ib. p. 13.

the joints, and that common to them all, another exquisite provision, manifestly adapted to their use, and concerning which there can, I think, be no dispute, namely, the regular supply of a *mucilage*, more emollient and slippery than oil itself, which is constantly softening and lubricating the parts that rub upon each other, and thereby diminishing the effect of attrition in the highest possible degree. For the continual secretion of this important liniment, and for the feeding of the cavities of the joint with it, glands are fixed near each joint; the excretory ducts of which glands, dripping with their balsamic contents, hang loose like fringes within the cavity of the joints. A late improvement in what are called friction wheels, which consists of a mechanism so ordered, as to be regularly dropping oil into a box, which incloses the axis, the nave, and certain balls upon which the nave revolves,* may be said, in some sort, to represent the contrivance in the animal joint; with this superiority, however, on the part of the joint, viz. that here, the oil is not only dropped, but *made*.

In considering the joints, there is nothing, perhaps, which ought to move our gratitude more than the reflection, *how well they wear*. A limb shall swing upon its hinge, or play in its socket, many hundred times in an hour, for sixty years together, without diminution of its agility: which is a long time for any thing to last; for any thing so much worked and exercised as the joints are. This durability I should attribute, in part, to the provision which is made for the preventing of wear and tear, first, by the polish of the cartilaginous surfaces, secondly, by the healing lubrication of the mucilage; and, in part, to that astonishing property of animal constitutions, assimilation, by which, in every portion of the body, let it consist of what it will, substance is restored, and waste repaired.

Moveable joints, I think, compose the curiosity of bones; but their union, even where no motion is intended or wanted, carries marks of mechanism and of mechanical wisdom. The teeth, especially the front teeth, are one bone fixed in another like a peg driven into a board. The sutures of the skull are like the edges of two saws clapped together, in such a manner as that the teeth of one enter the intervals of the other. We have sometimes one bone lapping over another, and planed down at the edges; sometimes also the thin lamella* of one bone received into a narrow furrow of another. In all which varieties we seem to discover the same design, viz. firmness of juncture, without clumsiness in the seam.

CHAPTER IX

OF THE MUSCLES

MUSCLES, with their tendons, are the instruments by which animal motion is performed. It will be our business to point out instances in which, and properties with respect to which, the disposition of these muscles is as strictly mechanical, as that of the wires and strings of a puppet.

I. We may observe, what I believe is universal, an exact relation between the joint and the muscles which move it. Whatever motion, the joint, by its mechanical construction, is capable of performing, that motion, the annexed muscles, by their position, are capable of producing. For example; if there be, as at the knee and elbow, a hinge joint, capable of motion only in the same plane, the leaders, as they are called, i. e. the muscular tendons, are placed in directions parallel to the bone, so as, by the contraction or relaxation of the muscles to which they belong, to produce that motion and no other. If these joints were capable of a freer motion, there are no muscles to produce it. Whereas at the shoulder and the hip, where the ball and socket joint allows by its construction of a rotatory or sweeping motion, tendons are placed in such a position, and pull in such a direction, as to produce the motion of which the joint admits. For instance, the sartorius* or taylor's muscle, rising from the spine, running diagonally across the thigh, and taking hold of the inside of the main bone of the leg a little below the knee, enables us, by its contraction, to throw one leg and thigh over the other; giving effect, at the same time, to the ball and socket joint at the hip, and the hinge joint at the knee. There is, as we have seen, a specific mechanism in the bones for the rotatory motions of the head and hands: there is, also, in the oblique direction of the muscles belonging to them, a specific provision for the putting of this mechanism of the bones into action. And mark the consent of uses. The oblique muscles would have been inefficient without the articulation: the articulation would have been lost, without the oblique muscles. It may be proper however to observe with respect to the *head*, although I think it does not vary the case, that its oblique motions and inclinations are often motions in a *diagonal*, produced by the joint action of muscles lying in straight directions. But, whether the pull be single or combined, the articulation is always such, as to be capable of obeying the action of the muscles. The oblique muscles attached to the head, are likewise so disposed, as to be capable of steadying the globe, as well as of moving it. The head of a new-born infant is often obliged to be filleted up. After death the head drops, and rolls in every direction. So that it is by the equilibre of the muscles, by the aid of a considerable and equipollent muscular force in constant exertion, that the head maintains its erect posture. The muscles here supply, what would otherwise be a great defect in the articulation: for the joint in the neck, although admirably adapted to the motion of the head, is insufficient for its support. It is not only by the means of a most curious structure of the bones that a man turns his head, but by virtue of an adjusted muscular power, that he even holds it up.

As another example of what we are illustrating, viz. conformity of use between the bones and the muscles, it has been observed of the different vertebræ, that their processes are exactly proportioned to the quantity of motion which the other bones allow of, and which the respective muscles are capable of producing.

II. A muscle acts only by contraction. Its force is exerted in no other way. When the exertion ceases it relaxes itself, that is, it returns by relaxation to its former state; but without energy. This is the nature of the muscular fibre: and being so, it is evident that the reciprocal *energetic* motion of the limbs, by which we mean motion *with force* in opposite directions, can only be produced by the instrumentality of opposite or antagonist muscles; of flexors and extensors answering to each other. For instance, the biceps and brachiæus *internus** muscles placed in the front part of the upper arm, by their contraction, bend the elbow; and with such degree of force, as the case requires, or the strength admits of. The relaxation of these muscles, after the effort, would merely let the fore arm drop down. For the *back stroke* therefore; and that the arm may not only bend at the elbow, but also extend and straighten itself with force, other muscles, the longus and brevis brachiæus *externus*, and the anconæus,* placed on the hinder part of the arm, by their contractile twitch fetch back the fore arm into a straight line with the cubit,* with no less force than that with which it was bent out of it. The same thing obtains in all the limbs, and in every moveable part of the body. A finger is not bent and straightened, without the contraction of two muscles taking place. It is evident therefore that the animal functions require that particular disposition of the muscles which we describe by the name of antagonist muscles. And they are accordingly so disposed. Every muscle is provided with an adversary. They act like two sawers in a pit by an opposite pull: and nothing surely can more strongly indicate design and attention to an end than their being thus stationed; than this collocation. The nature of the muscular fibre being what it is, the purposes of the animal could be answered by no other. And not only the capacity for motion, but the aspect and symmetry of the body is preserved by the muscles being marshalled according to this order, e. g. the mouth is held in the middle of the face, and its angles kept in a state of exact correspondency, by two muscles drawing against, and balancing, each other. In a hemiplegia, when the muscle on one side is weakened, the muscle on the other side draws the mouth awry.

III. Another property of the muscles, which could only be the result of care, is their being almost universally so disposed, as not to obstruct or interfere with one another's action. I know but one instance in which this impediment is perceived. We cannot easily swallow whilst we gape. This, I understand, is owing to the muscles employed in the act of deglutition* being so implicated with the muscles of the lower jaw, that, whilst these last are contracted, the former cannot act with freedom. The obstruction is, in this instance, attended with little inconveniency; but it shews what the effect is, where it does exist; and what loss of faculty there would be, if it were more frequent. Now when we reflect upon the number of muscles, not fewer than four hundred and forty-six in the human body,* known and named,[1] how contiguous they lie to each other, in layers, as it were, over one another, crossing one another, sometimes embedded in one another, sometimes perforating one another, an arrangement, which leaves to each its liberty and its full play, must necessarily require meditation and counsel.

IV. The following is oftentimes the case with the muscles. Their action is wanted where their situation would be inconvenient. In which case the body of the muscle is placed in some commodious position at a distance, and made to communicate with the point of action, by slender strings or wires. If the muscles, which move the

[1] Keill's Anat. p. 295, ed. 3d.

fingers, had been placed in the palm or back of the hand, they would have swelled that part to an awkward and clumsy thickness. The beauty, the proportions, of the part, would have been destroyed. They are therefore disposed in the arm, and even up to the elbow; and act by long tendons, strapped down at the wrist, and passing under the ligament to the fingers, and to the joints of the singers, which they are severally to move. In like manner, the muscles which move the toes, and many of the joints of the foot, how gracefully are they disposed in the calf of the leg, instead of forming an unwieldy tumefaction in the foot itself? The observation may be repeated of the muscle which draws the nictitating membrane over the eye. Its office is in the front of the eye; but its body is lodged in the back part of the globe, where it lies safe, and where it incumbers nothing.

V. The great mechanical variety in the figure of the muscles may be thus stated. It appears to be a fixed law, that the contraction of a muscle shall be towards its centre.* Therefore the subject for mechanism on each occasion is, so to modify the figure, and adjust the position, of the muscle, as to produce the motion required, agreeably with this law. This can only be done by giving to different muscles, a diversity of configuration, suited to their several offices, and to their situation with respect to the work which they have to perform. On which account we find them under a multiplicity of forms, and attitudes; sometimes with double, sometimes with treble tendons, sometimes with none; sometimes one tendon to several muscles, at other times one muscle to several tendons. The shape of the organ is susceptible of an incalculable variety, whilst the original property of the muscle, the law and line of its contraction, remains the same; and is simple. Herein the muscular system may be said to bear a perfect resemblance to our works of art. An artist does not alter the native quality of his materials, or their laws of action. He takes these as he finds them. His skill and ingenuity are employed in turning them, such as they are, to his account, by giving to the parts of his machine a form and relation, in which these unalterable properties may operate to the production of the effects intended.

VI. The ejaculations can never too often be repeated, How many things must go right for us to be an hour at ease! How many more, to be vigorous and active! Yet vigor and activity are, in a vast plurality of instances, preserved in human bodies, notwithstanding that they depend upon so great a number of instruments of motion, and notwithstanding that the defect or disorder sometimes of a very small instrument, of a single pair, for instance, out of the four hundred and forty-six muscles which are employed, may be attended with grievous inconveniency. There is piety and good sense in the following observation taken out of the *Religious Philosopher*. 'With much compassion,' says this writer,* 'as well as astonishment at the goodness of our loving Creator, have I considered the sad state of a certain gentleman, who, as to the rest, was in pretty good health, but only wanted the use of these *two little muscles* that serve to lift up the eyelids, and so had almost lost the use of his sight, being forced, as long as this defect lasted, to shove up his eyelids every moment with his own hands!' In general we may remark how little those, who enjoy the perfect use of their organs, know the comprehensiveness of the blessing, the variety of their obligation. They perceive a result, but they think little of the multitude of concurrences and rectitudes which go to form it.

BESIDE these observations, which belong to the muscular organ as such, we may notice some advantages of structure which are more conspicuous in muscles of a certain class or description than in others. Thus,

I. The variety, quickness, and precision, of which muscular motion is capable, are seen, I think, in no part so remarkably as in the *tongue*. It is worth any man's while to watch the agility of his tongue; the wonderful promptitude with which it executes changes of position, and the perfect exactness. Each syllable of articulated sound requires for its utterance a specific action of the tongue, and of the parts adjacent to it. The disposition and configuration of the mouth appertaining to every letter and word, is not only peculiar, but, if nicely and accurately attended to, perceptible to the fight; insomuch that curious persons have availed themselves of this circumstance to teach the deaf to speak, and to understand what is said by others. In the same person, and after his habit of speaking is formed, one, and only one, position of the parts, will produce a given articulate sound correctly. How instantaneously are these positions assumed and dismissed; how numerous are the permutations, how various, yet how infallible? Arbitrary and antic variety is not the thing we admire; but variety obeying a rule, conducing to an effect, and commensurate with exigencies infinitely diversified. I believe also that the anatomy of the tongue corresponds with these observations upon its activity.

The muscles of the tongue are so numerous, and so implicated with one another, that they cannot be traced by the nicest dissection; nevertheless, which is a great perfection of the organ, neither the number, nor the complexity, nor what might seem to be, the entanglement of its fibres, in any wise impede its motion, or render the determination or success of its efforts uncertain.

I here intreat the reader's permission to step a little out of my way to consider *the parts of the mouth* in some of their other properties. It has been said, and that by an eminent physiologist, that, whenever nature attempts to work two or more purposes by one instrument, she does both or all imperfectly. Is this true of the tongue regarded as an instrument of speech, and of taste; or regarded as an instrument of speech, of taste, and of deglutition? So much otherwise, that many persons, that is to say, nine hundred and ninety-nine persons out of a thousand,* by the instrumentality of this one organ, talk, and taste, and swallow, very well. In fact the constant warmth and moisture of the tongue, the thinness of the skin, the papillæ upon its surface, qualify this organ for its office of tasting, as much as its inextricable multiplicity of fibres do for the rapid movements which are necessary to speech. Animals which feed upon grass, have their tongues covered with a perforated skin, so as to admit the dissolved food to the papillæ underneath, which, in the mean time, remain defended from the rough action of the unbruised spiculæ.

There are brought together within the cavity of the mouth more distinct uses, and parts executing more distinct offices, than I think can be found lying so near to one another, or within the same compass, in any other portion of the body: viz. teeth of different shape, first for cutting, secondly for grinding: muscles, most artificially disposed for carrying on the compound motion of the lower jaw, half lateral and half vertical, by which the mill is worked: fountains of saliva, springing up in different parts of the cavity for the moistening of the food, whilst the mastication is going on: glands, to feed the fountains: a muscular constriction of a very peculiar kind in the back part of the cavity, for the guiding of the prepared aliment into its passage towards the stomach, and in many cases for carrying it along that passage: for, although we may imagine this to be done simply by the weight of the food itself, it in truth is not so, even in the upright posture of the human neck; and most evidently is not the case with quadrupeds, with a horse for instance, in which, when pasturing, the food is thrust upward by muscular strength, instead of descending of its own accord.

In the mean time, and within the same cavity, is going on another business, altogether different from what is here described, that of respiration and speech. In addition therefore to all that has been mentioned, we have a passage opened, from this cavity to the lungs, for the admission of air, exclusively of every other substance: we have muscles, some in the larynx, and without number in the tongue, for the purpose of modulating that air in its passage, with a variety, a compass, and precision, of which no other musical instrument is capable. And, lastly, which in my opinion crowns the whole as a piece of machinery, we have a specific contrivance for dividing the pneumatic part from the mechanical,* and for preventing one set of actions interfering with the other. Where various functions are united, the difficulty is to guard against the inconveniences of a too great complexity. In no apparatus put together by art, and for the purposes of art, do I know such multifarious uses so aptly combined as in the natural organization of the human mouth; or where the structure, compared with the uses, is so simple. The mouth, with all these intentions to serve, is a single cavity; is one machine; with its parts neither crowded nor confused, and each unembarrassed by the rest: each at least at liberty in a degree sufficient for the end to be attained. If we cannot eat and sing at the same moment, we can eat one moment and sing the next; the respiration proceeding freely all the while.

There is one case however of this double office, and that of the *earliest* necessity, which the mouth alone could not perform; and that is, carrying on together the two actions of sucking and breathing. Another rout therefore is opened for the air, namely, through the nose, which lets the breath pass backward and forward, whilst the lips, in the act of sucking, are necessarily shut close upon the body, from which the nutriment is drawn. This is a circumstance, which always appeared to me worthy of notice. The nose would have been necessary, although it had not been the organ of smelling. The making it the seat of a sense, was superadding a new use to a part already wanted: was taking a wise advantage of an antecedent and a constitutional necessity.

———————

But to return to that, which is the proper subject of the present section, the celerity and precision of muscular motion. These qualities may be particularly observed in the execution of many species of instrumental *music*, in which the changes produced by the hand of the musician are exceedingly rapid; are exactly measured, even when most minute; and display, on the part of the muscles, an obedience of action, alike wonderful for its quickness and its correctness.

Or let a person only observe his own hand whilst he is *writing*; the number of muscles, which are brought to bear upon the pen; how the joint and adjusted operation of several tendons is concerned in every stroke, yet that five hundred such strokes are drawn in a minute. Not a letter can be turned without more than one or two or three tendinous contractions, definite, both as to the choice of the tendon, and as to the space through which the contraction moves; yet how currently does the work proceed? and, when we look at it, how faithful have the muscles been to their duty, how true to the order which endeavour or habit hath inculcated? For let it be remembered, that, whilst a man's hand writing is the same, an exactitude of order is preserved, whether he write well or ill. These two instances of music and writing, shew not only the quickness and precision of muscular action, but the docility.

II. Regarding the particular configuration of muscles, *sphincter* or circular muscles appear to me admirable pieces of mechanism. It is the muscular power most happily applied; the same quality of the muscular substance, but under a new modification. The circular disposition of the fibres is strictly mechanical; but, though the most mechanical, is not the only thing in sphincters which deserves our notice. The regulated degree of contractile force with which they are endowed, sufficient for retention, yet vincible when requisite; together with their ordinary state of actual contraction, by means of which their dependence upon the will is not constant but occasional, gives to them a constitution of which the conveniency is inestimable. This their semi-voluntary character, is exactly such as suits with the wants and functions of the animal.

III. We may also, upon the subject of muscles, observe, that many of our most important actions are achieved by the combined help of different muscles. Frequently, a diagonal motion is produced, by the contraction of tendons, pulling in the direction of the sides of the parallelogram. This is the case, as hath been already noticed, with some of the oblique nutations of the head. Sometimes the number of cooperating muscles is very great. Dr Nieuentyt, in the Leipsic Transactions,* reckons up a hundred muscles that are employed every time we breathe: yet we take in, or let out, our breath, without reflecting what a work is thereby performed; what an apparatus is laid in of instruments for the service, and how many such contribute their assistance to the effect. Breathing with ease is a blessing of every moment: yet, of all others, it is that which we possess with the least consciousness. A man in an asthma is the only man who knows how to estimate it.

IV. Mr Home has observed,[1] that the most important and the most delicate actions are performed in the body by the smallest muscles: and he mentions, as his examples, the muscles which have been discovered in the iris of the eye and the drum of the ear. The tenuity of these muscles is astonishing. They are microscopic hairs; must be magnified to be visible; yet are they real effective muscles; and not only such, but the grandest and most precious of our faculties, sight and hearing, depend upon their health and action.

V. The muscles act in the limbs with what is called a mechanical disadvantage. The muscle at the shoulder, by which the arm is raised, is fixed nearly in the same manner, as the load is fixed upon a steelyard,* within a few decimals, we will say, of an inch, from the centre upon which the steelyard turns. In this situation, we find that a very heavy draught is no more than sufficient to countervail the force of a small lead plummet, placed upon the long arm of the steelyard, at the distance of perhaps fifteen or twenty inches from the centre, and on the other side of it. And this is the disadvantage which is meant. And an absolute disadvantage, no doubt, it would be, if the object were to spare the force of muscular contraction. But observe how conducive is this constitution to animal conveniency. Mechanism has always in view one or other of these two purposes; either to move a great weight slowly, and through a small space; or to move a light weight rapidly, through a considerable sweep. For the former of these purposes, a different species of lever,* and a different collocation of the muscles, might be better than the present: but for the second, the present structure is the true one. Now so it happens, that the second, and not the first, is that which the occasions of animal

[1] Phil. Trans. part i. 1800,* p. 8.

life principally call for. In what concerns the human body, it is of much more consequence to any man to be able to carry his hand to his head with due expedition, than it would be to have the power of raising from the ground a heavier load (of two or three more hundred weight, we will suppose,) than he can lift at present. This last is a faculty, which, upon some extraordinary occasions, he may desire to posses; but the other is what he wants and uses every hour or minute. In like manner, a husbandman* or a gardener will do more execution, by being able to carry his scythe, his rake, or his flail,* with a sufficient dispatch through a sufficient space, than if, with greater strength, his motions were proportionably more confined and slow. It is the same with a mechanic in the use of his tools. It is the same also with other animals in the use of their limbs. In general, the vivacity of their motions would be ill exchanged for greater force under a clumsier structure.

We have offered our observations upon the structure of muscles in general; we have also noticed certain species of muscles; but there are also *single* muscles, which bear marks of mechanical contrivance, appropriate as well as particular. Out of many instances of this kind we select the following.

I. Of muscular actions, even of those which are well understood, some of the most curious are incapable of popular explanation; at least without the aid of plates and figures. This is in a great measure the case, with a very familiar, but, at the same time, a very complicated motion, that of the lower jaw; and with the muscular structure by which it is produced. One of the muscles concerned, may, however, be described in such a manner, as to be, I think, sufficiently comprehended for our present purpose. The problem is to pull the lower jaw *down*. The obvious method should seem to be, to place a straight muscle, viz. to fix a string from the chin to the breast, the contraction of which would open the mouth, and produce the motion required at once. But it is evident that the form and liberty of the neck forbid a muscle being laid in such a position; and that, consistently with the preservation of this form, the motion, which we want, must be effectuated, by some muscular mechanism disposed further back in the jaw. The mechanism adopted is as follows. A certain muscle called the *digastric** rises on the side of the face, considerably *above* the insertion of the lower jaw; and comes down, being converted in its progress into a round tendon. Now it is evident, that the tendon, whilst it pursues a direction *descending* towards the jaw, must, by its contraction, pull the jaw up, instead of down. What then was to be done? This, we find, is done. The descending tendon, when it is got low enough, is passed through a loop, or ring, or pulley, in the os hyoides, and then made to ascend; and, having thus changed its line of direction, is inserted into the inner part of the chin: by which device, viz. the turn at the loop, the action of the muscle (which in all muscles is contraction) that before would have pulled the jaw up, now as necessarily draws it down. 'The mouth,' saith Heister, is opened by means of this trochlea in a most wonderful and elegant manner.'

II. What contrivance can be more mechanical than the following, viz. a slit in one tendon to let another tendon pass through it? This structure is found in the tendons which move the toes and fingers. The long tendon, as it is called, in the foot, which bends the first joint of the toe, passes *through* the short tendon which bends the second joint; which course allows to the sinew more liberty, and a more commodious action than it would otherwise have been capable of exerting.[1] There is nothing, I believe, in a silk or cotton mill; in the belts, or straps, or ropes, by which motion is communicated from one part of the machine to another, that is more artificial, or more evidently so, than this *perforation*.

III. The next circumstance which I shall mention, under this head of muscular arrangement, is so decisive a mark of intention, that it always appeared to me, to supersede, in some measure, the necessity of seeking for any other observation upon the subject: and that circumstance is, the tendons, which pass from the leg to the foot, being bound down by a ligament at the ancle. The foot is placed at a considerable angle with the leg. It is manifest, therefore, that flexible strings, passing along the interior of the angle, if left to themselves, would, when stretched, start from it. The obvious preventative is to tie them down. And this is done in fact. Across the instep, or rather just above it, the anatomist finds a strong ligament, *under* which the tendons pass to the foot. The effect of the ligament as a bandage, can be made evident to the senses; for, if it be cut, the tendons start up. The simplicity, yet the clearness of this contrivance, its exact resemblance to established resources of art, place it

[1] Ches. Anat. p. 93, 119.

amongst the most indubitable manifestations of design with which we are acquainted.

There is also a further use to be made of the present example, and that is, as it precisely contradicts the opinion, that the parts of animals may have been all formed by what is called appetency, i. e. endeavour, perpetuated, and imperceptibly working its effect, through an incalculable series of generations.* We have here no endeavour, but the reverse of it; a constant renitency* and reluctance. The endeavour is all the other way. The pressure of the ligament constrains the tendons; the tendons react upon the pressure of the ligament. It is impossible that the ligament should ever have been generated by the exercise of the tendon, or in the course of that exercise, forasmuch as the force of the tendon perpendicularly resists the fibre which confines it, and is constantly endeavouring, not to form, but to rupture and displace, the threads, of which the ligament is composed.

Keill has reckoned up, in the human body, four hundred and forty-six muscles, dissectible and describable; and hath assigned an use to every one of the number. This cannot be all imagination.

Bishop Wilkins* hath observed from Galen,* that there are, at least, ten several qualifications to be attended to in each particular muscle, viz. its proper figure, its just magnitude, its fulcrum, its point of action supposing the figure to be fixed, its collocation with respect to its two ends the upper and the lower, the place, the position of the whole muscle, the introduction into it of nerves, arteries, veins. How are things, including so many adjustments, to be made; or, when made, how are they to be put together, without intelligence?

I have sometimes wondered, why we are not struck with mechanism in animal bodies, as readily and as strongly as we are struck with it, at first sight, in a watch or a mill.* One reason of the difference may be, that animal bodies are, in a great measure, made up of soft, flabby, substances, such as muscles and membranes; whereas we have been accustomed to trace mechanism in sharp lines, in the configuration of hard materials, in the moulding, chiseling, and filing into shapes, such articles as metals or wood. There is something therefore of habit in the case: but it is sufficiently evident, that there can be no proper reason for any distinction of the sort. Mechanism may be displayed in the one kind of substance, as well as in the other.

Although the few instances we have selected, even as they stand in our description, are nothing short perhaps of logical proofs* of design, yet it must not be forgotten, that, in every part of anatomy, description is a poor substitute for inspection. It was well said by an able anatomist,[1]* and said in reference to the very part of the subject which we have been treating of, 'Imperfecta hæc musculorum descriptio, non minùs arida est legentibus, quàm inspectantibus fuerit jucunda eorundem præparatio. Elegantissima enim mechanicês artificia, creberrimè in illis obvia, verbis nonnisi obscurè exprimuntur; carnium autem ductu, tendinum colore, infertionum proportione, et trochlearium distributione, oculis exposita, omnem superant admirationem.'*

[1] Steno in Blas. Anat. Animal. p. 2. c. 4.

CHAPTER X

OF THE VESSELS OF ANIMAL BODIES

THE circulation of the *blood*, through the bodies of men and quad-rupeds, and the apparatus by which it is carried on, compose a system, and testify a contrivance, perhaps the best understood of any part of the animal frame. The lymphatic system, or the nervous system, may be more subtile and intricate; nay, it is possible that in their structure they be even more artificial than the sanguiferous; but we do not know so much about them.

The utility of the circulation of the blood, I assume as an acknowl-edged point. One grand purpose is plainly answered by it; the dis-tributing to every part, every extremity, every nook and corner, of the body, the nourishment which is received into it by one aperture. What enters at the mouth, finds its way to the fingers' ends. A more difficult mechanical problem could hardly I think be proposed, than to discover a method of constantly repairing the waste, and of sup-plying an accession of substance to every part, of a complicated machine at the same time.

This system presents itself under two views: first, the disposition of the blood vessels, i. e. the laying of the pipes; and, secondly, the construction of the engine* at the centre, viz. the heart, for driving the blood through them.

I. The disposition of the blood vessels, as far as regards the supply of the body, is like that of the water pipes in a city,* viz. large and main trunks branching off by smaller pipes (and these again by still narrower tubes) in every direction, and towards every part, in which the fluid, which they convey, can be wanted. So far, the water pipes, which serve a town, may represent the vessels, which carry the blood from the heart. But there is another thing necessary to the blood, which is not wanted for the water; and that is, the carrying of it back again to its source. For this office a reversed system of vessels is prepared, which, uniting at their extremities with the extremities of the first system, collects the divided and subdivided streamlets, first by capillary ramifications into larger branches, sec-ondly by these branches into trunks; and thus returns the blood (almost exactly inverting the order in which it went out) to the fountain from whence its motion proceeded. All which is evident mechanism.

The body, therefore, contains two systems of blood-vessels, arteries and veins. Between the constitution of the systems there are also two differences, suited to the functions which the systems have to execute. The blood, in going out, passing always from wider into narrower tubes; and, in coming back, from narrower into wider; it is evident, that the impulse and pressure upon the sides of the blood-vessels, will be much greater in one case than the other. Accordingly, the arteries which carry out the blood, are formed with much tougher and stronger coats, than the veins which bring it back. That is one difference: the other is still more artificial, or, if I may so speak, indicates, still more clearly, the care and anxiety of the artificer. Forasmuch as in the arteries, by reason of the greater force with which the blood is urged along them, a wound or rupture would be more dangerous, than in the veins, these vessels are defended from injury, not only by their texture, but by their situation; and by every advantage of situation which can be given to them. They are buried in sinuses, or they creep along grooves, made for them, in the bones; for instance, the under edge of the ribs is sloped and furrowed solely for the passage of these vessels. Sometimes they proceed in channels, protected by stout parapets on each side; which last description is remarkable in the bones of the fingers, these being hollowed out, on the under side, like a scoop, and with such a concavity that the finger may be cut across to the bone without hurting the artery which runs along it. At other times, the arteries pass in canals wrought in the substance, and in the very middle of the substance, of the bone: this takes place in the lower jaw; and is found where there would, otherwise, be danger of compression by sudden curvature. All this care is wonderful, yet not more than what the importance of the case required. To those, who venture their lives in a ship, it has been often said, that there is only an inch-board between them and death; but in the body itself, especially in the arterial system, there is, in many parts, only a membrane, a skin, a thread. For which reason this system lies deep under the integuments; whereas the veins, in which the mischief that ensues from injuring the coats is much less, lie in general above the arteries; come nearer to the surface; are more exposed.

It may be further observed concerning the two systems taken

together, that, though the arterial, with its trunk and branches and small twigs, may be imagined to issue or proceed, in other words to *grow* from the heart, like a plant from its root, or the fibres of a leaf from its foot stalk (which however, were it so, would be only to resolve one mechanism into another), yet the venal, the returning system, can never be formed in this manner. The arteries might go on shooting out from their extremities, i. e. lengthening and sub-dividing indefinitely; but an inverted system, continually uniting its streams, instead of dividing, and thus carrying back what the other system carried out, could not be referred to the same process.

II. The next thing to be considered is the engine which works this machinery, viz. the *heart*. For our purpose it is unnecessary to ascertain the principle upon which the heart acts.* Whether it be irritation excited by the contact of the blood, by the influx of the nervous fluid, or whatever else be the cause of its motion, it is something, which is capable of producing, in a living muscular fibre, reciprocal contraction and relaxation. This is the power we have to work with: and the enquiry is, how this power is applied in the instance before us. There is provided in the central part of the body a hollow muscle, invested with spiral fibres, running in both directions, the layers intersecting one another; in some animals, however, appearing to be semicircular rather than spiral. By the contraction of these fibres, the sides of the muscular cavities are necessarily squeezed together, so as to force out from them any fluid which they may at that time contain: by the relaxation of the same fibres, the cavities are in their turn dilated; and, of course, prepared to admit every fluid which may be poured into them. Into these cavities are inserted the great trunks, both of the arteries which carry out the blood, and of the veins which bring it back. This is a general account of the apparatus: and the simplest idea of its action is, that, by each contraction, a portion of blood is forced as by a syringe into the arteries; and, at each dilatation, an equal portion is received from the veins. This produces, at each pulse, a motion and change in the mass of blood, to the amount of what the cavity contains, which in a full grown human heart, I understand, is about an ounce, or two table-spoons full. How quickly these changes succeed one another, and by this succession how sufficient they are to support a stream or circulation throughout the system, may be understood by the following computation, abridged from Keill's Anatomy, p. 117. ed. 3. 'Each ventricle will at least contain one ounce of blood. The heart contracts four thousand times in one hour; from which it follows, that there passes through the heart, every hour, four thousand ounces, or three hundred and fifty pounds, of blood. Now the whole mass of blood is said to be about twenty-five pounds, so that a quantity of blood equal to the whole mass of blood passes through the heart fourteen times in one hour; which is about once every four minutes.' Consider what an affair this is, when we come to very large animals. The aorta of a whale is larger in the bore than the main pipe of the water-works at London Bridge; and the water roaring in its passage through that pipe, is inferior, in impetus and velocity, to the blood gushing from the whale's heart. Hear Dr Hunter's account of the dissection of a whale.* 'The aorta measured a foot diameter. Ten or fifteen gallons of blood is thrown out of the heart at a stroke with an immense velocity, through a tube of a foot diameter. The whole idea fills the mind with wonder.'[1]

The account which we have here stated, of the injection of blood into the arteries by the contraction, and of the corresponding reception of it from the veins by the dilatation, of the cavities of the heart, and of the circulation being thereby maintained through the blood-vessels of the body, is true, but imperfect. The heart performs this office, but it is in conjunction with another of equal curiosity and importance. It was necessary that the blood should be successively brought into contact, or contiguity, or proximity with the *air*. I do not know that the chymical reason, upon which this necessity is founded, has been yet sufficiently explored. It seems to be made appear, that the atmosphere which we breathe is a mixture of two kinds of air; one pure and vital, the other, for the purposes of life, effete, foul, and noxious: that when we have drawn in our breath, the blood in the lungs imbibes from the air, thus brought into contiguity with it, a portion of its pure ingredient; and, at the same time, gives out the effete or corrupt air which it contained, and which is carried away, along with the halitus, every time we expire. At least; by comparing the air which is breathed from the lungs, with the air before it enter the lungs, it is found to have lost some of its pure part, and to have brought away with it an addition of its impure part.* Whether these experiments satisfy the question, as to the need which the blood stands in, of being visited by continual accesses of air, is not for

---

[1] Dr Hunter's account of the dissection of a whale. Phil. Trans.

us to enquire into; nor material to our argument: it is sufficient to know, that, in the constitution of most animals such a necessity exists, and that the air, by some means or other, *must* be introduced into a near communication with the blood. The lungs of animals are constructed for this purpose. They consist of blood-vessels and air-vessels lying close to each other; and wherever there is a branch of the trachea or windpipe, there is a branch accompanying it of the vein and artery, and the air-vessel is always in the middle between the blood-vessels.[1] The internal surface of these vessels, upon which the application of the air to the blood depends, would, if collected and expanded, be, in a man, equal to a superficies of fifteen feet square. Now in order to give the blood in its course the benefit of this organization (and this is the part of the subject with which we are chiefly concerned), the following operation takes place. As soon as the blood is received by the heart from the veins of the body, and *before* that it is sent out again into its arteries, it is carried, by the force of the contraction of the heart, and by means of a separate and supplementary artery,* to the lungs, and made to enter the vessels of the lungs; from which, after it has undergone the action, whatever it be, of that viscus, it is brought back by a large vein* once more to the heart, in order, when thus concocted and prepared, to be from thence distributed anew into the system. This assigns to the heart a double office. The pulmonary circulation is a system within a system; and one action of the heart is the origin of both.

For this complicated function, four cavities become necessary; and four are accordingly provided: two, called ventricles, which *send out* the blood, viz. one into the lungs, in the first instance; the other into the mass, after it has returned from the lungs: two others also, called auricles, which *receive* the blood from the veins; viz. one, as it comes immediately from the body; the other, as the same blood comes a second time after its circulation through the lungs. So that there are two receiving cavities, and two forcing cavities. The structure of the heart has reference to the lungs, for without the lungs one of each would have been sufficient. The translation of the blood in the heart itself is after this manner. The receiving cavities respectively communicate with the forcing cavities, and, by their contraction, unload the received blood into them. The forcing cavities, when it is their

---

[1] Keill's Anat. p. 121.

turn to contract, compel the same blood into the mouths of the arteries.

The account here given will not convey to a reader ignorant of anatomy, any thing like an accurate notion of the form, action, or use of the parts (nor can any short and popular account do this*), but it is abundantly sufficient to testify contrivance; and, although imperfect, being true as far as it goes, may be relied upon for the only purpose for which we offer it, the purpose of this conclusion.

'The wisdom of the Creator,' saith Hamburger,* 'is in nothing seen more gloriously than in the heart.' And how well doth it execute its office! An anatomist, who understood the structure of the heart, might say beforehand that it would play: but he would expect, I think, from the complexity of its mechanism, and the delicacy of many of its parts, that it should always be liable to derangement, or that it would soon work itself out. Yet shall this wonderful machine go, night and day, for eighty years together, at the rate of a hundred thousand strokes every twenty-four hours, having, at every stroke, a great resistance to overcome; and shall continue this action for this length of time, without disorder, and without weariness.

But further; from the account, which has been given of the mechanism of the heart, it is evident that it must require the interposition of *valves*; that the success indeed of its action must depend upon these, for when any one of its cavities contracts, the necessary tendency of the force will be to drive the inclosed blood, not only into the mouth of the artery where it ought to go, but also back again into the mouth of the vein from which it flowed. In like manner, when by the relaxation of the fibres the same cavity is dilated, the blood would not only run into it from the vein, which was the course intended, but back from the artery, through which it ought to be moving forward. The way of preventing a reflux of the fluid, in both these cases, is to six valves; which, like flood-gates, may open a way to the stream in one direction, and shut up the passage against it in another. The heart, constituted as it is, can no more work without valves, than a pump can. When the piston descends in a pump, if it were not for the stoppage by the valve beneath, the motion would only thrust down the water which it had before drawn up. A similar consequence would frustrate the action of the heart. Valves therefore properly disposed, i. e. properly with respect to the course of the blood which it is necessary to promote, are essential to the contrivance. *And valves so disposed are, accordingly, provided.* A valve is placed in the communication between each auricle and its ventricle, left, when the ventricle contracts, part of the blood should get back again into the auricle, instead of the whole entering, as it ought to do, the mouth of the artery. A valve is also fixed at the mouth of each of the great arteries which take the blood from the heart; leaving the passage free, so long as the blood holds its proper course forward; closing it, whenever the blood, in consequence of the relaxation of the ventricle, would attempt to flow back. There is some variety in the construction of these valves, though all the valves of the body act nearly upon the same principle, and are destined to the same use. In general they consist of a thin membrane, lying close to the side of the vessel, and consequently allowing an open passage whilst the stream runs one way, but thrust out from the side by the fluid getting behind it, and opposing the passage of the blood, when it would flow the other way. Where more than one membrane is employed, the different membranes only compose one valve. Their joint action fulfills the office of a valve: for instance; over the entrance of the right auricle of the heart into the right ventricle, three of these skins or membranes are fixed; of a triangular figure; the bases of the triangles fastened to the flesh; the sides and summits loose; but, though loose, connected by threads of a determinate length with certain small fleshy prominences adjoining. The effect of this construction is, that, when the ventricle contracts, the blood endeavouring to escape in all directions, and amongst other directions pressing upwards, gets *between* these membranes and the sides of the passage; and thereby forces them up into such a position, as that, together, they constitute, when raised, a hollow cone (the strings, before spoken of, hindering them from proceeding or separating further); which cone, entirely occupying the passage, prevents the return of the blood into the auricle. A shorter account of the matter may be this: So long as the blood proceeds in its proper course, the membranes which compose the valve are pressed close to the side of the vessel, and occasion no impediment to the circulation; when the blood would regurgitate, they are raised from the side of the vessel, and meeting in the middle of its cavity, shut up the channel. Can any one doubt of contrivance here; or is it possible to shut our eyes against the proof of it?

This valve also, is not more curious in its structure, than it is important in its office. Upon the play of the valve, even upon the proportioned length of the strings or fibres which check the ascent of the membranes, depends, as it should seem, nothing less than the life itself of the animal. We may here likewise repeat, what we before observed concerning some of the ligaments of the body, that they could not be formed by any action of the parts themselves. There are cases, in which, although good uses appear to arise from the shape or configuration of a part, yet that shape and configuration itself may seem to be produced by the action of the part, or by the action or pressure of adjoining parts. Thus the bend, and the internal smooth concavity of the ribs, may be attributed to the equal pressure of the soft bowels; the particular shape of some bones and joints, to the traction of the annexed muscles, or to the position of contiguous muscles. But valves could not be so formed. Action and pressure are all against them. The blood, in its proper course, has no tendency to produce such things; and, in its improper or reflected current, has a tendency to prevent their production. Whilst we see therefore the use and necessity of this machinery, we can look to no other account of its origin or formation than the intending mind of a Creator. Nor can we without admiration reflect, that such thin membranes, such weak and tender instruments, as these valves are, should be able to hold out for seventy or eighty years.

Here also we cannot consider but with gratitude, how happy it is that our vital motions are *involuntary*. We should have enough to do, if we had to keep our hearts beating, and our stomachs at work. Did these things depend, we will not say upon our effort, but upon our bidding, our care, or our attention, they would leave us leisure for nothing else. We must have been continually upon the watch, and continually in fear: nor would this constitution have allowed of sleep.

It might perhaps be expected, that an organ so precious, of such central and primary importance, as the heart is, should be defended by *a case*. The fact is, that a membranous purse or bag, made of strong tough materials, is provided for it; holding the heart within its cavity; fitting loosely and easily about it; guarding its substance, without confining its motion; and containing likewise a spoonful or two of water, just sufficient to keep the surface of the heart in a state of suppleness and moisture. How should such a loose covering be generated by the action of the heart? Does not the inclosing of it in a sac, answering no other purpose but that inclosure, shew the care that has been taken of its preservation?

ONE USE of the circulation of the blood (probably amongst other uses) is to distribute nourishment to the different parts of the body. How minute and multiplied the ramifications of the blood-vessels, for that purpose, are; and how thickly spread, over at least the superficies of the body, is proved by the single observation, that we cannot prick the point of a pin into the flesh, without drawing blood, i. e. without finding a blood-vessel. Nor, internally, is their diffusion less universal. Blood-vessels run along the surface of membranes, pervade the substance of muscles, penetrate the bones. Even into every tooth, we trace, through a small hole in the root, an artery to feed the bone, as well as a vein to bring back the spare blood from it; both which, with the addition of an accompanying nerve, form a thread only a little thicker than a horse-hair.

WHEREFORE, when the nourishment taken in at the mouth, has once reached, and mixed itself with, the blood, every part of the body is in the way of being supplied with it. And this introduces another grand topic, namely, the manner in which the aliment gets into the *blood*; which is a subject distinct from the preceding, and brings us to the consideration of another entire system of vessels.

II. For this necessary part of the animal œconomy, an apparatus is provided, in a great measure, capable of being, what anatomists call, demonstrated, that is, shewn in the dead body; and a line or course of conveyance, which we can pursue by our examinations.

First, the food descends by wide passages into the intestines, undergoing two great preparations on its way, one, in the mouth by mastication and moisture, (can it be doubted with what design the teeth were placed in the road to the stomach, or that there was choice in fixing them in this situation?) the other, by digestion in the stomach itself. Of this last surprising dissolution I say nothing; because it is chymistry, and I am endeavouring to display mechanism.* The figure and position of the stomach (I speak all along with a reference to the human organ) are calculated for detaining the food long enough for the action of its digestive juice. It has the shape of the pouch of a bagpipe; lies across the body; and the pylorus, or passage by which the food leaves it, is somewhat higher in the body, than the cardia or orifice by which it enters; so that it is by the contraction of the muscular coat of the stomach, that the contents, after having undergone the application of the gastric menstruum, are gradually pressed out. In dogs and cats, this action of the coats of the stomach has been displayed to the eye. It is a slow and gentle undulation, propagated from one orifice of the stomach to the other. For the same reason that I omitted, for the present, offering any observation upon the digestive fluid, I shall say nothing concerning the bile or the pancreatic juice, further than to observe upon the mechanism, viz. that from the glands in which these secretions are elaborated, pipes are laid into the first of the intestines, through which pipes the product of each gland flows into that bowel, and is there mixed with the aliment, as soon almost as it passes the stomach: adding also as a remark, how grievously this same bile offends the stomach itself, yet cherishes the vessel that lies next to it.

Secondly, We have now the aliment in the intestines, converted into pulp, and, though lately consisting of perhaps ten different viands, reduced to nearly an uniform substance, and to a state fitted for yielding its essence, which is called chyle, but which is milk, or more nearly resembling milk than any other liquor with which it can be compared. For the straining off of this fluid from the digested aliment in the course of its long progress through the body, myriads of capillary tubes, i. e. pipes as small as hairs, open their orifices into the cavity of every part of the intestines. These tubes, which are so fine and slender as not to be visible unless when distended with chyle, soon unite into larger branches. The pipes, formed by this union, terminate in glands, from which other pipes of a still larger diameter arising, carry the chyle, from all parts, into a common reservoir or *receptacle*. This receptacle is a bag, large enough to hold about two table spoonfulls; and from this vessel a duct or main pipe proceeds, climbing up the back part of the chest, and then creeping along the gullet till it reach the neck. Here it meets the river. Here it discharges itself into a large vein, which soon conveys the chyle, now flowing along with the old blood, to the heart. This whole route can be exhibited to the eye. Nothing is left to be supplied by imagination or conjecture. Now, beside the subserviency of this whole structure to a manifest and necessary purpose, we may remark two or three separate particulars in it, which shew, not only the contrivance, but the perfection of it. We may remark, first, the length of the intestines, which, in the human subject, is six times that of the body. Simply for a passage, these voluminous bowels, this prolixity of gut, seems in no wise necessary; but, in order to allow time and space for the successive extraction of the chyle from the digested aliment, namely, that the chyle, which escapes the lactals of one part of the guts, may be taken up by those of some other part, the length of the canal is of evident use and conduciveness. Secondly, we must also remark their peristaltic motion; which is made up of contractions, following one another like waves upon the surface of a fluid, and not unlike what we observe in the body of an earthworm crawling along the ground; and which is effected by the joint action of longitudinal and of spiral, or rather perhaps of a great number of separate semi-circular fibres. This curious action pushes forward the grosser part of the aliment, at the same time that the more subtile parts, which we call chyle, are, by a series of gentle compressions, squeezed into the narrow orifices of the lacteal veins. Thirdly, It was necessary that these tubes, which we denominate lacteals, or their mouths at least, should be made as narrow as possible, in order to deny admission into the blood to any particle, which is of size enough to make a lodgement afterwards in the small arteries, and thereby to obstruct the circulation: and it was also necessary that this extreme tenuity should be compensated by multitude; for a large quantity of chyle (in ordinary constitutions, not less, it has been computed, than two or three quarts in a day) is, by some means or other, to be passed through them. Accordingly, we find the number of the lacteals exceeding all powers of computation; and their pipes so fine and slender, as not to be visible, unless filled, to the naked eye; and their orifices, which open into the intestines, so small, as not to be discernible even by the best microscope. Fourthly, The main pipe which carries the chyle from the reservoir to the blood, viz. the thoracic duct, being fixed in an almost upright position, and wanting that advantage of propulsion which the arteries possess, is furnished with a succession of valves to check the ascending fluid, when once it has passed them, from falling back. These valves look upward, so as to leave the ascent free, but to prevent the return of the chyle, if, for want of sufficient force to push it on, its weight should at any time cause it to descend. Fifthly, The chyle enters the blood in an odd place, but perhaps the most commodious place possible, viz. at a large vein in the neck, so situated with respect to the circulation, as speedily to bring the mixture to the heart. And this seems to be a circumstance of great moment; for had the chyle entered the blood at an artery, or at a distant vein, the fluid, composed of the old and the new materials, must have performed a considerable part of the circulation, before it received that churning in the lungs, which is, probably, necessary for the intimate and perfect union of the old blood with the recent chyle. Who could have dreamt of a communication between the cavity of the intestines and the left great vein of the *neck*? Who could have suspected that this communication should be the medium through which all nourishment is derived to the body? Or this the place, where, by a side inlet, the important junction is formed between the blood and the material which feeds it?

We postponed the consideration of *digestion*, lest it should interrupt us in tracing the course of the food to the blood; but, in treating of the alimentary system,* so principal a part of the process cannot be omitted.

Of the gastric juice, the immediate agent, by which that change which food undergoes in our stomachs is effected, we shall take our account, from the numerous, careful, and varied experiments, of the Abbé Spallanzani.*

1. It is not a simple diluent, but a real solvent.* A quarter of an ounce of beef had scarce touched the stomach of a crow, when the solution began.

2. It has not the nature of saliva: it has not the nature of bile; but is distinct from both. By experiments out of the body* it appears, that neither of these secretions acts upon alimentary substances, in the same manner as the gastric juice acts.

3. Digestion is not *putrefaction*; for it resists putrefaction most pertinaciously; nay, not only checks its further progress, but restores putrid substances.

4. It is not a *fermentative* process; for the solution begins at the surface, and proceeds towards the centre, contrary to the order in which fermentation acts and spreads.

5. It is not the *digestion of heat*;* for the cold maw of a cod or sturgeon will dissolve the shells of crabs and lobsters, harder than the sides of the stomach which contains them.

In a word, animal digestion carries about it the marks of being a power and a process, completely *sui generis*; distinct from every other; at least from every chymical process with which we are acquainted. And the most wonderful thing about it is its appropriation; its subserviency to the particular œconomy of each animal. The gastric juice of an owl, falcon, or kite, will not touch grain; no not even to finish the macerated and half digested pulse, which is left in the crops of the sparrows that the bird devours. In poultry, the trituration* of the gizzard, and the gastric juice, conspire in the work of digestion. The gastric juice will not dissolve the grain whilst it is whole. Grains of barley inclosed in tubes or spherules are not affected by it. But if the same grain be by any means broken or ground, the gastric juice immediately lays hold of it. Here then is wanted, and here we find, a combination of mechanism and chymistry. For the preparatory grinding, the gizzard lends its mill. And, as all mill work should be strong, its structure is so, beyond that of any other muscle belonging to the animal. The internal coat also, or lining of the gizzard, is, for the same purpose, hard and cartilaginous. But, forasmuch as this is not the sort of animal substance suited for the reception of glands, or for secretion, the gastric juice, in this family, is not supplied, as in membranous stomachs, by the stomach itself, but by the gullet, in which the feeding glands are placed, and from which it trickles down into the stomach.

In sheep, the gastric fluid has no effect in digesting plants, *unless they have been previously masticated*. It only produces a slight maceration; nearly such as common water would produce, in a degree of heat somewhat exceeding the medium temperature of the atmosphere. But, provided that the plant has been reduced to pieces by chewing, the gastric juice then proceeds with it, first by softening its substance; next by destroying its natural consistency; and, lastly, by dissolving it so completely, as not even to spare the toughest and most stringy parts, such as the nerves of the leaves.

So far our accurate and indefatigable Abbé, Dr Stevens* of Edinburgh, in 1777, found by experiments tried with perforated balls, that the gastric juice of the sheep and the ox speedily dissolved vegetables, but made no impression upon beef, mutton, and other animal bodies. Dr Hunter* discovered a property of this fluid, of a most curious kind; viz. that, in the stomachs of animals which feed upon flesh, irresistibly as this fluid acts upon animal substances, it is only upon the *dead* substance, that it operates at all. The *living* fibre suffers no injury from lying in contact with it. Worms and insects are found alive in the stomachs of such animals. The coats of the human stomach, in a healthy state, are insensible to its presence: yet, in cases of sudden death, (wherein the gastric juice, not having been weakened by disease, retains its activity,) it has been known to eat a hole

through the bowel which contains it.[1] How nice is this discrimination of action, yet how necessary?

But to return to our hydraulics.

III. The gall bladder is a very remarkable contrivance. It is the
reservoir of a canal. It does not form the channel itself, i. e. the direct
communication between the liver and the intestine, which is by
another passage, viz. the ductus hepaticus,* continued under the
name of the ductus communis; but it lies adjacent to this channel,
joining it by a duct of its own, the ductus cysticus: by which structure it is enabled, as occasions may require, to add its contents to,
and increase, the flow of bile into the duodenum. And the position of
the gall bladder is such as to apply this structure to the best advantage. In its natural situation it touches the exterior surface of the
stomach, and consequently is compressed by the distension of that
vessel: the effect of which compression is, to force out from the bag,
and send into the duodenum, an extraordinary quantity of bile, to
meet the extraordinary demand which the repletion of the stomach
by food is about to occasion.[2] Cheselden describes[3] the gall bladder as
seated against the duodenum, and thereby liable to have its fluid
pressed out, by the passage of the aliment through that cavity; which
likewise will have the effect of causing it to be received into the
intestine, at a right time, and in a due proportion.

There may be other purposes answered by this contrivance; and it
is probable, that there are. The contents of the gall bladder are not
exactly of the same kind as what passes from the liver through the
direct passage.[4] It is possible that the gall may be changed, and for
some purposes meliorated, by keeping.

The entrance of the gall duct into the duodenum furnishes
another observation. Whenever either smaller tubes are inserted into
larger tubes, or tubes into vessels and cavities, such receiving tubes,
vessels, or cavities, being subject to muscular constriction, we always
find a contrivance to prevent *regurgitation*. In some cases valves are
used; in other cases, amongst which is that now before us, a different
expedient is resorted to: which may be thus described. The gall duct
enters the duodenum obliquely: after it has pierced the first coat, it

---

[1] Phil. Transac. vol. lxii.* p. 447.
[2] Keill's Anat. p. 64.
[3] Anat. p. 164.
[4] Keill from Malpighius,* p. 63.

runs near two fingers breadth *between* the coats, before it open into
the cavity of the intestine.[1] The same contrivance is used in another
part, where there is exactly the same occasion for it, viz. in the
insertion of the ureters into the bladder. These enter the bladder
near its neck, running obliquely for the space of an inch between its
coats.[2] It is, in both cases, sufficiently evident, that this structure has
a necessary mechanical tendency to resist regurgitation; for whatever
force acts in such a direction as to urge the fluid back into the orifices
of the tubes, must, at the same time, stretch the coats of the vessels,
and, thereby, compress that part of the tube, which is included
between them.

IV. Amongst the *vessels* of the human body, the pipe which conveys the saliva from the place where it is made, to the place where it
is wanted, deserves to be reckoned amongst the most intelligible
pieces of mechanism with which we are acquainted. The saliva, we
all know, is used in the mouth; but much of it is manufactured on the
outside of the cheek, by the parotid gland,* which lies between the ear
and the angle of the lower jaw. In order to carry the secreted juice to
its destination, there is laid from the gland on the outside, a pipe,
about the thickness of a wheat straw, and about three fingers breadth
in length; which, after riding over the masseter muscle, bores for
itself a hole through the very middle of the cheek; enters by that
hole, which is a complete perforation of the buccinator muscle,* into
the mouth; and there discharges its fluid very copiously.

V. Another exquisite structure, differing indeed from the four
preceding instances, in that it does not relate to the conveyance of
fluids, but still belonging, like these, to the class of pipes or conduits
of the body, is seen in the *larynx*. We all know, that there go down the
throat two pipes, one leading to the stomach, the other to the lungs;
the one being the passage for the food, the other for the breath and
voice: we know also that both these passages open into the bottom of
the mouth; the gullet, necessarily, for the conveyance of food; and
the windpipe, for speech and the modulation of sound, not much less
so: therefore the difficulty was, the passages being so contiguous, to
prevent the food, especially the liquids, which we swallow into the
stomach, from entering the windpipe, i. e. the road to the lungs; the

[1]　Keill's Anat. p. 62.
[2]　Ches. Anat. p. 260.

consequence of which error, when it does happen, is perceived by the convulsive throes that are instantly produced. This business, which is very nice, is managed in this manner. The gullet (the passage for food) opens into the mouth like the cone or upper part of a funnel, the capacity of which forms indeed the bottom of the mouth. Into the side of this funnel, at the part which lies the lowest, enters the windpipe, by a chink or slit, with a lid or flap, like a little tongue, accurately fitted to the orifice. The solids or liquids which we swallow, pass over this lid or flap, as they descend by the funnel into the gullet. Both the weight of the food, and the action of the muscles concerned in swallowing, contribute to keep the lid close down upon the aperture, whilst any thing is passing; whereas, by means of its natural cartilaginous spring, it raises itself a little, as soon as the food is passed, thereby allowing a free inlet and outlet for the respiration of air by the lungs. And we may here remark the almost complete success of the expedient, viz. how seldom it fails of its purpose, compared with the number of instances in which it fulfills it. Reflect, how frequently we swallow, how constantly we breathe. In a city feast, for example, what deglutition, what anhelation! yet does this little cartilage, the epiglottis, so effectually interpose its office, so securely guard the entrance of the windpipe, that, whilst morsel after morsel, draught after draught, are coursing one another over it, an accident of a crumb or a drop slipping into this passage, (which nevertheless must be opened for the breath every second of time,) excites, in the whole company, not only alarm by its danger, but surprise by its novelty. Not two guests are choked in a century.

There is no room for pretending, that the action of the parts may have gradually formed the epiglottis: I do not mean in the same individual, but in a succession of generations. Not only the action of the parts has no such tendency, but the animal could not live, nor consequently the parts act, either without it, or with it in a half formed state. The species was not to wait for the gradual formation or expansion of a part, which was, from the first, necessary to the life of the individual.

Not only is the larynx curious, but the whole windpipe possesses a structure, adapted to its peculiar office. It is made up (as any one may perceive by putting his fingers to his throat) of stout cartilaginous ringlets, placed at small and equal distances from one another. Now this is not the case with any other of the numerous conduits of the body. The use of these cartilages is to keep the passage for the air *constantly* open; which they do mechanically. A pipe with soft membranous coats, liable to collapse and close when empty, would not have answered here; although this be the general vascular structure, and a structure which serves very well for those tubes, which are kept in a state of perpetual distension by the fluid they inclose, or which afford a passage to solid and protruding substances.

Nevertheless, (which is another particularity well worthy of notice,) these rings are not complete, that is, are not cartilaginous and stiff all round; but their hinder part, which is contiguous to the gullet, is membranous and soft, easily yielding to the distensions of that organ occasioned by the descent of solid food. The same rings are also bevelled off at the upper and lower edges, the better to close upon one another, when the trachea is compressed or shortened.

The constitution of the trachea may suggest likewise another reflection. The membrane which lines its inside, is, perhaps, the most sensible, irritable, membrane of the body. It rejects the touch of a crumb of bread, or a drop of water, with a spasm which convulses the whole frame; yet, left to itself, and its proper office, the intromission of air alone, nothing can be so quiet. It does not even make itself felt: a man does not know that he has a trachea. This capacity of perceiving with such acuteness; this impatience of offence, yet perfect rest and ease when let alone; are properties, one would have thought, not likely to reside in the same subject. It is to the junction however of these almost inconsistent qualities, in this as well as in some other delicate parts of the body, that we owe our safety and our comfort; our safety to their sensibility, our comfort to their repose.

The larynx, or rather the whole windpipe taken together, (for the larynx is only the upper part of the windpipe,) beside its other uses, is also a musical instrument, that is to say, it is *mechanism* expressly adapted to the modulation of sound; for it has been found upon trial, that, by relaxing or tightening the tendinous bands at the extremity of the windpipe, and blowing in at the other end, all the cries and notes might be produced, of which the living animal was capable. It can be sounded, just as a pipe or flute is sounded. Birds, says Bonnet,* have, at the lower end of the windpipe, a conformation like the reed of a hautboy for the modulation of their notes. A tuneful bird is a ventriloquist. The seat of the song is in the breast.

The use of the lungs in the system has been said to be obscure:

one use however is plain, though, in some sense, external to the system, and that is, the formation, in conjunction with the larynx, of voice and speech. They are, to animal utterance, what the bellows are to the organ.

For the sake of method, we have considered animal bodies under three divisions, their bones, their muscles, and their vessels: and we have stated our observations upon these parts separately. But this is to diminish the strength of the argument. The wisdom of the Creator is seen, not in their separate but their collective action; in their mutual subserviency and dependence; in their contributing *together* to one effect, and one use. It has been said, that a man cannot lift his hand to his head without finding enough to convince him of the existence of a God. And it is well said; for he has only to reflect, familiar as this action is, and simple as it seems to be, how many things are requisite for the performing of it; how many things which we understand, to say nothing of many more, probably, which we do not; viz. first, a long, hard, strong cylinder, in order to give to the arm its firmness and tension; but which, being rigid and, in its substance, inflexible, can only turn upon joints: secondly therefore; joints for this purpose, one at the shoulder to raise the arm, another at the elbow to bend it; these joints continually fed with a soft mucilage to make the parts slip easily upon one another, and held together by strong braces to keep them in their position: then, thirdly, strings and wires, i. e. muscles and tendons, artificially inserted for the purpose of drawing the bones in the directions in which the joints allow them to move. Hitherto we seem to understand the mechanism pretty well; and understanding this, we possess enough for our conclusion: nevertheless we have hitherto only a machine standing still; a dead organization; an apparatus. To put the system in a state of activity (to set it at work) a further provision is necessary, viz. a communication with the brain by means of nerves. We know the existence of this communication, because we can see the communicating threads, and can trace them to the brain: its necessity we also know, because, if the thread be cut, if the communication be intercepted, the muscle becomes paralytic: but beyond this we know little; the organization being too minute and subtile for our inspection.

To what has been enumerated, as officiating in the single act of a man's raising his hand to his head, must be added likewise, all that is necessary, and all that contributes, to the growth, nourishment, and sustentation of the limb, the repair of its waste, the preservation of its health: such as the circulation of the blood through every part of it; its lymphatics, exhalants, absorbents; its excretions and integuments.* All these share in the result; join in the effect: and how all these, or any of them, come together without a designing, disposing intelligence, it is impossible to conceive.

CHAPTER XI

OF THE ANIMAL STRUCTURE REGARDED AS A MASS

CONTEMPLATING an animal body in its collective capacity, we cannot forget to notice, what a number of instruments are brought together, and often within how small a compass. In a canary bird, for instance, and in the ounce of matter which composes its body (but which seems to be all employed), we have instruments, for eating, for digesting, for nourishment, for breathing, for generation, for running, for flying, for seeing, for hearing, for smelling; each appropriate; each entirely different from all the rest.

I. The human, or indeed the animal frame, considered as a mass or assemblage, exhibits in its composition three properties, which have long struck my mind, as indubitable evidences, not only of design, but of a great deal of attention and accuracy in prosecuting the design.

1. The first is, the exact correspondency of the two sides of the same animal; the right hand answering to the left, leg to leg, eye to eye, one side of the countenance to the other; and with a precision, to imitate which in any tolerable degree forms one of the difficulties of statuary, and requires, on the part of the artist, a constant attention to this property of his work, distinct from every other.

It is the most difficult thing that can be to get a wig made even; yet how seldom is the *face* awry? And what care is taken that it should not be so, the anatomy of its bones demonstrates. The upper part of the face is composed of thirteen bones, six on each side, answering each to each, and the thirteenth, without a fellow, in the middle: the lower part of the face is in like manner composed of six bones, three on each side, respectively corresponding, and the lower jaw in the centre. In building an arch could more be done in order to make the curve *true*, i. e. the parts equi-distant from the middle, alike in figure and position?

The exact resemblance of the *eyes*, considering how compounded this organ is in its structure, how various and how delicate are the shades of colour with which its iris is tinged, how differently, as to effect upon appearance, the eye may be mounted in its socket, and how differently in different heads eyes actually are set, is a property of animal bodies much to be admired. Of ten thousand eyes, I don't know that it would be possible to match one, except with its own fellow; or to distribute them into suitable pairs by any other selection than that which obtains.

This regularity of the animal structure is rendered more remarkable by the three following considerations. First, the limbs, separately taken, have not this correlation of parts; but the contrary of it. A knife drawn down the chine cuts the human body into two parts, externally equal and alike; you cannot draw a straight line which will divide a hand, a foot, the leg, the thigh, the cheek, the eye, the ear, into two parts equal and alike. Those parts which are placed upon the middle or partition line of the body, or which traverse that line, as the nose, the tongue, the lips, may be so divided, or, more properly speaking, are double organs; but other parts cannot. This shews that the correspondency which we have been describing does not arise by any necessity in the nature of the subject; for, if necessary, it would be universal, whereas it is observed only in the system or assemblage: it is not true of the separate parts: that is to say, it is found where it conduces to beauty or utility;* it is not found, where it would subsist at the expence of both. The two wings of a bird always correspond; the two sides of a feather frequently do not. In centipedes, millepedes, and that whole tribe of insects, no two legs on the same side are alike; yet there is the most exact parity between the legs opposite to one another.

2. The next circumstance to be remarked, is, that, whilst the cavities of the body are so configurated, as, *externally*, to exhibit the most exact correspondency of the opposite sides, the contents of these cavities have no such correspondency. A line drawn down the middle of the breast divides the thorax into two sides exactly similar; yet these two sides inclose very different contents. The heart lies on the left side; a lobe of the lungs on the right; balancing each other, neither in size nor shape. The same thing holds of the abdomen. The liver lies on the right side, without any similar viscus opposed to it on the left. The spleen indeed is situated over against the liver; but agreeing with the liver, neither in bulk nor form. There is no equipollency between these. The stomach is a vessel, both irregular in its shape, and oblique in its position. The foldings and doublings of the intestines do not present a parity of sides. Yet that symmetry which depends upon the correlation of the sides, is externally preserved

throughout the whole trunk: and is the more remarkable in the lower parts of it, as the integuments are soft; and the shape, consequently, is not, as the thorax is by its ribs, reduced by natural stays. It is evident, therefore, that the external proportion does not arise from any equality in the shape or pressure of the internal contents. What is it indeed but a correction of inequalities? an adjustment, by mutual compensation of anomalous forms into a regular congeries? the effect, in a word, of artful, and, if we might be permitted so to speak, of studied collocation?

3. Similar also to this, is the third observation; that an internal inequality in the feeding vessels is so managed, as to produce no inequality in parts which were intended to correspond. The right arm answers accurately to the left, both in size and shape; but the arterial branches, which supply the two arms, do not go off from their trunk, in a pair, in the same manner, at the same place, or at the same angle. Under which want of similitude, it is very difficult to conceive how the same quantity of blood should be pushed through each artery: yet the result is right; the two limbs, which are nourished by them, perceive no difference of supply, no effects of excess or deficiency.

Concerning the difference of manner, in which the subclavian and carotid arteries, upon the different sides of the body, separate themselves from the aorta, Cheselden seems to have thought, that the advantage which the left gain by going off at a much acuter angle than the right, is made up to the right by their going off together in one branch.[1] It is very possible that this may be the compensating contrivance: and, if it be so, how curious, how hydrostatical!*

II. ANOTHER perfection of the animal mass is the *package*. I know nothing which is so surprising. Examine the contents of the trunk of any large animal. Take notice how soft, how tender, how intricate they are; how constantly in action, how necessary to life. Reflect upon the danger of any injury to their substance, any derangement of their position, any obstruction to their office. Observe the heart pumping at the centre, at the rate of eighty strokes in a minute: one set of pipes carrying the stream away from it, another set, bringing, in its course, the fluid back to it again: the lungs performing their elaborate office, viz. distending and contracting their many thousand

[1] Ches. Anat. p. 184. ed. 7.

vesicles, by a reciprocation which cannot cease for a minute: the stomach exercising its powerful chymistry; the bowels silently propelling the changed aliment; collecting from it, as it proceeds, and transmitting to the blood an incessant supply of prepared and assimilated nourishment: that blood pursuing its course; the liver, the kidneys, the pancreas, the parotid, with many other known and distinguishable glands, drawing off from it, all the while, their proper secretions. These several operations, together with others more subtile but less capable of being investigated, are going on within us, at one and the same time. Think of this; and then observe how the body itself, the case which holds this machinery, is rolled, and jolted, and tossed about, the mechanism remaining unhurt, and with very little molestation even of its nicest motions. Observe a rope dancer, a tumbler, or a monkey; the sudden inversions and contortions which the internal parts sustain by the postures into which their bodies are thrown; or rather observe the shocks, which these parts, even in ordinary subjects, sometimes receive from falls and bruises, or by abrupt jerks and twists, without sensible, or with soon recovered damage. Observe this, and then reflect how firmly every part must be secured, how carefully surrounded, how well tied down and packed together.

This property of animal bodies has never, I think, been considered under a distinct head, or so fully as it deserves. I may be allowed therefore, in order to verify my observation concerning it, to set forth a short anatomical detail, though it oblige me to use more technical language, than I should wish to introduce into a work of this kind.

1. The *heart* (such care is taken of the centre of life) is placed between two soft lobes of the lungs; is *tied* to the mediastinum and to the pericardium, which pericardium is not only itself an exceedingly strong membrane, but *adheres* firmly to the duplicature of the mediastinum, and, by its point, to the middle tendon of the diaphragm. The heart is also *sustained* in its place by the great blood-vessels which issue from it.[1]

2. The *lungs* are *tied* to the sternum by the mediastinum, before; to the vertebræ by the pleura, behind. It seems indeed to be the very use of the mediastinum (which is a membrane that goes, straight

---

[1] Keill's Anat. p. 107, ed. 3.

through the middle of the thorax, from the breast to the back) to keep the contents of the thorax in their places; in particular to hinder one lobe of the lungs from incommoding another, or the parts of the lungs from pressing upon each other when we lie on one side.[1]

3. The *liver* is fastened in the body by two ligaments; the first, which is large and strong, comes from the covering of the diaphragm, and penetrates the substance of the liver; the second is the umbilical vein, which, after birth, degenerates into a ligament.* The first, which is the principal, fixes the liver in its situation, whilst the body holds an erect posture; the second prevents it from pressing upon the diaphragm when we lie down; and both together sling or suspend the liver when we lie upon our backs, so that it may not compress or obstruct the ascending vena cava,[2] to which belongs the important office of returning the blood from the body to the heart.

4. The *bladder* is tied to the navel by the urachus transformed into a ligament:* thus, what was a passage for urine to the fœtus becomes, after birth, a support or stay to the bladder. The peritonæum also keeps the viscera from confounding themselves with, or pressing irregularly upon, the bladder: for the kidneys and bladder are contained in a distinct duplicature of that membrane, being thereby partitioned off from the other contents of the abdomen.

5. The *kidneys* are lodged in a bed of fat.

6. The *pancreas* or sweetbread is strongly tied to the peritonæum, which is the great wrapping sheet, that incloses all the bowels contained in the lower belly.[3]

7. The *spleen* also is confined to its place by an adhesion to the peritonæum and diaphragm, and by a connection with the omentum.[4]* It is possible, in my opinion, that the spleen may be merely a *stuffing*, a soft cushion to fill up a vacancy or hollow, which unless occupied, would leave the package loose and unsteady: for supposing that it answers no other purpose than this, it must be vascular, and admit of a circulation through it, in order to be kept alive, or be a part of a living body.

8. The *omentum*, epiploon, or cawl, is an apron, tucked up, or doubling upon itself, at its lowest part. The upper edge is tied to the

[1] Ib. 119.
[2] Ches. Anat. p. 162.
[3] Keill's Anat. p. 57.
[4] Ches. Anat. p. 167.

bottom of the stomach, to the spleen, as hath already been observed, and to part of the duodenum. The reflected edge, also, after forming the doubling, comes up behind the front flap, and is tied to the colon and adjoining viscera.[1]

9. The septa of the brain, probably, prevent one part of that organ from pressing with too great a weight upon another part. The processes of the dura mater divide the cavity of the skull, like so many inner partition walls; and, thereby, confine each hemisphere and lobe of the brain to the chamber which is assigned to it, without its being liable to rest upon, or intermix with, the neighbouring parts. The great art and caution of packing, is to prevent one thing hurting another. This, in the head, the chest, and the abdomen, of an animal body, is, amongst other methods, provided for, by membranous partitions and wrappings, which keep the parts separate.

THE ABOVE may serve as a short account of the manner, in which the principal viscera are sustained in their places. But, of the provisions for this purpose, by far, in my opinion, the most curious, and where also such a provision was most wanted, is in the *guts*. It is pretty evident, that a long narrow tube (in man about five times the length of the body) laid from side to side in folds upon one another, winding in oblique and circuitous directions, composed also of a soft and yielding substance, must, without some extraordinary precaution for its safety, be continually displaced by the various, sudden, and abrupt motions of the body which contains it. I should expect, that, if not bruised or wounded by every fall, or leap, or twist, it would be entangled, or be involved with itself; or, at the least, slipped and shaken out of the order in which it is disposed, and which order is necessary to be preserved for the carrying on of the important functions, which it has to execute in the animal œconomy. Let us see therefore how a danger so serious, and yet so natural to the length, narrowness, and tubular form of the part, is provided against. The expedient is admirable, and it is this. The intestinal canal, throughout its whole process, is knit to the edge of a broad fat membrane, called the mesentery. It forms the margin of this mesentery, being stitched and fastened to it like the edging of a ruffle; being four times as long as the mesentery itself, it is, what a sempstress would call, 'gathered on' to it. This is the nature of the connection of the gut

[1] Ib. p. 149.

with the mesentery; and, being thus joined to, or rather made a part of the mesentery, it is folded and wrapped up together with it. Now the mesentery, having a considerable dimension in breadth, being in its substance, withal, both thick and suety, is capable of a close and safe folding, in comparison of what the intestinal tube would admit of, if it had remained loose. The mesentery likewise not only keeps the intestinal canal in its proper place and position under all the turns and windings of its course, but sustains the numberless small vessels, the arteries, the veins, the lympheducts, and, above all, the lacteals, which lead from or to almost every point of its coats and cavity. This membrane, which appears to be the great support and security of the alimentary apparatus, is itself strongly tied to the first three vertebræ of the loins.[1]

III. A third general property of animal forms is *beauty*. I do not mean relative beauty, or that of one individual above another of the same species, or of one species compared with another species; but I mean, generally, the provision which is made, in the body of almost every animal, to adapt its appearance to the perception of the animals with which it converses. In our own species, for example, only consider what the parts and materials are, of which the fairest body is composed; and no further observation will be necessary to shew, how well these things are wrapped up, so as to form a mass, which shall be capable of symmetry in its proportion, and of beauty in its aspect; how the bones are covered, the bowels concealed, the roughnesses of the muscles smoothed and softened; and how over the whole is drawn an integument, which converts the disgusting materials of a dissecting-room into an object of attraction to the sight, or one, upon which it rests, at least, with ease and satisfaction. Much of this effect is to be attributed to the intervention of the cellular or adipose membrane,* which lies immediately under the skin; is a kind of lining to it; is moist, soft, slippery, and compressible; every where filling up the interstices of the muscles, and forming thereby their roundness and flowing line, as well as the evenness and polish of the whole surface.

All which seems to be a strong indication of design, and of a design studiously directed, to this purpose. And it being once allowed, that such a purpose existed with respect to any of the

[1] Keill's Anat. p. 45.

productions of nature, we may refer, with a considerable degree of probability, other particulars to the same intention; such as the teints of flowers, the plumage of birds, the furs of beasts, the bright scales of fishes, the painted wings of butterflies and beetles, the rich colours and spotted lustre of many tribes of insects.

There are parts also of animals ornamental,* and the properties by which they are so, not subservient, that we know of, to any other purpose. The *irides** of most animals are very beautiful, without conducing at all, by their beauty, to the perfection of vision; and nature could in no part have employed her pencil to so much advantage, because no part presents itself so conspicuously to the observer, or communicates so great an effect to the whole aspect.

In plants, especially in the flowers of plants, the principle of beauty holds a still more considerable place in their composition; is still more confessed than in animals. Why, for one instance out of a thousand, does the corolla* of the tulip, when advanced to its size and maturity, change its colour? The purposes, so far as we can see, of vegetable nutrition, might have been carried on as well by its continuing green. Or, if this could not be, consistently with the progress of vegetable life, why break into such a variety of colours? This is no proper effect of age, or of declension in the ascent of the sap;* for that, like the autumnal teints, would have produced one colour in one leaf, with marks of fading and withering. It seems a lame account, to call it, as it has been called, a disease of the plant. Is it not more probable, that this property, which is independent, as it should seem, of the wants and utilities of the plant, was calculated for beauty, intended for display?

A ground, I know, of objection, has been taken against this whole topic of argument, namely, that there is no such thing as beauty at all: in other words, that whatever is useful and familiar comes of course to be thought beautiful; and that things appear to be so, only by their alliance with these qualities. Our idea of beauty is capable of being so modified by habit, by fashion, by the experience of advantage or pleasure, and by associations arising out of that experience, that a question has been made, whether it be not altogether generated by these causes, or would have any proper existence without them. It seems however a carrying of the conclusion too far, to deny the existence of the principle, viz. a native capacity of perceiving beauty, on account of the influence, or the varieties proceeding from that influence, to which it is subject: seeing that principles the most acknowledged, are liable to be affected in the same manner. I should rather argue thus. The question respects objects of sight. Now every other sense hath its distinction of agreeable and disagreeable. Some tastes offend the palate, others gratify it. In brutes and insects, this distinction is stronger, and more regular, than in man. Every horse, ox, sheep, swine, when at liberty to choose, and when in a natural state, that is, when not vitiated by habits forced upon it,* eats and rejects the same plants. Many insects which feed upon particular plants, will rather die than change their appropriate leaf. All this looks like a determination in the sense itself to particular tastes. In like manner, smells affect the nose with sensations pleasurable or disgusting. Some sounds, or compositions of sound, delight the ear, others torture it. Habit can do much in all these cases, (and it is well for us that it can; for it is this power which reconciles us to many necessities,) but has the distinction, in the mean time, of agreeable and disagreeable, no foundation in the sense itself? What is true of the other senses is most probably true of the eye, (the analogy is irresistible) viz. that there belongs to it an original constitution, fitted to perceive pleasure from some impressions, and pain from others.

I do not however know that the argument which alledges beauty as a final cause, rests upon this concession. We possess a sense of beauty, however we come by it. It in fact exists. Things are not indifferent to this sense: all objects do not suit it: many, which we see, are agreeable to it; many others disagreeable. It is certainly not the effect of habit upon the particular object, because the most agreeable objects are often the most rare; many, which are very common, continue to be offensive. If they be made supportable by habit, it is all which habit can do; they never become agreeable. If this sense, therefore, be acquired, it is a result; the produce of numerous and complicated actions of external objects upon the senses, and of the mind upon its sensations.* With this *result* there must be a certain congruity to enable any particular object to please: and that congruity, we contend, is consulted in the *aspect* which is given to animal and vegetable bodies.

IV. The skin and covering of animals is that upon which their appearance chiefly depends, and it is that part which, perhaps, in all animals is most decorated; and most free from impurities. But were beauty, or agreeableness of aspect, entirely out of the question, there is another purpose answered by this integument, and by the collocation of the parts of the body beneath it, which is of still greater importance; and that purpose is *concealment*. Were it possible to view through the skin the mechanism of our bodies, the sight would frighten us out of our wits. 'Durst we make a single movement,' asks a lively French writer, 'or stir a step from the place we were in, if we *saw* our blood circulating, the tendons pulling, the lungs blowing, the humours filtrating, and all the incomprehensible assemblage of fibres, tubes, pumps, valves, currents, pivots, which sustain an existence, at once so frail, and so presumptuous?'

V. Of animal bodies, considered as masses, there is another property, more curious than it is generally thought to be; which is the faculty of *standing*: and it is more remarkable in two-legged animals than in quadrupeds, and, most of all, as being the tallest, and resting upon the smallest base, in man. There is more, I think, in the matter than we are aware of. The statue of a man, placed loose upon its pedestal, would not be secure of standing half an hour. You are obliged to fix its feet to the block by bolts and solder, or the first shake, the first gust of wind, is sure to throw it down. Yet this statue shall express all the mechanical proportions of a living model. It is not therefore the mere figure, or merely placing the centre of gravity within the base, that is sufficient. Either the law of gravitation is suspended in favor of living substances, or something more is done for them, in order to enable them to uphold their posture. There is no reason whatever to doubt, but that their parts descend by gravitation in the same manner as those of dead matter. The gift therefore appears to me to consist in a faculty of perpetually shifting the centre of gravity, by a set, of obscure indeed, but of quick balancing actions, so as to keep the line of direction, which is a line drawn from that centre to the ground, within its prescribed limits. Of these actions it may be observed, first, that they in part constitute what we call strength. The dead body drops down. The mere adjustment therefore of weight and pressure, which may be the same the moment after death as the moment before, does not support the column. In cases also of extreme weakness the patient cannot stand upright. Secondly, that these actions are only in a small degree voluntary. A man is seldom conscious of his voluntary powers in keeping himself upon his legs. A child learning to walk is the greatest posture-master in the world: but art, if it may be so called, sinks into habit; and he is soon able to poise himself in a great variety of attitudes without being sensible either of caution or effort. But still there must be an aptitude of parts upon which habit can thus attach; a previous capacity of motions which the animal is thus taught to exercise: and the facility, with which this exercise is acquired, forms one object of our admiration. What parts are principally employed, or in what manner each contributes its office, is, as hath already been confessed, difficult to explain. Perhaps the obscure motion of the bones of the feet may have their share in this effect. They are put in action by every slip or vacillation of the body, and seem to assist in restoring its balance. Certain it is, that this circumstance in the structure of the foot, viz. its being composed of many small bones, applied to, and articulating with, one another, by diversely shaped surfaces, instead of being made of one piece, like the last of a shoe, is very remarkable. I suppose also that it would be difficult to stand firm upon stilts or wooden legs, though their base exactly imitated the figure and dimensions of the sole of the foot. The alternation of the joints, the knee joint bending backward, the hip joint forward; the flexibility, in every direction, of the spine, especially in the loins and neck, appear to be of great moment in preserving the equilibrium of the body. With respect to this last circumstance it is observable, that the vertebræ are so confined by ligaments as to allow no more slipping upon their bases, than what is just sufficient to break the shock which any violent motion may occasion to the body. A certain degree also of tension of the sinews appears to be essential to an erect posture; for it is by the loss of this, that the dead or paralytic body drops down. The whole is a wonderful result of combined powers, and of very complicated operations.

We have said that this property is the most worthy of observation in the *human* body: but a *bird*, resting upon its perch, or hopping upon a spray, affords no mean specimen of the same faculty. A chicken runs off as soon as it is hatched from the egg; yet a chicken, considered geometrically, and with relation to its centre of gravity, its line of direction, and its equilibrium, is a very irregular solid. Is this gift, therefore, or instruction? May it not be said to be with great attention, that nature hath balanced the body upon its pivots?*

I observe also in the same *bird* a piece of useful mechanism of this kind. In the trussing of a fowl, upon bending the legs and thighs up towards the body, the cook finds that the claws close of their own accord. Now let it be remembered, that this is the position of the limbs, in which the bird rests upon its perch. And in this position it sleeps in safety; for the claws do their office in keeping hold of the support, not by any exertion of voluntary power, which sleep might suspend, but by the traction of the tendons, in consequence of the attitude which the legs and thighs take by the bird sitting down, and to which the mere weight of the body gives the force that is necessary.

VI. Regarding the human body as a mass; regarding the general conformations which obtain in it; regarding also particular parts in respect to those conformations; we shall be led to observe what I call 'interrupted analogies.' The following are examples of what I mean by these terms: and I don't know, how such critical deviations can, by any possible hypothesis, be accounted for, without design.

1. All the bones of the body are covered with a *periosteum*,* except the teeth; where it ceases, and an enamel of ivory, which saws and files will hardly touch, comes into its place. No one can doubt of the use and propriety of this difference; of the 'analogy' being thus 'interrupted;' of the rule, which belongs to the conformation of the bones, stopping where it does stop: for, had so exquisitely sensible a membrane as the periosteum, invested the teeth, as it invests every other bone of the body, their action, necessary exposure, and irritation, would have subjected the animal to continual pain. General as it is, it was not the sort of integument which suited the teeth. What they stood in need of, was a strong, hard, insensible, defensive coat: and exactly such a covering is given to them, in the ivory enamel which adheres to their surface.

2. The scarf-skin, which clothes all the rest of the body, gives way, at the extremities of the toes and fingers, to *nails*. A man has only to look at his hand, to observe with what nicety and precision, that covering, which extends over every other part, is here superseded by a different substance and a different texture. Now, if either the rule had been necessary, or the deviation from it accidental, this effect would not be seen. When I speak of the rule being necessary, I mean the formation of the skin upon the surface being produced by a set of causes constituted without design, and acting, as all ignorant causes must act, by a general operation. Were this the case, no account could be given of the operation being suspended at the fingers' ends,

or on the back part of the fingers, and not on the fore part. On the other hand; if the deviation were accidental, an error, an anomalism; were it any thing else than settled by intention; we should meet with nails upon other parts of the body. They would be scattered over the surface, like warts or pimples.

3. All the great cavities of the body are inclosed by membranes except the *skull*. Why should not the brain be content with the same covering as that which serves for the other principal organs of the body? The heart, the lungs, the liver, the stomach, the bowels, have all soft integuments, and nothing else. The muscular coats are all soft and membranous. I can see a reason for this distinction in the final cause, but in no other. The importance of the brain to life, (which experience proves to be immediate,) and the extreme tenderness of its substance, make a solid case more necessary for it, than for any other part: and such a case the hardness of the skull supplies. When the smallest portion of this natural casquet is lost, how carefully, yet how imperfectly, is it replaced by a plate of metal? If an anatomist should say, that this bony protection is not confined to the brain, but is extended along the course of the spine, I answer, that he adds strength to the argument. If he remark, that the chest also is fortified by bones, I reply, that I should have alledged this instance myself, if the ribs had not appeared subservient to the purpose of motion, as well as of defence. What distinguishes the skull from every other cavity is, that the bony covering completely surrounds its contents, and is calculated, not for motion, but solely for defence. Those hollows likewise and inequalities, which we observe in the inside of the skull, and which exactly sit the folds of the brain, answer the important design of keeping the substance of the brain steady, and of guarding it against concussions.

CHAPTER XII

COMPARATIVE ANATOMY

WHENEVER we find a general plan pursued, yet with such variations in it, as are, in each case, required by the particular exigency of the subject to which it is applied, we possess, in such plan and such adaptation, the strongest evidence, that can be afforded, of intelligence and design; and evidence, which the most completely excludes every other hypothesis.* If the general plan proceeded from any fixed necessity in the nature of things, how could it accommodate itself to the various wants and uses which it had to serve, under different circumstances, and on different occasions? Arkwright's mill* was invented for the spinning of cotton. We see it employed for the spinning of wool, flax, and hemp, with such modifications of the original principle, such variety in the same plan, as the texture of those different materials rendered necessary. Of the machine's being put together with design, if it were possible to doubt, whilst we saw it only under one mode, and in one form; when we came to observe it in its different applications, with such changes of structure, such additions, and supplements, as the special and particular use in each case demanded, we could not refuse any longer our assent to the proposition, 'that intelligence, properly and strictly so called (including under that name, foresight, consideration, reference to utility,) had been employed, as well in the primitive plan, as in the several changes and accommodations which it is made to undergo.'

Very much of this reasoning is applicable to what has been called *Comparative Anatomy*. In their general œconomy, in the outlines of the plan, in the construction as well as offices of their principal parts, there exists, between all large terrestrial animals, a close resemblance. In all, life is sustained, and the body nourished, by nearly the same apparatus. The heart, the lungs, the stomach, the liver, the kidneys, are much alike in all. The same fluid (for no distinction of blood has been observed) circulates through their vessels, and nearly in the same order. The same cause, therefore, whatever that cause was, has been concerned in the origin; has governed the production of these different animal forms.

When we pass on to smaller animals, or to the inhabitants of a different element, the resemblance becomes more distant and more obscure, but still the plan accompanies us.

And what we can never enough commend, and which it is our business at present to exemplify, the plan is attended through all its varieties and deflections, by subserviences to special occasions and utilities.

I. The *covering* of different animals (though, whether I am correct in classing this under their anatomy, I don't know) is the first thing which presents itself to our observation; and is, in truth, both for its variety, and its suitableness to their several natures, as much to be admired as any part of their structure. We have bristles, hair, wool, furs, feathers, quills, prickles, scales; yet in this diversity both of material and form, we cannot change one animal's coat for another, without evidently changing it for the worse: taking care however to remark, that these coverings are, in many cases, armour as well as clothing; intended for protection, as well as warmth.

The *human* animal is the only one which is naked, and the only one which can clothe itself. This is one of the properties which renders him an animal of all climates, and of all seasons. He can adapt the warmth or lightness of his covering to the temperature of his habitation. Had he been born with a fleece upon his back, although he might have been comforted by its warmth in high latitudes, it would have oppressed him by its weight and heat, as the species spread towards the equator.

What art, however, does for men, nature has, in many instances, done for those animals which are incapable of art. Their clothing, of its own accord, changes with their necessities. This is particularly the case with that large tribe* of quadrupeds which are covered with *furs*. Every dealer in hare-skins, and rabbit-skins, knows how much the fur is thickened by the approach of winter. It seems to be a part of the same constitution and the same design, that wool, in hot countries, degenerates, as it is called, but in truth (most happily for the animal's ease) passes into hair; whilst, on the contrary, that hair, in the dogs of the polar regions, is turned into wool, or something very like it. To which may be referred, what naturalists have remarked, that bears, wolves, foxes, hares, which do not take the water, have the fur much thicker on the back than the belly: whereas in the beaver it is the thickest upon the belly; as are the feathers in water fowl. We know the final cause of all this; and we know no other.

The *covering of birds* cannot escape the most vulgar observation. Its lightness, its smoothness, its warmth; the disposition of the feathers all inclined backward, the down about their stem, the overlapping of their tips, their different configuration in different parts, not to mention the variety of their colours, constitute a vestment for the body, so beautiful, and so appropriate to the life which the animal is to lead, as that, I think, we should have had no conception of any thing equally perfect, if we had never seen it, or can now imagine any thing more so. Let us suppose (what is possible only in supposition) a person who had never seen a bird, to be presented with a plucked pheasant, and bid to set his wits to work, how to contrive for it a covering which shall unite the qualities of warmth, levity, and least resistance to the air, and the highest degree of each; giving it also as much of beauty and ornament as he could afford. He is the person to behold the work of the Deity, in this part of his creation, with the sentiments which are due to it.

The commendation, which the general aspect of the feathered world seldom fails of exciting, will be increased by further examination. It is one of those cases in which the philosopher has more to admire, than the common observer. Every *feather* is a mechanical wonder. If we look at the quill, we find properties not easily brought together, strength and lightness. I know few things more remarkable, than the strength and lightness of the very pen, with which I am writing. If we cast our eye to the upper part of the stem, we see a material, made for the purpose, used in no other class of animals, and in no other part of birds; tough, light, pliant, elastic. The pith, also, which feeds the feather, is, amongst animal substances, sui generis; neither bone, flesh, membrane, nor tendon.

But the artificial part of a feather is the *beard*, or, as it is sometimes I believe called, the vane. By the beards are meant, what are fastened on each side the stem, and what constitute the breadth of the feather; what we usually strip off, from one side or both, when we make a pen. The separate pieces, or laminæ, of which the beard is composed, are called threads, sometimes filaments, or rays. Now the first thing which an attentive observer will remark is, how much stronger the beard of the feather shews itself to be, when pressed in a direction perpendicular to its plane, than when rubbed, either up or down, in the line of the stem; and he will soon discover the structure which occasions this difference, viz. that the laminæ whereof these beards are composed, are flat, and placed with their flat sides towards each other; by which means, whilst they *easily* bend for the approaching of each other, as any one may perceive by drawing his finger ever so lightly upwards, they are much harder to bend out of their plane, which is the direction in which they have to encounter the impulse and pressure of the air; and in which their strength is wanted, and put to the trial.

This is one particularity in the structure of a feather: a second is still more extraordinary. Whoever examines a feather, cannot help taking notice, that the threads or laminæ of which we have been speaking, in their natural state *unite*; that their union is something more than the mere apposition of loose surfaces; that they are not parted asunder without some degree of force; that nevertheless there is no glutinous* cohesion between them; that, therefore, by some mechanical means or other, they catch or clasp among themselves, thereby giving to the beard or vane its closeness and compactness of texture. Nor is this all: when two laminæ, which have been separated by accident or force, are brought together again, they immediately *reclasp*: the connection, whatever it was, is perfectly recovered, and the beard of the feather becomes as smooth and firm as if nothing had happened to it. Draw your finger down the feather, which is against the grain, and you break, probably, the junction of some of the contiguous threads; draw your finger up the feather, and you restore all things to their former state. This is no common contrivance; and now for the mechanism by which it is effected. The threads or laminæ above mentioned are *interlaced* with one another; and the interlacing is performed by means of an infinite number of fibres or teeth, which the laminæ shoot forth *on each side*, and which hook and grapple together. A friend of mine counted fifty of these fibres in one twentieth of an inch. These fibres are crooked; but curved after a different manner; for those, which proceed from the thread on the side towards the extremity of the feather, are longer, more flexible, and bent downward: whereas those which proceed from the side towards the beginning or quill end of the feather are shorter, firmer, and turn upwards. The process then which takes place is as follows. When two laminæ are pressed together, so that these long fibres are forced far enough over the short ones, *their* crooked parts fall into the cavity made by the crooked parts of the others: just as the latch that is fastened to a door, enters into the cavity of the catch fixed to the door post, and, there hooking itself, *fastens* the door; for it is properly in this manner, that one thread of a feather is fastened to the other.

This admirable structure of the feather, which it is easy to see with the microscope,* succeeds perfectly for the use to which nature has designed it, which use was, not only that the laminæ might be united, but that when one thread or lamina has been separated from another by some external violence, it might be reclasped with sufficient facility and expedition.[1]

In the *ostrich*, this apparatus of crotchets and fibres, of hooks and teeth, is wanting; and we see the consequence of the want. The filaments hang loose and separate from one another, forming only a kind of down; which constitution of the feathers, however it may fit them for the flowing honours of a lady's head-dress, may be reckoned an imperfection in the bird, inasmuch as wings, composed of these feathers, although they may greatly assist it in running, do not serve for flight.

But under the present division of our subject, our business with feathers is, as they are the *covering* of the bird. And herein a singular circumstance occurs. In the small order of birds which winter with us, from a snipe downwards, let the external colour of the feathers be what it will, their Creator has universally given them a bed of *black* down next their bodies. Black, we know, is the warmest colour: and the purpose here is, to *keep in* the heat, arising from the heart and circulation of the blood. It is further likewise remarkable, that this is not found in larger birds; for which there is also a reason. Small birds are much more exposed to the cold than large ones; forasmuch as they present, in proportion to their bulk, a much larger surface to the air. If a turkey was divided into a number of wrens, supposing the shape of the turkey and the wren to be similar, the surface of all the wrens would exceed the surface of the turkey, in the proportion of the length, breadth, (or, of any homologous line) of a turkey to that of a wren; which would be perhaps a proportion of ten to one. It was necessary therefore that small birds should be warmer clad than large ones; and this seems to be the expedient, by which that exigency is provided for.

---

[1] The above account is taken from Memoirs for a Natural History of Animals by the Royal Academy of Paris, published 1701, p. 219.

II. In comparing different animals, I know no part of their structure which exhibits greater variety, or, in that variety, a nicer accommodation to their respective conveniency, than that which is seen in the different formations of their *mouths*. Whether the purpose be the reception of aliment merely, or the catching of prey, the picking up of seeds, the cropping of herbage, the extraction of juices, the suction of liquids, the breaking and grinding of food, the taste of that food, together with the respiration of air, and in conjunction with the utterance of sound; these various offices are assigned to this one part, and, in different species, provided for, as they are wanted, by its different constitution. In the human species, forasmuch as there are hands to convey the food to the mouth, the mouth is flat, and by reason of its flatness fitted only for *reception*: whereas the projecting jaws, the wide rictus,* the pointed teeth, of the dog and his affinities, enable them to apply their mouths to *snatch and seize* the objects of their pursuit. The full lips, the rough tongue, the corrugated cartilaginous palate, the broad cutting teeth, of the ox, the deer, the horse and the sheep, qualify this tribe for *browsing* upon their pasture; either gathering large mouthfulls at once, where the grass is long, which is the case with the ox in particular; or biting close, where it is short, which the horse and the sheep are able to do, in a degree that one could hardly expect. The retired under jaw of a swine *works in the ground*, after the protruding snout, like a prong or ploughshare, has made its way to the roots upon which it feeds. A conformation so happy was not the gift of chance.

In *birds* this organ assumes a new character; new both in substance and in form, but, in both, wonderfully adapted to the wants and uses of a distinct mode of existence. We have, no longer, the fleshy lips, the teeth of enamelled bone; but we have, in the place of these two parts, and to perform the office of both, a hard substance (of the same nature with that which composes the nails, claws, and hoofs of quadrupeds) cut out into proper shapes, and mechanically suited to the actions which are wanted. The sharp edge and tempered point of the *sparrow's* bill, picks almost every kind of seed from its concealment in the plant; and not only so, but hulls the grain, breaks and shatters the coats of the seed, in order to get at the kernel. The *hooked* beak of the hawk tribe, separates the flesh from the bones of the animals which it feeds upon, almost with the cleanness and precision of a dissector's knife. (The butcher bird, transfixes its prey

upon the spike of a thorn, whilst it picks its bones.) In some birds of
this class, we have the *cross* bill, i. e. both the upper and lower bill
hooked, and their tips crossing. The *spoon* bill, enables the goose to
graze, to collect its food from the bottom of pools, or to seek it
amidst the soft or liquid substances with which it is mixed. The *long*
tapering bill of the snipe and woodcock, penetrates still deeper into
moist earth, which is the bed in which the food of that species is
lodged. This is exactly the instrument which the animal wanted. It
did not want strength in its bill, which was inconsistent with the
slender form of the animal's neck, as well as unnecessary for the kind
of aliment upon which it subsists; but it wanted length to reach its
object.

But the species of bill which belongs to birds that live by *suction*,
deserves to be described in its particular relation to that office. They
are what naturalists call serrated or dentated bills; the inside of
them, towards the edge, being thickly set with parallel or concentric
rows, of short, strong, sharp-pointed prickles. These, though they
should be called teeth, are not for the purpose of mastication, like
the teeth of quadrupeds; nor yet, as in fish, for the seizing and
retaining of their prey; but for a quite different use. They form a
filter. The *duck* by means of them discusses the mud; examining,
with great accuracy, the puddle, the brake, every mixture which is
likely to contain her food. The operation is thus carried on. The
liquid or semiliquid substances, in which the animal has plunged her
bill, she draws, by the action of her lungs, through the narrow interstices which lie between these teeth; catching, as the stream passes
across her beak, whatever it may happen to bring along with it, that
proves agreeable to her choice, and easily dismissing all the rest.
Now suppose the purpose to have been, out of a mass of confused
and heterogeneous substances, to separate for the use of the animal,
or rather to enable the animal to separate for its own, those few
particles which suited its taste and digestion, what more artificial, or
more commodious, instrument of selection, could have been given to
it, than this natural filter? It has been observed also, what must
enable the bird to choose and distinguish with greater acuteness, as
well, probably, as what increases its gratification and its luxury, that
the bills of this species are furnished with large nerves, that they are
covered with a skin, and that the nerves run down to the very
extremity. In the curlew, woodcock, and snipe, there are *three pairs* of

nerves, equal almost to the optic nerve in thickness, which pass first along the roof of the mouth, and then along the upper chap, down to the point of the bill, long as the bill is.

But to return to the train of our observations. The similitude* between the bills of birds and the mouths of quadrupeds, is exactly such, as, for the sake of the argument, might be wished for. It is near enough to shew the continuation of the same plan: it is remote enough to exclude the supposition of the difference being produced by action or use. A more prominent contour, or a wider gape, might be resolved into the effect of continued efforts, on the part of the species, to thrust out the mouth, or open it to the stretch. But by what course of action, or exercise, or endeavour, shall we get rid of the lips, the gums, the teeth; and acquire, in the place of them, pincers of horn? By what habit shall we so completely change, not only the shape of the part, but the substance of which it is composed?* The truth is, if we had seen no other than the mouths of quadrupeds, we should have thought no other could have been formed: little could we have supposed, that all the purposes of a mouth, furnished with lips, and armed with teeth, could be answered by an instrument which had none of these; could be supplied, and that with many additional advantages, by the hardness, and sharpness, and figure, of the bills of birds.

Every thing about the animal *mouth* is mechanical. The teeth of fish, have their points turned backwards, like the teeth of a wool or cotton-card.* The teeth of lobsters, work one against another, like the sides of a pair of shears. In many insects, the mouth is converted into a pump or sucker, fitted at the end sometimes with a wimble, sometimes with a forceps; by which double provision, viz. of the tube and the penetrating form of the point, the insect first bores through the integuments of its prey, and then extracts the juices. And, what is most extraordinary of all, one sort of mouth, as the occasion requires, shall be changed into another sort. The caterpillar could not live without teeth; in several species, the butterfly formed from it, could not use them. The old teeth therefore are cast off with the exuviæ* of the grub; a new and totally different apparatus assumes their place in the fly. Amidst these novelties of form, we sometimes forget that it is, all the while, the animal's *mouth*; that, whether it be lips, or teeth, or bill, or beak, or shears, or pump, it is the same part diversified: and it is also remarkable, that under all the varieties of configuration with which we are acquainted, and which are very great, the organs of taste and smelling are situated near each other.

III. To the mouth adjoins the *gullet*: in this part also, comparative anatomy discovers a difference of structure adapted to the different necessities of the animal. In brutes, because the posture of their neck conduces little to the passage of the aliments, the fibres of the gullet, which act in this business, run in two close spiral lines, crossing each other: in men, these fibres run only a little obliquely from the upper end of the œsophagus to the stomach, into which, by a gentle contraction, they easily transmit the descending morsels; that is to say, for the more laborious deglutition of animals, which thrust their food *up* instead of *down*, and also through a longer passage, a proportionably more powerful apparatus of muscles is provided; more powerful, not merely by the strength of the fibres, which might be attributed to the greater exercise of their force, but in their collocation, which is a determinate circumstance, and must have been original.

IV. The gullet leads to the *intestines*: here, likewise, as before, comparing quadrupeds with man, under a general similitude we meet with appropriate differences. The valvulæ conniventes,* or, as they are by some called, the semilunar valves, found in the human intestine, are wanting in that of brutes. These are wrinkles or plaits of the innermost coat of the guts, the effect of which is to retard the progress of the food through the alimentary canal. It is easy to understand how much more necessary such a provision may be to the body of an animal of an erect posture, and in which, consequently, the weight of the food is added to the action of the intestine, than in that of a quadruped, in which the course of the food, from its entrance to its exit, is nearly horizontal: but it is impossible to assign any cause, except the final cause, for this distinction actually taking place. So far as depends upon the action of the part, this structure was more to be expected in a quadruped than a man. In truth, it must, in both, have been formed, not by action, but in direct opposition to action, and to pressure: but the opposition, which would arise from pressure, is greater in the upright trunk than in any other. That theory therefore is pointedly contradicted by the example before us. The structure is found, where its generation, according to the method by which the theorist would have it generated, is the most difficult; but (*observe*) it is found, where its effect is most useful.

The different length of the intestines in carnivorous and herbivorous animals has been noticed on a former occasion. The shortest, I believe, is that of some birds of prey, in which the intestinal canal is little more than a straight passage from the mouth to the vent. The longest is in the deer kind. The intestines of a Canadian stag, four feet high, measured ninety-six feet.[1] The intestine of a sheep, unravelled, measures thirty times the length of the body. The intestine of a wild cat is only three times the length of the body. Universally, where the substance upon which the animal feeds, is of slow concoction, or yields its chyle with more difficulty, there the passage is circuitous and dilatory, that time and space may be allowed for the change and the absorption which are necessary. Where the food is soon dissolved, or already half assimilated, an unnecessary, or, perhaps, hurtful detention is avoided, by giving to it a shorter and a readier route.

V. In comparing the *bones* of different animals, we are struck, in the bones of birds, with a *propriety*, which could only proceed from the wisdom of an intelligent and designing Creator. In the bones of an animal which is to fly, the two qualities required, are strength and lightness. Wherein, therefore, do the bones of birds (I speak of the cylindrical bones) differ, in these respects, from the bones of quadrupeds? In three properties; first, their cavities are much larger in proportion to the weight of the bone, than in those of quadrupeds: secondly, these cavities are empty: thirdly, the shell is of a firmer texture, that is the substance of other bones. It is easy to observe these particulars, even in picking the wing or leg of a chicken. Now, the weight being the same, the diameter, it is evident, will be greater in a hollow bone than a solid one; and, with the diameter, as every mathematician can prove, is increased, cæteris paribus,* the strength of the cylinder, or its resistance to breaking. In a word; a bone of the *same weight* would not have been so strong in any other form; and, to have made it heavier, would have incommoded the animal's flight. Yet this form could not be acquired by use, or the bone become hollow and tubular by exercise. What appetency could excavate a bone?

VI. The *lungs* also of birds, as compared with the lungs of quadrupeds, contain in them a provision, distinguishingly calculated for

[1] Mem. of Acad. Paris, 1701, p. 170.

this same purpose of levitation; namely, a communication (not found in other kinds of animals) between the air-vessels of the lungs and the cavities of the body: so that by the intromission of air from one to the other, at the will, as it should seem, of the animal, its body can be occasionally puffed out, and its tendency to descend in the air, or its specific gravity,* made less. The bodies of birds are blown up from their lungs, which no other animal bodies are; and thus rendered buoyant.

VII. All birds are *oviparous*. This, likewise, carries on the work of gestation, with as little increase as possible of the weight of the body. A gravid uterus* would have been a troublesome burthen to a bird in its flight. The advantage, in this respect, of an oviparous procreation is, that, whilst the whole brood are hatched together, the eggs are excluded singly, and at considerable intervals. Ten, fifteen, or twenty young birds may be produced in one cletch or covey,* yet the parent bird have never been encumbered by the load of more than one full grown egg at one time.

VIII. A principal topic of comparison between animals, is in their *instruments of motion*. These come before us under three divisions, feet, wings, and fins. I desire any man to say, which of the three is best fitted for its use: or whether the same consummate art be not conspicuous in them all. The constitution of the elements, in which the motion is to be performed, is very different. The animal action must necessarily follow that constitution. The Creator therefore, if we might so speak, had to prepare for different situations, for different difficulties: yet the purpose is accomplished not less successfully, in one case than the other. And, as between *wings* and the corresponding limbs of quadrupeds, it is accomplished without deserting the general idea. The idea is modified, not deserted. Strip a wing of its feathers, and it bears no obscure resemblance to the fore-leg of a quadruped. The articulations at the shoulder and the cubitus* are much alike; and, what is a closer circumstance, in both cases the upper part of the limb consists of a single bone, the lower part of two.

But, fitted up with its furniture of feathers and quills, it becomes a wonderful instrument; more artificial than its first appearance indicates, though that be very striking: at least, the use, which the bird makes of its wings in flying, is more complicated, and more curious, than is generally known. One thing is certain; that, if the flapping of the wings in flight were no more than the reciprocal motion of the same surface in opposite directions, either upwards and downwards, or estimated in any oblique line, the bird would lose as much by one motion, as she gained by another. The skylark could never ascend by such an action as this; for, though the stroke upon the air by the under side of her wing would carry her up, the stroke from the upper side, when she raised her wing again, would bring her down. In order, therefore, to account for the advantage which the bird derives from her wings, it is necessary to suppose, that the surface of the wing, measured upon the same plane, is contracted, whilst the wing is drawn up; and let out to its full expansion, when it descends upon the air for the purpose of moving the body by the reaction of that element. Now the form and structure of the wing, its external convexity, the disposition, and particularly the overlapping, of its larger feathers, the action of the muscles and joints of the pinions, are all adapted to this alternate adjustment of its shape and dimensions. Such a twist, for instance, or semirotatory motion, is given to the great feathers of the wing, that they strike the air with their flat side, but rise from the stroke slantwise. The turning of the oar in rowing, whilst the rower advances his hand for a new stroke, is a similar operation to that of the feather, and takes its name from the resemblance. I believe that this faculty is not found in the great feathers of the tail. This is the place also for observing, that the pinions are so set on upon the body, as to bring down the wings, not vertically, but in a direction obliquely tending towards the tail: which motion, by virtue of the common resolution of forces, does two things at the same time; supports the body in the air, and carries it forward.

The *steerage* of a bird in its flight is effected partly by the wings, but, in a principal degree, by the tail. And herein we meet with a circumstance not a little remarkable. Birds with long legs have short tails; and, in their flight, place their legs close to their bodies, at the same time stretching them out backwards as far as they can. In this position the legs extend beyond the rump, and become the rudder; supplying that steerage which the tail could not.

From the *wings* of birds, the transition is easy to the *fins* of fish. They are both, to their respective tribes, the instruments of their motion; but, in the work which they have to do, there is a considerable difference, founded in this circumstance. Fish, unlike birds, have very nearly the same specific gravity with the element in which they move. In the case of fish, therefore, there is little or no weight to bear up: what is wanted, is only an impulse sufficient to carry the body through a resisting medium, or to maintain the posture, or to support or restore the balance of the body, which is always the most unsteady where there is no weight to sink it. For these offices the fins are as large as necessary, though much smaller than wings, their action mechanical, their position, and the muscles by which they are moved, in the highest degree, convenient. The following short account of some experiments upon fish, made for the purpose of ascertaining the use of their fins, will be the best confirmation of what we assert. In most fish, beside the great fin the tail, we find two pair of fins upon the sides, two single fins upon the back, and one upon the belly, or rather between the belly and the tail. The *balancing* use of these organs is proved in this manner. Of the large-headed fish, if you cut off the pectoral fins, i. e. the pair which lies close behind the gills, the head falls prone to the bottom: if the right pectoral fin only be cut off, the fish leans to that side; if the ventral fin on the same side be cut away, then it loses its equilibrium entirely: if the dorsal and ventral fins be cut off, the fish reels to the right and left. When the fish dies, that is, when the fins cease to play, the belly turns upwards. The use of the same parts for *motion* is seen in the following observation upon them when put in action. The pectoral, and more particularly the ventral fins, serve to *raise and depress* the fish: when the fish desires to have a *retrograde* motion, a stroke forward with the pectoral fin effectually produces it: if the fish desire to *turn* either way, a single blow with the tail the opposite way, sends it round at once: if the tail strike both ways, the motion produced by the double lash is *progressive*; and enables the fish to dart forwards with an astonishing velocity.[1] The result is, not only, in some cases, the most rapid, but, in all cases, the most gentle, pliant, easy, animal motion, with which we are acquainted. However, when the tail is cut off, the fish loses all motion, and gives itself up to where the water impels it. The rest of the fins, therefore, so far as respects motion, seem to be merely subsidiary to this. In their mechanical use, the anal fin may be reckoned the keel, the ventral fins, out-riggers; the pectoral muscles, the oars:* and if there be any similitude between these parts of a boat and a fish, observe, that it is not the resemblance of

[1] Goldsmith's Hist. of An. Nat.* vol. vi. p. 154.

imitation, but the likeness which arises from applying similar mechanical means to the same purpose.

We have seen that the *tail* in the fish is the great instrument of motion. Now, in cetaceous or warm-blooded fish, which are obliged to rise every two or three minutes to the surface to take breath, the tail, unlike what it is in other fish, is horizontal; its stroke, consequently, perpendicular to the horizon, which is the right direction for sending the fish to the top, or carrying it down to the bottom.

REGARDING animals in their instruments of motion, we have only followed the comparison through the first great division of animals into beasts, birds, and fish. If it were our intention to pursue the consideration further, I should take in that generic distinction amongst birds, the *web foot* of water fowl. It is an instance which may be pointed out to a child. The utility of the web to water fowl, the inutility to land fowl, are so obvious, that it seems impossible to notice the difference without acknowledging the design. I am at a loss to know, how those who deny the agency of an intelligent Creator, dispose of this example. There is nothing in the action of swimming, as carried on by a bird upon the surface of the water, that should generate a membrane between the toes. As to that membrane, it is an exercise of constant resistance. The only supposition I can think of is, that all birds have been originally water fowl, and web footed; that sparrows, hawks, linnets, etc. which frequent the land, have, in process of time, and in the course of many generations, had this part worn away by treading upon hard ground.* To such evasive assumptions must atheism always have recourse; and, after all, it confesses that the structure of the feet of birds, in their original form, was critically adapted to their original destination. The web feet of amphibious quadrupeds, seals, otters, etc. fall under the same observation.

IX. The *five senses* are common to most large animals: nor have we much difference to remark in their constitution; or much however which is referable to mechanism.

The superior sagacity of animals which hunt their prey, and which, consequently, depend for their livelihood upon their *nose*, is well known, in its use; but not at all known in the organization which produces it.

The external *ears* of beasts of prey, of lions, tigers, wolves, have their trumpet part or concavity standing forwards, to seize the sounds which are before them, viz. the sounds of the animals, which they pursue or watch. The ears of animals of flight are turned backward, to give notice of the approach of their enemy from behind, when he may steal upon them unseen. This is a critical distinction; and is mechanical: but it may be suggested, and, I think, not without probability, that it is the effect of continued habit.

The *eyes* of animals which follow their prey by night, as cats, owls, etc. possess a faculty, not given to those of other species, namely, of closing the pupil *entirely*. The final cause of which seems to be this. It was necessary for such animals to be able to descry objects with very small degrees of light. This capacity depended upon the superior sensibility of the retina; that is, upon its being affected by the most feeble impulses. But that tenderness of structure, which rendered the membrane thus exquisitely sensible, rendered it also liable to be offended by the access of stronger degrees of light. The contractile range therefore of the pupil is increased in these animals, so as to enable them to close the aperture entirely; which includes the power of diminishing it in every degree; whereby at all times such portions, and only such portions of light are admitted, as may be received without injury to the sense.

There appears to be also in the figure, and in some properties of the pupil of the eye, an appropriate relation to the wants of different animals. In horses, oxen, goats, sheep, the pupil of the eye is elliptical; the transverse axis being horizontal: by which structure, although the eye be placed on the side of the head, the anterior elongation of the pupil catches the forward rays, or those which come from objects immediately in front of the animal's face.

CHAPTER XIII

PECULIAR ORGANIZATIONS

I BELIEVE that all the instances which I shall collect under this title, might, consistently enough with technical language, have been placed under the head of *Comparative Anatomy*. But there appears to me an impropriety in the use which that term hath obtained: it being, in some sort, absurd, to call that a case of comparative anatomy, in which there is nothing to 'compare;' in which a conformation is found in one animal, which hath nothing properly answering to it in another. Of this kind are the examples which I have to propose in the present chapter; and the reader will see that, though some of them be the strongest, perhaps, he will meet with under any division of our subject, they must necessarily be of an unconnected and miscellaneous nature. To dispose them however into some sort of order, we will notice, first, particularities of structure which belong to quadrupeds, birds, and fish, as such, or to many of the kinds included in these classes of animals; and then, such particularities as are confined to one or two species.

I. Along each side of the neck of large *quadrupeds*, runs a stiff robust cartilage, which butchers call the pax wax. No person can carve the upper end of a crop of beef without driving his knife against it. It is a tough, strong, tendinous substance, braced from the head to the middle of the back: its office is to assist in supporting the weight of the head. It is a mechanical provision, of which this is the undisputed use; and it is sufficient, and not more than sufficient, for the purpose which it has to execute. The head of an ox or a horse is a heavy weight, acting at the end of a long lever, (consequently with a great purchase,) and in a direction nearly perpendicular to the joints of the supporting neck. From such a force, so advantageously applied, the bones of the neck would be in constant danger of dislocation, if they were not fortified by this strong tape. No such organ is found in the human subject, because, from the erect position of the head, (the pressure of it acting nearly in the direction of the spine,) the junction of the vertebræ appears to be sufficiently secure without it. The care of the Creator is seen where it is wanted. This cautionary expedient is limited to quadrupeds.

II. The oil with which *birds* prune their feathers, and the organ which supplies it, is a specific provision for the winged creation. On each side of the rump of birds is observed a small nipple, yielding upon pressure a butter-like substance, which the bird extracts by pinching the pap with its bill. With this oil or ointment, thus procured, the bird dresses its coat; and repeats the action as often as its own sensations teach it that it is in any part wanted, or as the excretion may be sufficient for the expense. The gland, the pap, the nature and quality of the excreted substance, the manner of obtaining it from its lodgment in the body, the application of it when obtained, form, collectively, an evidence of intention, which it is not easy to withstand. Nothing similar to it is found in unfeathered animals. What blind conatus* of nature should produce it in birds; should not produce it in beasts?

III. The air bladder also of a *fish*, affords a plain and direct instance, not only of contrivance, but strictly of that species of contrivance, which we denominate mechanical. It is a philosophical apparatus in the body of an animal. The principle of the contrivance is clear: the application of the principle is also clear. The use of the organ to sustain, and, at will, also to elevate, the body of the fish in the water, is proved by observing, what has been tried, that, when the bladder is burst, the fish grovels at the bottom; and also, that flounders, soles, skaits, which are without the air bladder, seldom rise in the water, and that with effort. The manner in which the purpose is attained, and the suitableness of the means to the end, are not difficult to be apprehended. The rising and sinking of a fish in water, so far as it is independent of the stroke of the fins and tail, can only be regulated by the specific gravity of the body. When the bladder, contained in the body of the fish, is contracted, which the fish probably possesses a muscular power of doing, the bulk of the fish is contracted along with it; whereby, since the absolute weight remains the same, the specific gravity, which is the sinking force, is increased, and the fish descends: on the contrary, when, in consequence of the relaxation of the muscles, the elasticity of the inclosed, and now compressed air, restores the dimensions of the bladder, the tendency downwards becomes proportionably less than it was before, or is turned into a contrary tendency. These are known properties of bodies immersed in a fluid. The enamelled figures, or little glass bubbles, in a jar of water, are made to rise and fall by the same artifice. A diving machine* might be made to ascend and descend upon the like principle; namely, by introducing into the inside of it an air vessel, which by its contraction would diminish, and by its distension enlarge, the bulk of the machine itself, and thus render it specifically heavier, or specifically lighter, than the water which surrounds it. Suppose this to be done; and the artist to solicit a patent for his invention. The inspectors of the model, whatever they might think of the use or value of the contrivance, could, by no possibility, entertain a question in their minds, whether it were a contrivance or not. No reason has ever been assigned, no reason can be assigned, why the conclusion is not as certain in the fish, as in the machine; why the argument is not as firm, in one case as the other.

It would be very worthy of enquiry, if it were possible to discover, by what method an animal, which lives constantly in water, is able to supply a repository of air. The expedient, whatever it be, forms part, and perhaps the most curious part, of the provision. Nothing similar to the air bladder is found in land animals; and a life in the water has no natural tendency to produce a bag of air. Nothing can be further from an acquired organization than this is.

THESE examples mark the attention of the Creator to three great kingdoms* of his animal creation, and to their constitution as such. The example which stands next in point of generality, belonging to a large tribe of animals, or rather to various species of that tribe, is the poisonous tooth of serpents.

I. The *fang of a viper* is a clear and curious example of mechanical contrivance. It is a perforated tooth, loose at the root; in its quiet state lying down flat upon the jaw, but furnished with a muscle, which, with a jerk, and by the pluck as it were of a string, suddenly erects it. Under the tooth, close to its root, and communicating with the perforation, lies a small bag containing the venom. When the fang is raised, the closing of the jaw presses its root against the bag underneath; and the force of this compression sends out the fluid, and with a considerable impetus, through the tube in the middle of the tooth. What more unequivocal or effectual apparatus could be devised, for the double purpose of at once inflicting the wound and injecting the poison? Yet, though lodged in the mouth, it is so constituted, as, in its inoffensive and quiescent state, not to interfere with the animal's ordinary office of receiving its food. It has been observed also, that none of the harmless serpents, the black snake, the blind worm, etc. have these fangs, but teeth of an equal size; not moveable, as this is, but fixed into the jaw.

II. In being the property of several different species, the preceding example is resembled by that which I shall next mention, which is the *bag of the opossum*. This is a mechanical contrivance, most properly so called. The simplicity of the expedient renders the contrivance more obvious than many others; and, by no means, less certain. A false skin under the belly of the animal, forms a pouch, into which the young litter are received at their birth; where they have an easy and constant access to the teats; in which they are transported by the dam from place to place; where they are at liberty to run in and out, and where they find a refuge from surprise and danger. It is their cradle, their conveyance, and their asylum. Can the use of this structure be doubted of? Nor is it a mere doubling of the skin, but it is a new organ, furnished with bones and muscles of its own. Two bones are placed before the os pubis,* and joined to that bone as their base. These support, and give a fixture to, the muscles, which serve to open the bag. To these muscles there are antagonists, which serve in the same manner to shut it: and this office they perform so exactly, that, in the living animal, the opening can scarcely be discerned, except when the sides are forcibly drawn asunder.[1] Is there any action in this part of the animal, any process arising from that action, by which these members could be formed? any account to be given of the formation, except design?

III. As a particularity, yet appertaining to more species than one; and also as strictly mechanical; we may notice a circumstance in the structure of the *claws* of certain birds. The middle claw of the heron and cormorant is toothed and notched like a saw. These birds are great fishers, and these notches assist them in holding their slippery prey. The use is evident; but the structure such, as cannot at all be accounted for by the effort of the animal, or the exercise of the part. Some other fishing birds have these notches in their *bills*; and for the same purpose. The gannet, or Soland goose, has the side of its bill irregularly jagged, that it may hold its prey the faster. Nor can the structure in this, more than in the former case, arise from the man-

---

[1] Goldsmith's Nat. Hist. vol. iv. p. 244.

ner of employing the part. The smooth surfaces, and soft flesh of fish, were less likely to notch the bills of birds, than the hard bodies upon which many other species feed.

We now come to particularities strictly so called, as being limited to a single species of animal. Of these I shall take one from a quadruped, and one from a bird.

I. The *stomach of the camel* is well known to retain large quantities of water, and to retain it unchanged for a considerable length of time. This property qualifies it for living in the desart. Let us see therefore what is the internal organization, upon which a faculty, so rare and so beneficial, depends. A number of distinct sacs or bags (in a dromedary thirty of these have been counted) and observed to lie between the membranes of the second stomach, and to open into the stomach near the top by small square apertures. Through these orifices, after the stomach is full, the annexed bags are filled from it. And the water so deposited, is, in the first place, not liable to pass into the intestines; in the second place, is kept separate from the solid aliment; and, in the third place, is out of the reach of the digestive action of the stomach, or of mixture with the gastric juice. It appears probable, or rather certain, that the animal, by the conformation of its muscles, possesses the power of squeezing back this water from the adjacent bags into the stomach, whenever thirst excites it to put this power in action.

II. The *tongue of the wood-pecker*, is one of those singularities, which nature presents us with, when a singular purpose is to be answered. It is a particular instrument for a particular use; and what else but design ever produces such? The woodpecker lives chiefly upon insects, lodged in the bodies of decayed or decaying trees. For the purpose of boring into the wood, it is furnished with a bill, straight, hard, angular, and sharp. When, by means of this piercer, it has reached the cells of the insects, then comes the office of its tongue; which tongue is first, of such a length that the bird can dart it out three or four inches from the bill, in this respect differing greatly from every other species of bird; in the second place, it is tipped with a stiff, sharp, bony thorn; and, in the third place, which appears to me the most remarkable property of all, this tip is dentated on both sides, like the beard of an arrow or the barb of a hook. The description of the part declares its use. The bird, having exposed the retreats of the insects by the assistance of its bill, with a motion inconceivably quick lanches out at them this long tongue; transfixes them upon the barbed needle at the end of it; and thus draws its prey within its mouth. If this be not mechanism, what is? Should it be said, that, by continual endeavours to shoot out the tongue to the stretch, the woodpecker species may by degrees have lengthened the organ itself, beyond that of other birds, what account can be given of its form; of its tip? How, in particular, did it get its barbs, its dentition? These barbs, in my opinion, wherever they occur, are decisive proofs of mechanical contrivance.

III. I shall add one more example for the sake of its novelty. It is always an agreeable discovery, when, having remarked in an animal an extraordinary structure, we come at length to find out an unexpected use for it. The following narrative, which Goldsmith has taken from Buffon,* furnishes an instance of this kind. The babyrou-essa, or Indian hog,* a species of wild boar found in the East Indies, has two *bent* teeth, more than half a yard long, growing upwards, and, which is the singularity, from the upper jaw. These instruments are not wanted for defence, that service being provided for by two tusks issuing from the under jaw, and resembling those of the common boar. Nor does the animal use them for defence. They might seem therefore to be both a superfluity and an incumbrance. But observe the event. The animal hitches one of these bent upper teeth upon the branch of a tree, and then suffers its whole body to swing from it. This is its manner of taking repose, and of consulting for its safety. It continues the whole night suspended by its tooth, both easy in its posture, and secure; being out of the reach of animals which hunt it for prey.[1]

---

[1] Goldsmith's Nat. Hist. vol. iii. p. 195.

CHAPTER XIV

PROSPECTIVE CONTRIVANCES

I CAN hardly imagine to myself a more distinguishing mark, and, consequently, a more certain proof of design, than *preparation*, i. e. the providing of things beforehand, which are not to be used until a considerable time afterwards; for this implies a contemplation of the future, which belongs only to intelligence.

Of these prospective contrivances the bodies of animals furnish various examples.

I. The human teeth afford an instance, not only of prospective contrivance, but of the completion of the contrivance being designedly suspended. They are formed within the gums, and there they stop: the fact being, that their further advance to maturity would not only be useless to the new-born animal, but extremely in its way; as it is evident that the act of *sucking*, by which it is for some time to be nourished, will be performed with more ease both to the nurse and to the infant, whilst the inside of the mouth, and edges of the gums, are smooth and soft, than if set with hard pointed bones. By the time they are wanted, the teeth are ready. They have been lodged within the gums for some months past, but detained, as it were, in their sockets, so long as their further protrusion would interfere with the office to which the mouth is destined. Nature, namely, that intelligence which was employed in creation, looked beyond the first year of the infant's life; yet, whilst she was providing for functions which were after that term to become necessary, was careful not to incommode those which preceded them. What renders it more probable that this is the effect of design is, that the teeth are imperfect, whilst all other parts of the mouth are perfect. The lips are perfect, the tongue is perfect; the cheeks, the jaws, the palate, the pharynx, the larynx, are all perfect. The teeth alone are not so. This is the fact with respect to the human mouth: the fact also is, that the parts above enumerated, are called into use from the beginning; whereas the teeth would be only so many obstacles and annoyances, if they were there. When a contrary order is necessary, a contrary order prevails. In the worm of the beetle, as hatched from the egg, the teeth are the first things which arrive at perfection. The insect begins to gnaw as soon as it escapes from the shell, though its other parts be only gradually advancing to their maturity.

What has been observed of the teeth, is true of the *horns* of animals; and for the same reason. The horn of a calf or lamb does not bud, or at least does not sprout to any considerable length, until the animal be capable of browsing upon its pasture; because such a substance upon the forehead of the young animal, would very much incommode the teat of the dam in the office of giving suck.

But in the case of the *teeth*, of the human teeth at least, the prospective contrivance looks still further. A succession of crops is provided, and provided from the beginning; a second tier being originally formed beneath the first, which do not come into use till several years afterwards. And this double or suppletory provision meets a difficulty in the mechanism of the mouth, which would have appeared almost unsurmountable. The expansion of the jaw, (the consequence of the proportionable growth of the animal, and of its skull) necessarily separates the teeth of the first set, however compactly disposed, to a distance from one another, which would be very inconvenient. In due time therefore, i. e. when the jaw has attained a great part of its dimensions, a new set of teeth springs up, (loosening and pushing out the old ones before them) more exactly fitted to the space which they are to occupy, and rising also in such close ranks, as to allow for any extension of line which the subsequent enlargement of the head may occasion.

II. It is not very easy to conceive a more evidently prospective contrivance, than that which, in all viviparous animals,* is found in the *milk* of the female parent. At the moment the young animal enters the world, there is its maintenance ready for it. The particulars to be remarked in this œconomy are neither few nor slight. We have, first, the nutritious quality of the fluid, unlike, in this respect, every other excretion of the body; and in which nature hitherto remains unimitated, neither cookery nor chymistry having been able to make milk out of grass: we have, secondly, the organ for its reception and retention: we have, thirdly, the excretory duct, annexed to it: and we have, lastly, the determination of the milk to the breast, at the particular juncture when it is about to be wanted. We have all these properties in the subject before us; and they are all indications of design. The last circumstance is the strongest of any. If I had been to guess beforehand, I should have conjectured, that, at the time when there was an extraordinary demand of nourishment in one part of the system, there would be the least likelihood of a redundancy to supply another part. The advanced pregnancy of the female has no intelligible tendency to fill the breasts with milk. The lacteal system* is a constant wonder: and it adds to other causes of our admiration, that the number of the teats or paps in each species is found to bear a proportion to the number of the young. In the sow, the bitch, the rabbit, the cat, the rat, which have numerous litters, the paps are numerous and are disposed along the whole length of the belly: in the cow and mare they are few. The most simple account of this, is to refer it to a designing Creator.

But, in the argument before us, we are entitled to consider not only animal bodies when framed, but the circumstances under which they are framed. And, in this view of the subject, the constitution of many of their parts, is, most strictly, prospective.

III. The eye is of no use, at the time when it is formed. It is an optical instrument made in a dungeon; constructed for the refraction of light to a focus, and perfect for its purpose, before a ray of light has had access to it; geometrically adapted to the properties and action of an element, with which it has no communication. It is about indeed to enter into that communication; and this is precisely the thing which evidences intention. It is *providing* for the *future* in the closest sense which can be given to these terms; for it is providing for a future change: not for the then subsisting condition of the animal; not for any gradual progress or advance in that same condition; but for a new state, the consequence of a great and sudden alteration, which the animal is to undergo at its birth. Is it to be believed that the eye was formed, or, which is the same thing, that the series of causes was fixed by which the eye is formed, without a view to this change; without a prospect of that condition, in which its fabric, of no use at present, is about to be of the greatest; without a consideration of the qualities of that element, hitherto entirely excluded, but with which it was hereafter to hold so intimate a relation? A young man makes a pair of spectacles for himself against he grows old: for which spectacles he has no want or use whatever at the time he makes them. Could this be done without knowing and considering the defect of vision to which advanced age is subject? Would not the precise suitableness of the instrument to its purpose, of the remedy to the defect, of the convex lens to the flattened eye, establish the certainty of the conclusion, that the case, afterwards to arise, had been considered beforehand, speculated upon, provided for? all which are exclusively the acts of a reasoning mind. The eye formed in one state, for use only in another state, and in a different state, affords a proof no less clear of destination to a future purpose; and a proof proportionably stronger, as the machinery is more complicated, and the adaptation more exact.

IV. What has been said of the eye, holds equally true of the *lungs*. Composed of air vessels, where there is no air; elaborately constructed for the alternate admission and exclusion of an elastic fluid, where no such fluid exists; this great organ, with the whole apparatus belonging to it, lies collapsed in the fœtal thorax,* yet in order, and in readiness for action, the first moment that the occasion requires its service. This is having a machine locked up in store for future use; which incontestably proves, that the case was expected to occur, in which this use might be experienced: but expectation is the proper act of intelligence. Considering the state in which an animal exists *before* its birth, I should look for nothing less in its body than a system of lungs. It is like finding a pair of bellows in the bottom of the sea; of no sort of use in the situation in which they are found; formed for an action which was impossible to be exerted; holding no relation or fitness to the element which surrounds them, but both to another element in another place.

As part and parcel of the same plan, ought to be mentioned, in speaking of the lungs, the provisionary contrivances of the foramen ovale and ductus arteriosus.* In the fœtus, pipes are laid for the passage of the blood through the lungs; but, until the lungs be inflated by the inspiration of air, that passage is impervious, or in a great degree obstructed. What then is to be done? What would an artist, what would a master, do upon the occasion? He would endeavour, most probably, to provide a *temporary* passage, which might carry on the communication required, until the other was open. Now this is the thing, which is, actually, done in the heart. Instead of the circuitous rout through the lungs, which the blood afterwards takes, before it get from one auricle of the heart to the other; a portion of the blood passes immediately from the right auricle to the left, through a hole, placed in the partition, which separates these cavities. This hole anatomists call the *foramen ovale*. There is likewise another cross cut, answering the same purpose, by what is called the *ductus* *arteriosus*, lying between the pulmonary artery and the aorta. But both expedients are so strictly temporary, that, after birth, the one passage is closed, and the tube which forms the other shrivelled up into a ligament. If this be not contrivance, what is?

But, forasmuch as the action of the air upon the blood in the lungs, appears to be necessary to the perfect concoction of that fluid, i. e. to the life and health of the animal, (otherwise the shortest rout might still be the best,) how comes it to pass that the fœtus lives, and grows, and thrives, without it? The answer is, that the blood of the fœtus is the mother's; that it has undergone that action in her habit; that one pair of lungs serves for both. When the animals are separated, a new necessity arises; and to meet this necessity as soon as it occurs, an organization is prepared. It is ready for its purpose: it only waits for the atmosphere: it begins to play, the moment the air is admitted to it.

CHAPTER XV

RELATIONS

WHEN several different parts contribute to one effect; or, which is the same thing, when an effect is produced by the joint action of different instruments; the fitness of such parts or instruments to one another, for the purpose of producing, by their united action, the effect, is what I call *relation*:* and wherever this is observed in the works of nature or of man, it appears to me to carry along with it decisive evidence of understanding, intention, art. In examining, for instance, the several parts of a *watch*, the spring, the barrel, the chain, the fusee, the balance, the wheels of various sizes, forms, and positions, what is it which would take the observer's attention, as most plainly evincing a construction, directed by thought, deliberation, and contrivance? It is the suitableness of these parts to one another, first, in the succession and order in which they act; and, secondly, with a view to the effect finally produced. Thus, referring the spring to the wheels, he sees, in it, that which originates and upholds *their* motion; in the chain, that which transmits the motion to the fusee; in the fusee, that which communicates it to the wheels; in the conical figure of the fusee, if he refer back again to the spring, he sees that which corrects the inequality of its force. Referring the wheels to one another, he notices, first, their teeth, which would have been without use or meaning, if there had been only one wheel, or if the wheels had had no connection between themselves, or common bearing upon some joint effect; secondly, the correspondency of their position, so that the teeth of one wheel catch into the teeth of another; thirdly, the proportion observed in the number of teeth of each wheel, which determines the rate of going. Referring the balance to the rest of the works, he saw, when he came to understand its action, that which rendered their motions equable. Lastly, in looking upon the index and face of the watch, he saw the use and conclusion of the mechanism, viz. marking the succession of minutes and hours; but all depending upon the motions within, all upon the system of intermediate actions between the spring and the pointer. What thus struck his attention in the several parts of the watch he might probably designate by one general name of 'relation:'

and observing, with respect to all the cases whatever, in which the origin and formation of a thing could be ascertained by evidence, that these relations were found in things produced by art and design, and in no other things, he would rightly deem of them as characteristic of such productions. To apply the reasoning here described to the works of nature.

The animal œconomy is full; is made up of these *relations*.

I. There are first, what, in one form or other, belong to all animals, the parts and powers which successively act upon their *food*. Compare this action with the process of a manufactory. In man and quadrupeds, the aliment is, first, broken and bruised by mechanical instruments of mastication, viz. sharp spikes or hard knobs, pressing against, or rubbing upon, one another: thus ground and comminuted, it is carried by a pipe into the stomach, where it waits to undergo a great chymical action, which we call digestion: when digested, it is delivered through an orifice, which opens and shuts as there is occasion, into the first intestine: there, after being mixed with certain proper ingredients, poured through a hole in the side of the vessel, it is further dissolved: in this state, the milk, chyle, or part which is wanted, and which is suited for animal nourishment, is strained off by the mouths of very small tubes, opening into the cavity of the intestines: thus freed from its grosser parts, the percolated fluid is carried by a long, winding, but traceable course, into the main stream of the old circulation; which conveys it, in its progress, to every part of the body. Now I say again, compare this with the process of a manufactory; with the making of cyder, for example, the bruising of the apples in the mill, the squeezing of them when so bruised in the press, the fermentation in the vat, the bestowing of the liquor thus fermented in the 'hogsheads, the drawing off into bottles, the pouring out for use into the glass. Let any one shew me any difference between these two cases, as to the point of contrivance. That which is at present under our consideration, the 'relation' of the parts successively employed, is not more clear in the last case, than in the first. The aptness of the jaws and teeth to prepare the food for the stomach, is, at least, as manifest, as that of the cyder-mill to crush the apples for the press. The concoction of the food in the stomach is as necessary for its future use, as the fermentation of the stum* in the vat is to the perfection of the liquor. The disposal of the aliment afterwards; the action and change which it undergoes;

the rout which it is made to take, in order that, and until that, it arrive at its destination, is more complex indeed and intricate, but, in the midst of complication and intricacy, as evident and certain, as is the apparatus of cocks, pipes, tunnels, for transferring the cyder from one vessel to another, of barrels and bottles for preserving it till fit for use, or of cups and glasses for bringing it, when wanted, to the lip of the consumer. The character of the machinery is in both cases this, that one part answers to another part, and every part to the final result.

This parallel between the alimentary operation and some of the processes of art, might be carried further into detail. Spallanzani has remarked[1] a circumstantial resemblance between the stomachs of gallinaceous* fowls and the structure of *corn-mills*. Whilst the two sides of the gizzard perform the office of the mill-stones, the craw or crop* supplies the place of the hopper.* When our fowls are abundantly supplied with meat they soon fill their craw; but it does not immediately pass thence into the gizzard. It always enters in very small quantities, in proportion to the progress of trituration: in like manner as in a mill a receiver is fixed above the two large stones which serve for grinding the corn; which receiver, although the corn be put into it by bushels, allows the grain to dribble only in small quantities into the central hole in the upper mill-stone.

But we have not done with the alimentary history. There subsists a general *relation* between the external organs of an animal by which it procures its food, and the internal powers by which it digests it. Birds of prey, by their talons and beaks, are qualified to seize and devour many species, both of other birds, and of quadrupeds. The constitution of the stomach agrees exactly with the form of the members. The gastric juice of a bird of prey, of an owl, a falcon, or a kite, acts upon the animal fibre alone; will not act upon seeds or grasses at all. On the other hand, the conformation of the mouth of the sheep or the ox is suited for browsing upon herbage. Nothing about these animals is fitted for the pursuit of living prey. Accordingly it has been found by experiments, tried not many years ago with perforated balls,* that the gastric juice of ruminating animals, such as the sheep and the ox, speedily dissolves vegetables, but makes no impression upon animal bodies. This accordancy is still

---

[1] Diss. I. sec. liv.

more particular. The gastric juice even of graminivorous* birds, will not act upon the grain, whilst whole and entire. In performing the experiment of digestion with the gastric juice in vessels, the grain must be crushed and bruised, before it be submitted to the menstruum, that is to say, must undergo by art, without the body, the preparatory action which the gizzard exerts upon it within the body, or no digestion will take place. So strict is the relation between the offices assigned to the digestive organ; between the mechanical operation, and the chymical process.

II. The relation of the kidneys to the bladder, and of the ureters to both, i. e. of the secreting organ to the vessel receiving the secreted liquor, and the pipe laid from one to the other for the purpose of conveying it from one to the other, is as manifest as it is amongst the different vessels employed in a distillery, or in the communications between them. The animal structure, in this case, being simple, and the parts easily separated, it forms an instance of correlation which may be presented by dissection to every eye, or which, indeed, without dissection is capable of being apprehended by every understanding. This correlation of instruments to one another fixes intention somewhere.

Especially when every other solution is negatived by the conformation. If the bladder had been merely an expansion of the ureter, produced by retention of the fluid, there ought to have been a bladder for each ureter. One receptacle, fed by two pipes, issuing from different sides of the body, yet from both conveying the same fluid, is not to be accounted for by any such supposition as this.

III. Relation of parts to one another accompanies us throughout the whole animal œconomy. Can any relation be more simple, yet more convincing, than this, that the eyes are so placed as to look in the direction in which the legs move and the hands work? It might have happened very differently, if it had been left to chance. There were, at least, three quarters of the compass out of four to have erred in. Any considerable alteration in the position of the eye, or the figure of the joints, would have disturbed the line, and destroyed the alliance between the sense and the limbs.

IV. But relation perhaps is never so striking, as when it subsists, not between different parts of the same thing, but between different things. The relation between a lock and a key is more obvious, than it is between different parts of the lock. A bow was designed for an arrow, and an arrow for a bow; and the design is more evident for their being separate implements.

Nor do the works of the Deity want this clearest species of relation. The *sexes* are manifestly made for each other.* They form the grand relation of animated nature; universal, organic, mechanical; subsisting, like the clearest relations of art, in different individuals; unequivocal, inexplicable without design:

So much so, that, were every other proof of contrivance in nature dubious or obscure, this alone would be sufficient. The example is complete. Nothing is wanting to the argument. I see no way whatever of getting over it.

V. The teats of animals, which give suck, bear a relation to the mouth of the suckling progeny; particularly to the lips and tongue. Here also, as before, is a correspondency of parts; which parts subsist in different individuals.

THESE are *general* relations, or the relations of parts which are found, either in all animals, or in large classes and descriptions of animals. *Particular* relations, or the relations which subsist between the particular configuration of one or more parts of certain species of animals, and the particular configuration of one or more other parts of the same animal, (which is the sort of relation, that is, perhaps, most striking,) are such as the following.

I. In the *swan*; the web foot, the spoon bill, the long neck, the thick down, the graminivorous stomach, bear all a relation to one another, inasmuch as they all concur in one design, that of supplying the occasions of an aquatic fowl, floating upon the surface of shallow pools of water, and seeking its food at the bottom. Begin with any one of these particularities of structure, and observe how the rest follow it. The web foot qualifies the bird for swimming; the spoon bill enables it to graze. But how is an animal, floating upon the surface of pools of water, to graze at the bottom, except by the mediation of a long neck? A long neck accordingly is given to it. Again, a warm-blooded animal, which was to pass its life upon water, required a defence against the coldness of that element. Such a defence is furnished to the swan, in the muff in which its body is wrapped. But all this outward apparatus would have been in vain, if the intestinal system had not been suited to the digestion of vegetable substances. I say suited to the digestion of vegetable substances: for it is well known, that there are two intestinal systems found in birds, one with a membranous stomach and a gastric juice, capable of dissolving animal substances alone; the other with a crop and gizzard, calculated for the moistening, bruising, and afterwards digesting, of vegetable aliment.

Or set off with any other distinctive part in the body of the swan; for instance, with the long neck. The long neck, without the web foot, would have been an incumbrance to the bird; yet there is no necessary connection between a long neck and a web foot. In fact they do not usually go together. How happens it, therefore, that they meet, only when a particular design demands the aid of both?

II. This mutual relation, arising from a subserviency to a common purpose, is very observable also in the parts of a *mole*. The strong short legs of that animal, the palmated* feet armed with sharp nails, the piglike nose, the teeth, the velvet coat, the small external ear, the sagacious smell, the sunk protected eye, all conduce to the utilities, or to the safety, of its underground life. It is a special purpose, specially consulted throughout. The form of the feet fixes the character of the animal. They are so many shovels: they determine its action to that of rooting in the ground; and every thing about its body agrees with this destination. The cylindrical figure of the mole, as well as the compactness of its form, arising from the terseness of its limbs, proportionally lessens its labour; because, according to its bulk, it thereby requires the least possible quantity of earth to be removed for its progress. It has nearly the same structure of the face and jaws as a swine, and the same office for them. The nose is sharp, slender, tendinous, strong; with a pair of nerves going down to the end of it. The plush covering, which, by the smoothness, closeness, and polish of the short piles that compose it, rejects the adhesion of almost every species of earth, defends the animal from cold and wet, and from the impediment, which it would experience by the mold sticking to its body. From soils of all kinds the little pioneer comes forth bright and clean. Inhabiting dirt, it is, of all animals, the neatest.

But what I have always most admired in the mole is its *eyes*. This animal occasionally visiting the surface, and wanting, for its safety and direction, to be informed when it does so, or when it approaches it, a perception of light was necessary. I do not know that the clearness of sight depends at all upon the size of the organ. What is gained by the largeness or prominence of the globe of the eye is width in the field of vision. Such a capacity would be of no use to an animal which was to seek its food in the dark. The mole did not want to look about it; nor would a large advanced eye have been easily defended from the annoyance, to which the life of the animal must constantly expose it. How indeed was the mole, working its way under ground, to guard its eyes at all? In order to meet this difficulty, the eyes are made scarcely larger than the head of a corking pin;* and these minute globules are sunk so deep in the skull, and lie so sheltered within the velvet of its covering, as that any contraction of what may be called the eyebrows, not only closes up the apertures which lead to the eyes, but presents a cushion, as it were, to any sharp or protruding substance, which might push against them. This aperture even in its ordinary state is like a pin hole in a piece of velvet, scarcely pervious to loose particles of earth.

Observe then, in this structure, that which we call relation. There is no natural connection between a small sunk eye and a shovel palmated foot. Palmated feet might have been joined with goggle eyes; or small eyes might have been joined with feet of any other form. What was it therefore which brought them together in the mole? That which brought together the barrel, the chain, and the fusee, in a watch: design; and design, in both cases, inferred, from the relation which the parts bear to one another in the prosecution of a common purpose. As hath already been observed, there are different ways of stating the relation, according as we set out from a different part. In the instance before us, we may either consider the shape of the feet, as qualifying the animal for that mode of life and inhabitation, to which the structure of its eye confines it; or we may consider the structure of the eye, as the only one which would have suited with the action to which the feet are adapted. The relation is manifest, whichever of the parts related we place first in the order of our consideration. In a word: the feet of the mole are made for digging; the neck, nose, eyes, ears and skin, are peculiarly adapted to an underground life: and this is what I call relation.

CHAPTER XVI

COMPENSATION

COMPENSATION is a species of relation. It is relation, when the *defects* of one part, or of one organ, are supplied by the structure of another part, or of another organ. Thus,

I. The short, unbending neck of the *elephant*, is compensated by the length and flexibility of his *proboscis*.* He could not have reached the ground without it: or, if it be supposed that he might have fed upon the fruit, leaves, or branches of trees, how was he to drink? Should it be asked, Why is the elephant's neck so short? it may be answered that the weight of a head so heavy could not have been supported at the end of a longer lever. To a form therefore, in some respects necessary, but in some respects also inadequate to the occasions of the animal, a supplement is added, which exactly makes up the deficiency under which he laboured.

If it be suggested, that this proboscis may have been produced in a long course of generations, by the constant endeavour of the elephant to thrust out his nose, (which is the general hypothesis by which it has lately been attempted to account for the forms of animated nature,)* I would ask, how was the animal to subsist in the mean time; during the process; *until* this prolongation of snout were completed? What was to become of the individual, whilst the species was perfecting?

Our business at present is, simply to point out the relation, which this organ bears to the peculiar figure of the animal, to which it belongs. And, herein, all things correspond. The necessity of the elephant's proboscis arises from the shortness of his neck; the shortness of the neck is rendered necessary by the weight of the head. Were we to enter into an examination of the structure and anatomy of the proboscis itself, we should see in it one of the most curious of all examples of animal mechanism. The disposition of the ringlets and fibres, for the purpose, first, of forming a long cartilaginous pipe; secondly, of contracting and lengthening that pipe; thirdly, of turning it in every direction at the will of the animal; with the superaddition, at the end, of a fleshy production, of about the length and thickness of a finger, and performing the office of a finger, so as to pick up a straw from the ground; these properties of the same organ, taken together, exhibit a specimen, not only of design, (which is attested by the advantage,) but of consummate art, and, as I may say, of elaborate preparation, in accomplishing that design.

II. The hook in the wing of a *bat*, is strictly a mechanical, and, also, a *compensating* contrivance. At the angle of its wing there is a bent claw, exactly in the form of a hook, by which the bat attaches itself to the sides of rocks, caves, and buildings, laying hold of crevices, joinings, chinks, and roughnesses. It hooks itself by this claw; remains suspended by this hold; takes its flight from this position: which operations compensate for the decrepitude of its legs and feet. Without her hook, the bat would be the most helpless of all animals. She can neither run upon her feet, nor raise herself from the ground. These inabilities are made up to her by the contrivance in her wing: and in placing a claw on that part, the Creator has deviated from the analogy observed in winged animals. A singular defect required a singular substitute.

III. The *crane* kind are to live and seek their food amongst the waters; yet, having no web feet, are incapable of swimming. To make up for this deficiency, they are furnished with long legs for wading, or long bills for groping; or usually with both. This is *compensation*. But I think the true reflection upon the present instance is, how every part of nature is tenanted by appropriate inhabitants. Not only is the surface of deep waters peopled by numerous tribes* of birds that swim, but marshes and shallow pools are furnished with hardly less numerous tribes of birds that wade.

IV. The common *parrot* has, in the structure of its beak, both an inconveniency, and a *compensation* for it. When I speak of an inconveniency, I have a view to a dilemma which frequently occurs in the works of nature, viz. that the peculiarity of structure by which an organ is made to answer one purpose, necessarily unfits it for some other purpose. This is the case before us. The upper bill of the parrot is so much hooked, and so much overlaps the lower, that, if, as in other birds, the lower chap alone had motion, the bird could scarcely gape wide enough to receive its food: yet this hook and overlapping of the bill could not be spared, for it forms the very instrument by which the bird climbs: to say nothing of the use which it makes of it in breaking nuts, and the hard substances upon which it feeds. How therefore has nature provided for the opening of this occluded mouth? By making the upper chap moveable, as well as the lower. In most birds the upper chap is connected, and makes but one piece, with the skull; but, in the parrot, the upper chap is joined to the bone of the head by a strong membrane, placed on each side of it, which lifts and depresses it at pleasure.[1]

V. The *spider's web* is a *compensating* contrivance. The spider lives upon flies, without wings to pursue them; a case, one would have thought, of great difficulty, yet provided for; and provided for by a resource, which no stratagem, no effort of the animal, could have produced, had not both its external and internal structure been specifically adapted to the operation.

VI. In many species of insects the eye is fixed; and consequently without the power of turning the pupil to the object. This great defect is however perfectly *compensated*; and by a mechanism which we should not suspect. The eye is a multiplying glass; with a lense looking in every direction, and catching every object. By which means, although the orb of the eye be stationary, the field of vision is as ample as that of other animals; and is commanded on every side. When this lattice work was first observed,* the multiplicity and minuteness of the surfaces must have added to the surprise of the discovery. Adams tells us,* that fourteen hundred of these reticulations have been counted in the two eyes of a drone bee.

In other cases, the *compensation* is effected, by the number and position of the eyes themselves. The spider has eight eyes, mounted upon different parts of the head, two in front, two in the top of the head, two on each side. These eyes are without motion; but, by their situation, suited to comprehend every view, which the wants or safety of the animal render it necessary for it to take.

VII. The Memoirs for the Natural History of Animals, published by the French Academy, A. D. 1687, furnish us with some curious particulars in the eye of a camelion. Instead of two eyelids it is covered by an eyelid with a hole in it. This singular structure appears to be *compensatory*, and to answer to some other singularities in the shape of the animal. The neck of the camelion is inflexible. To make up for this, the eye is so prominent, as that more than half of the ball stands out of the head. By means of which extraordinary projection, the pupil of the eye can be carried by the muscles in every direction,

[1] Goldsmith's Nat. Hist. vol. v. p. 274.

and is capable of being pointed towards every object. But then so unusual an exposure of the globe of the eye, requires for its lubricity and defence, a more than ordinary protection of eyelid, as well as more than ordinary supply of moisture; yet the motion of an eyelid, formed according to the common construction, would be impeded, as it should seem, by the convexity of the organ. The aperture in the lid meets this difficulty. It enables the animal to keep the principal part of the surface of the eye under cover, and to preserve it in a due state of humidity, without shutting out the light; or without performing every moment a nictitation,* which, it is probable, would be more laborious to this animal than to others.

VIII. In another animal, and in another part of the animal œconomy, the same Memoirs describe a most remarkable *substitution*. The reader will remember what we have already observed concerning the *intestinal* canal; that its length, so many times exceeding that of the body, promotes the extraction of the chyle from the aliment, by giving room for the lacteal vessels to act upon it through a greater space. This long intestine, wherever it occurs, is, in other animals, disposed in the abdomen from side to side in returning folds. But, in the animal now under our notice, the matter is managed otherwise. The same intention is mechanically effectuated; but by a mechanism of a different kind. The animal of which I speak, is an amphibious quadruped, which our authors call the alopecias, or sea fox. The intestine is straight from one end to the other: but in this straight, and consequently short intestine, is a winding, corkscrew, spiral passage, through which, the food, not without several circumvolutions, and in fact by a long rout, is conducted to its exit. Here the shortness of the gut is *compensated* by the obliquity of the perforation.

IX. But the works of the Deity are known by expedients. Where we should look for absolute destitution; where we can reckon up nothing but wants; some contrivance always comes in to supply the privation. A *snail*, without wings, feet, or thread, climbs up the stalks of plants, by the sole aid of a viscid humour discharged from her skin. She adheres to the stems, leaves, and fruits of plants, by means of a sticking plaister. A *muscle*, which might seem, by its helplessness, to lie at the mercy of every wave that went over it, has the singular power of spinning, strong, tendinous threads, by which she moors her shell to rocks and timbers. A *cockle*, on the contrary, by means of its stiff tongue, works for itself a shelter in the sand. The provisions of nature extend to cases the most desperate. A *lobster* has a difficulty in its constitution so great, that one could hardly conjecture before hand how nature would dispose of it. In most animals, the skin grows with their growth. If, instead of a soft skin, there be a shell, still it admits of a gradual enlargement. If the shell, as in the tortoise, consist of several pieces, the accession* of substance is made at the sutures. Bivalve shells grow bigger by receiving an accretion at their edge: it is the same with spiral shells at their mouth. The simplicity of their form admits of this. But the lobster's shell being applied to the limbs of the body, as well as to the body itself, allows not of either of the modes of growth which are observed to take place in other shells. Its hardness resists expansion; and its complexity renders it incapable of increasing its size by addition of substance to its edge. How then was the growth of the lobster to be provided for? Was room to be made for it in the old shell, or was it to be successively fitted with new ones? If a change of shell became necessary, how was the lobster to extricate himself from his present confinement? How was he to uncase his buckler, or draw his legs out of his boots? The process, which fishermen have observed to take place, is as follows. At certain seasons, the shell of the lobster grows soft; the animal swells its body; the seams open, and the claws burst at the joints. When the shell is thus become loose upon the body, the animal makes a second effort, and by a tremulous, spasmodic motion, casts it off. In this state the liberated, but defenceless, fish, retires into holes in the rock. The released body now suddenly pushes its growth. In about eight-and-forty hours, a fresh concretion of humour upon the surface, i. e. a new shell, is formed, adapted in every part to the increased dimensions of the animal. This wonderful mutation* is repeated every year.

If there be imputed defects without compensation, I should suspect that they were defects only in appearance. Thus, the body of the *sloth* has often been reproached for the slowness of its motions, which has been attributed to an imperfection in the formation of its limbs. But it ought to be observed, that it is this slowness, which alone suspends the voracity of the animal. He fasts during his migration from one tree to another; and this fast may be necessary for the relief of his overcharged vessels, as well as to allow time for

the concoction of the mass of coarse and hard food which he has
taken into his stomach. The tardiness of his pace seems to have
reference to the capacity of his organs, and to his propensities with
respect to food; h. e.* is calculated to counteract the effects of
repletion.

Or there may be cases, in which a defect is artificial, and compensated by the very cause which produces it. Thus the *sheep*, in the
domesticated state in which we see it, is destitute of the ordinary
means of defence or escape; is incapable either of resistance or flight.
But this is not so with the wild animal. The natural sheep is swift
and active: and, if it lose these qualities when it comes under the
subjection of man, the loss is compensated by his protection. Perhaps there is no species of quadruped whatever, which suffers so
little as this does, from the depredation of animals of prey.

FOR THE SAKE of making our meaning better understood, we
have considered this business of compensation under certain *particularities* of constitution, in which it appears to be most conspicuous.
This view of the subject necessarily limits the instances to single
species of animals. But there are compensations, perhaps, not less
certain, which extend over large classes, and to large portions, of
living nature.

I. In quadrupeds, the deficiency of teeth is usually *compensated* by
the faculty of rumination. The sheep, deer, and ox tribe, are without
fore teeth in the upper jaw. These ruminate. The horse and ass are
furnished with teeth in the upper jaw, and do not ruminate. In the
former class the grass and hay descend into the stomach, nearly in
the state in which they are cropped from the pasture, or gathered
from the bundle. In the stomach they are softened by the gastric
juice, which in these animals is unusually copious. Thus softened,
and rendered tender, they are returned a second time to the action of
the mouth, where the grinding teeth complete at their leisure the
trituration which is necessary, but which was before left imperfect. I
say the trituration which is necessary; for it appears from experiments that the gastric fluid of sheep, for example, has no effect in
digesting plants, unless they have been previously masticated; that it
only produces a slight maceration, nearly as common water would
do in a like degree of heat: but that, when once vegetables are
reduced to pieces by mastication, the fluid then exerts upon them its
specific operation. Its first effect is to soften them, and to destroy

their natural consistency: it then goes on to dissolve them; not sparing even the toughest parts, such as the nerves of the leaves.[1]

I think it very probable that the gratification also of the animal is renewed and prolonged by this faculty. Sheep, deer, and oxen, appear to be in a state of enjoyment whilst they are chewing the cud. It is then, perhaps, that they best relish their food.

II. In birds, the *compensation* is still more striking. They have no teeth at all. What have they then to make up for this severe want? I speak of graminivorous and herbivorous* birds; such as common fowls, turkeys, ducks, geese, pigeons, etc. for it is concerning these alone that the question need be asked. All these are furnished with a peculiar and most powerful muscle, called the *gizzard*; the inner coat of which is fitted up with rough plaits, which, by a strong friction against one another, break and grind the hard aliment, as effectually, and by the same mechanical action, as a coffee-mill would do. It has been proved by the most correct experiments, that the gastric juice of these birds will not operate upon the *entire* grain; not even when softened by water or macerated in the crop. Therefore without a grinding machine within its body; without the trituration of the gizzard; a chicken would have starved upon a heap of corn. Yet why should a bill and a gizzard go together? Why should a gizzard never be found where there are teeth?

Nor does the gizzard belong to birds as such. A gizzard is not found in birds of prey. *Their* food requires not to be ground down in a mill. The compensatory contrivance goes no further than the necessity. In both classes of birds however, the digestive organ within the body, bears a strict and mechanical relation to the external instruments for procuring food. The soft membranous stomach, accompanies the hooked, notched, beak; the short, muscular legs; the strong, sharp, crooked talons: the cartilaginous stomach, attends that conformation of bill and toes, which restrains the bird to the picking of seeds or the cropping of plants.

III. But to proceed with our *compensations*. A very numerous and comprehensive tribe of terrestrial animals are entirely without feet; yet locomotive; and, in a very considerable degree, swift in their motion. How is the *want of feet* compensated? It is done by the disposition of the muscles and fibres of the trunk. In consequence of

[1] Spal. Diss.* III. sec. cxl.

the just collocation, and by means of the joint action of longitudinal and annular fibres, that is to say, of strings and rings, the body and train of reptiles are capable of being reciprocally shortened and lengthened, drawn up and stretched out. The result of this action is a progressive, and, in some cases, a rapid movement of the whole body, in any direction to which the will of the animal determines it. The meanest creature is a collection of wonders. The play of the rings in an *earth-worm*, as it crawls; the undulatory motion propagated along the body; the beards or prickles, with which the annuli* are armed, and which the animal can either shut up close to its body, or let out to lay hold of the roughnesses of the surface upon which it creeps; and, the power arising from all these, of changing its place and position, affords, when compared with the provisions for motion in other animals, proofs of new and appropriate mechanism. Suppose that we had never seen an animal move upon the ground without feet, and that the problem was, muscular action, i. e. reciprocal contraction and relaxation being given, to describe how such an animal might be constructed, capable of voluntarily changing place. Something, perhaps, like the organization of reptiles, might have been hit upon by the ingenuity of an artist; or might have been exhibited in an automaton, by the combination of springs, spiral wires, and ringlets: but to the solution of the problem would not be denied, surely, the praise of invention and of successful thought; least of all could it ever be questioned, whether intelligence had been employed about it, or not.

CHAPTER XVII

THE RELATION OF ANIMATED BODIES TO
INANIMATE NATURE

WE have already confidered *relation*, and under different views; but it was the relation of parts to parts, of the parts of an animal to other parts of the same animal, or of another individual of the same species.

But the bodies of animals hold, in their constitution and properties, a close and important relation to natures altogether external to their own; to inanimate substances, and to the specific qualities of these, e. g. *they hold a strict relation to the elements by which they are surrounded.**

I. Can it be doubted, whether the *wings of birds* bear a relation to air, and the *fins of fish* to water? They are instruments of motion, severally suited to the properties of the medium in which the motion is to be performed: which properties are different. Was not this difference contemplated, when the instruments were differently constituted?

II. The structure of the animal *ear* depends for its use not simply upon being surrounded by a fluid, but upon the specific nature of that fluid. Every fluid would not serve: its particles must repel one another; it must form an elastic medium: for it is by the successive pulses of *such* a medium, that the undulations excited by the founding body are carried to the organ; that a communication is formed between the object and the sense; which must be done, before the internal machinery of the ear, subtile as it is, can act at all.

III. The *organs* of speech, and voice, and respiration, are, no less than the ear, indebted, for the success of their operation, to the peculiar qualities of the fluid, in which the animal is immersed. They, therefore, as well as the ear, are constituted upon the supposition of such a fluid, i. e. of a fluid with such particular properties, being always present. Change the properties of the fluid, and the organ cannot act: change the organ, and the properties of the fluid would be lost. The structure therefore of our organs, and the properties of our atmosphere, are made for one another. Nor does it alter the relation, whether you alledge the organ to be made for the element, (which seems the most natural way of considering it,) or the element as prepared for the organ.

IV. But there is another fluid with which we have to do; with properties of its own; with laws of acting, and of being acted upon, totally different from those of air or water:—and that is *light*. To this new, this singular element; to qualities perfectly peculiar, perfectly distinct and remote from the qualities of any other substance with which we are acquainted, an organ is adapted, an instrument is correctly adjusted, not less peculiar amongst the parts of the body, not less singular in its form, and, in the substance of which it is composed, not less remote from the materials, the model, and the analogy of any other part of the animal frame, than the element, to which it relates, is specific amidst the substances with which we converse. If this does not prove appropriation, I desire to know what would prove it.

Yet the element of light and the organ of vision, however related in their office and use, have no connection whatever in their original. The action of rays of light upon the surfaces of animals has no tendency to breed eyes in their heads. The sun might shine for ever upon living bodies without the smallest approach towards producing the sense of sight. On the other hand also, the animal eye does *not* generate or emit light.

V. Throughout the universe there is a wonderful *proportioning* of one thing to another. The size of animals, of the human animal especially, when considered with respect to other animals, or to the plants which grow around him, is such, as a regard to his conveniency would have pointed out. A giant or a pigmy could not have milked goats, reaped corn, or mowed grass; we may add, could not have rode a horse, trained a vine, shorn a sheep, with the same bodily ease as we do, if at all. A pigmy would have been lost amongst rushes, or carried off by birds of prey.

It may be mentioned likewise, that, the model and the materials of the human body being what they are, a much greater bulk would have broken down by its own weight. The persons of men, who much exceed the ordinary stature, betray this tendency.

VI. Again; and which includes a vast variety of particulars, and those of the greatest importance, how close is the *suitableness* of the earth and sea to their several inhabitants; and of these inhabitants to the places of their appointed residence?

Take the *earth* as it is; and consider the correspondency of the powers of its inhabitants with the properties and condition of the soil which they tread. Take the inhabitants as they are; and consider the substances which the earth yields for their use. They can scratch its surface, and its surface supplies all which they want. This is the length of their faculties; and such is the constitution of the globe, and their own, that this is sufficient for all their occasions.

When we pass from the earth to the *sea*, from land to water, we pass through a great change; but an adequate change accompanies us of animal forms and functions, of animal capacities and wants, so that *correspondency* remains. The earth in its nature is very different from the sea, and the sea from the earth;* but one accords with its inhabitants, as exactly as the other.

VII. The last relation of this kind which I shall mention is that of *sleep* to *night*. And it appears to me to be a relation which was expressly intended. Two points are manifest: first, that the animal frame requires sleep; secondly, that night brings with it a silence, and a cessation of activity, which allows of sleep being taken without interruption, and without loss. Animal existence is made up of action and slumber: nature has provided a season for each. An animal, which stood not in need of rest, would always live in daylight. An animal, which, though made for action, and delighting in action, must have its strength repaired by sleep, meets by its constitution the returns of day and night. In the human species for instance, were the bustle, the labour, the motion of life, upheld by the constant prefence of light, sleep could not be enjoyed without being disturbed by noise, and without expence of that time, which the eagerness of private interest would not contentedly resign. It is happy therefore for this part of the creation, I mean that it is conformable to the frame and wants of their constitution, that nature, by the very disposition of her elements, has commanded, as it were, and imposed upon them, at moderate intervals, a general intermission of their toils, their occupations, and pursuits.

But it is not for man, either solely or principally, that night is made. Inferior, but less perverted natures, taste its solace, and expect its return, with greater exactness and advantage than he does. I have often observed, and never observed but to admire, the satisfaction no less than the regularity, with which the greatest part of the irrational world yield to this soft necessity, this grateful vicissitude; how comfortably, the birds of the air, for example, address themselves to the repose of the evening; with what alertness they resume the activity of the day.

Nor does it disturb our argument to confess, that certain species of animals are in motion during the night, and at rest in the day. With respect even to them it is still true, that there is a change of condition in the animal, and an external change corresponding with it. There is still the relation, though inverted. The fact is, that the repose of other animals sets these at liberty, and invites them to their food or their sport.

If the relation of *sleep* to *night*, and, in some instances, its converse, be real, we cannot reflect without amazement upon the extent to which it carries us. Day and night are things close to us: the change applies immediately to our sensations: of all the phænomena of nature, it is the most familiar to our experience: but, in its cause, it belongs to the great motions which are passing in the heavens. Whilst the earth glides round her axle, she ministers to the alternate necessities of the animals dwelling upon her surface, at the same time that she obeys the influence of those attractions, which regulate the order of many thousand worlds. The relation therefore of sleep to night, is the relation of the inhabitants of the earth to the rotation of their globe; probably it is more: it is a relation to the system, of which that globe is a part; and, still further, to the congregation of systems, of which theirs is only one. If this account be true, it connects the meanest individual with the universe itself; a chicken roosting upon its perch, with the spheres revolving in the firmament.

But if any one object to our representation, that the succession of day and night, or the rotation of the earth upon which it depends, is not resolvible into central attraction, we will refer him to that which certainly is,—to the change of the seasons. Now the constitution of animals susceptible of torpor, bears a relation to winter, similar to that which sleep bears to night. Against not only the cold, but the want of food, which the approach of winter induces, the preserver of the world has provided, in many animals by migration, in many others by torpor. As one example out of a thousand, the bat, if it did not sleep through the winter, must have starved, as the moths and flying insects, upon which it feeds, disappear. But the transition from summer to winter carries us into the very midst of physical astronomy, that is to say, into the midst of those laws which govern the solar system at least, and probably all the heavenly bodies.

CHAPTER XVIII

INSTINCTS

THE order may not be very obvious, by which I place *instincts* next to relations. But I consider them as a species of relation. They contribute, along with the animal organization, to a joint effect, in which view they are related to that organization. In many cases they refer from one animal to another animal; and, when this is the case, become strictly relations in a second point of view.

An INSTINCT is a propensity, prior to experience, and independent of instruction.* We contend, that it is by instinct that the sexes of animals seek each other; that animals cherish their offspring; that the young quadruped is directed to the teat of its dam; that birds build their nests, and brood with so much patience upon their eggs; that insects, which do not sit upon their eggs; deposit them in those particular situations, in which the young, when hatched, find their appropriate food; that it is instinct, which carries the salmon, and some other fish, out of the sea into rivers, for the purpose of shedding their spawn in fresh water.

We may select out of *this* catalogue the incubation of eggs. I entertain no doubt, but that a couple of sparrows hatched in an oven, and kept separate from the rest of their species, would proceed as other sparrows do, in every office which related to the production and preservation of their brood. Assuming this fact, the thing is inexplicable upon any other hypothesis, than that of an instinct, impressed upon the constitution of the animal. For, first, what should induce the female bird to prepare a nest before she lays her eggs? It is in vain to suppose her to be possessed of the faculty of reasoning; for no reasoning will reach the case. The fullness or distension which she might feel in a particular part of her body, from the growth and solidity of the egg within her, could not possibly inform her, that she was about to produce something, which, when produced, was to be preserved and taken care of. Prior to experience, there was nothing to lead to this inference, or to this suspicion. The analogy was *all* against it; for, in every other instance, what issued from the body was cast out and rejected.

But, secondly, let us suppose the egg to be produced into day:

How should birds know that their eggs contain their young? There is nothing either in the aspect, or in the internal composition of an egg, which could lead even the most daring imagination to a conjecture, that it was hereafter to turn out, from under its shell, a living perfect bird. The form of the egg bears not the rudiments of a resemblance to that of the bird. Inspecting its contents, we find still less reason, if possible, to look for the result which actually takes place. If we should go so far, as, from the appearance of order and distinction in the disposition of the liquid substances which we noticed in the egg, to guess that it might be designed for the abode and nutriment of an animal, (which would be a very bold hypothesis,) we should expect a tadpole dabbling in the slime, much rather than a dry, winged, feathered creature; a compound of parts and properties impossible to be used in a state of confinement in the egg, and bearing no conceivable relation, either in quality or material, to any thing observed in it. From the white of an egg, would any one look for the feather of a goldfinch? or expect from a simple uniform mucilage, the most complicated of all machines, the most diversified of all collections of substances? Nor would the process of incubation, for some time at least, lead us to suspect the event. Who that saw red streaks, shooting in the fine membrane which divides the white from the yolk, would suppose that these were about to become bones and limbs? Who, that espied two discoloured points first making their appearance in the cicatrix,* would have had the courage to predict, that these points were to grow into the heart and head of a bird? It is difficult to strip the mind of its experience. It is difficult to resuscitate surprise, when familiarity has once laid the sentiment asleep. But could we forget all that we know, and which *our* sparrows never knew, about oviparous generation; could we divest ourselves of every information, but what we derived from reasoning upon the appearances or quality discovered in the objects presented to us, I am convinced that Harlequin coming out of an egg upon the stage, is not more astonishing to a child, than the hatching of a chicken both would be, and ought to be, to a philosopher.

But admit the sparrow by some means to know, that within that egg was concealed the principle of a future bird, from what chymist was she to learn, that *warmth* was necessary to bring it to maturity, or that the degree of warmth, imparted by the temperature of her own body, was the degree required?

To suppose, therefore, that the female bird acts in this process from a sagacity and reason of her own, is to suppose her to arrive at conclusions, which there are no premises to justify. If our sparrow, sitting upon her eggs, expect young sparrows to come out of them, she forms, I will venture to say, a wild and extravagant expectation, in opposition to present appearances, and to probability. She must have penetrated into the order of nature, further than any faculties of ours will carry us: and it hath been well observed, that this deep sagacity, if it be sagacity, subsists in conjunction with great stupidity, even in relation to the same subject. 'A chymical operation,' says Addison,* 'could not be followed with greater art or diligence, than is seen in hatching a chicken: yet is the process carried on without the least glimmering of thought or common sense. The hen will mistake a piece of chalk for an egg; is insensible of the increase or diminution of their number; does not distinguish between her own, and those of another species; is frightened when her supposititious breed of ducklings take the water.'

But it will be said, that what reason could not do for the bird, observation, or instruction, or tradition might. Now if it be true, that a couple of sparrows brought up from the first in a state of separation from all other birds, would build their nest, and brood upon their eggs, then there is an end of this solution. What can be the traditionary knowledge of a chicken hatched in an oven?

Of young birds taken in their nests, a few species breed, when kept in cages; and they which do so, build their nests nearly in the same manner as in the wild state, and sit upon their eggs. This is sufficient to prove an instinct, without having recourse to experiments upon birds, hatched by artificial heat, and deprived, from their birth, of all communication with their species: for we can hardly bring ourselves to believe, that the parent bird informed her unfledged pupil of the history of her gestation, her timely preparation of a nest, her exclusion of the eggs, her long incubation, and of the joyful eruption at last of her expected offspring: all which the bird in the cage must have learnt in her infancy, if we resolve her conduct into *institution*.

Unless we will rather suppose that she remembers her own escape from the egg; had attentively observed the conformation of the nest in which she was nurtured; and had treasured up her remarks for future imitation. Which is not only extremely improbable, (for who that sees a brood of callow birds in their nest, can believe that they are taking a plan of their habitation?) but leaves unaccounted for, one principal part of the difficulty, 'the preparation of the nest before the laying of the egg.' This she could not gain from observation in her infancy.

It is remarkable also, that the hen sits upon eggs, which she has laid without any communication with the male; and which are therefore necessarily unfruitful. That secret she is not let into. Yet, if incubation had been a subject of instruction or of tradition, it should seem that this distinction would have formed part of the lesson: whereas the instinct of nature is calculated for a state of nature; the exception, here alluded to, taking place, chiefly, if not solely, amongst domesticated fowls, in which nature is forced out of her course.

There is another case of oviparous œconomy, which is still less likely to be the effect of education, than it is even in birds, namely, that of *moths* and *butterflies*, which deposit their eggs in the precise substance, that of a cabbage for example, from which, not the butterfly herself, but the caterpillar which is to issue from her egg, draws its appropriate food. The butterfly cannot taste the cabbage. Cabbage is no food for her: yet in the cabbage, not by chance,* but studiously and electively, she lays her egg. There are, amongst many other kinds, the willow caterpillar, and the cabbage caterpillar; but we never find upon a willow, the caterpillar which eats the cabbage; nor the converse. This choice, as appears to me, cannot in the butterfly proceed from instruction. She had no teacher in her caterpillar state. She never knew her parent. I do not see, therefore, how knowledge acquired by experience, if it ever were such, could be transmitted from one generation to another. There is no opportunity either for instruction or imitation. The parent race is gone before the new brood is hatched. And, if it be original reasoning in the butterfly, it is profound reasoning indeed. She must remember her caterpillar state, its tastes and habits; of which memory she shews no signs whatever.* She must conclude from analogy, for here her recollection cannot serve her, that the little round body, which drops from her abdomen, will at a future period produce a living creature, not like herself, but like the caterpillar which she remembers herself once to have been. Under the influence of these reflections she goes about to make provision for an order of things, which, she concludes, will, some time or other, take place. And it is to be observed, that not a few out of many, but that all butterflies argue thus, all draw this conclusion, all act upon it.

But suppose the address, and the selection, and the plan, which we perceive in the preparations which many irrational animals make for their young, to be traced to some probable origin; still there is left to be accounted for, that which is the source and foundation of these phænomena, that which sets the whole at work, the στοργη, the parental affection, which I contend to be inexplicable upon any other hypothesis than that of instinct.

For we shall, hardly, I imagine, in brutes, refer their conduct towards their offspring to a sense of duty, or of decency, a care of reputation, a compliance with public manners, with public laws, or with rules of life built upon a long experience of their utility. And all attempts to account for the parental affection from association, I think, fail. With what is it associated? Most immediately with the throes of parturition,* that is, with pain, and terror, and disease. The more remote, but not less strong association, that which depends upon analogy, is all against it. Every thing else, which proceeds from the body, is cast away and rejected.

In birds, is it the egg which the hen loves? or is it the expectation which she cherishes of a future progeny, that keeps her upon her nest? What cause has she to expect delight from her progeny? Can any rational answer be given to the question, why, prior to experience, the brooding hen should look for pleasure from her chickens? It does not, I think, appear, that the cuckoo ever knows her young: yet, in her way, she is as careful in making provision for them, as any other bird. She does not leave her egg in every hole.

The salmon suffers no surmountable obstacle to oppose her progress up the stream of fresh rivers. And what does she do there? She sheds a spawn, which she immediately quits, in order to return to the sea; and this issue of her body she never afterwards recognizes in any shape whatever. Where shall we find a motive for her efforts, and her perseverance? Shall we seek it in argumentation, or in instinct? The violet crab of Jamaica performs a fatiguing march, of some months continuance, from the mountains to the sea side. When she reaches the coast, she casts her spawn into the open sea; and sets out upon her return home.

Moths and butterflies, as hath already been observed, seek out for their eggs, those precise situations and substances, in which the offspring caterpillar will find its appropriate food. That dear cater-pillar the parent butterfly must never see. There are no experiments to prove that she would retain any knowledge of it, if she did. How shall we account for her conduct? I do not mean for her art and judgment in selecting and securing a maintenance for her young, but for the impulse upon which she acts. What should induce her to exert any art, or judgment, or choice, about the matter? The undisclosed grub, the animal, which she is destined not to know, can hardly be the object of a particular affection, if we deny the influence of instinct. There is nothing, therefore, left to her, but that, of which her nature seems incapable, an abstract anxiety for the general pre-servation of the species; a kind of patriotism; a solicitude lest the butterfly race should cease from the creation.*

Lastly; the principle of association will not explain the discontinu-ance of the affection when the young animal is grown up. Associ-ation, operating in its usual way, would rather produce a contrary effect. The object would become more necessary by habits of society: whereas birds and beasts, after a certain time, banish their offspring; disown their acquaintance; seem to have even no knowledge of the objects which so lately engrossed the attention of their minds, and occupied the industry and labour of their bodies. This change, in different animals, takes place at different distances of time from the birth; but the time always corresponds with the ability of the young animal to maintain itself; never anticipates it. In the sparrow tribe, when it is perceived that the young brood can fly and shift for them-selves, then the parents forsake them for ever; and, though they continue to live together, pay them no more attention than they do to other birds in the same flock.[1] I believe the same thing is true of all gregarious quadrupeds.

In this part of the case the variety of resources, expedients, and materials, which animals of the same species are said to have recourse to, under different circumstances and when differently supplied, makes nothing against the doctrine of instincts. The thing which we want to account for is the propensity. The propensity being there, it is probable enough that it may put the animal upon different actions according to different exigences. And this adapta-tion of resources may look like the effect of art and consideration,

[1] Goldsmith's Nat. Hist. vol. iv. p. 244.

rather than of instinct; but still the propensity is instinctive. For instance, suppose what is related of the woodpecker to be true, that, in Europe, she deposits her eggs in cavities, which she scoops out in the trunks of soft or decayed trees, and in which cavities the eggs lie concealed from the eye, and in some sort safe from the hand, of man; but that, in the forests of Guinea and the Brasils, which man seldom frequents, the same bird hangs her nest to the twigs of tall trees; thereby placing them out of the reach of *monkeys* and *snakes*, i. e. that in each situation she prepares against the danger which she has most occasion to apprehend: suppose I say this to be true, and to be alledged, on the part of the bird that builds these nests, as evidence of a reasoning and distinguishing precaution, still the question returns, whence the propensity to build at all?

Nor does parental affection accompany generation by any universal law of animal organization, if such a thing were intelligible.* Some animals cherish their progeny with the most ardent fondness, and the most assiduous attention; others entirely neglect them: and this distinction always meets the constitution of the young animal, with respect to its wants and capacities. In many, the parental care extends to the young animal; in others, as in all oviparous fish, it is confined to the egg, and even, as to that, to the disposal of it in its proper element. Also, as there is generation without parental affection, so is there parental instinct, or what exactly resembles it, without generation. In the bee tribe, the grub is nurtured neither by the father nor the mother, but by the neutral bee. Probably the case is the same with ants.

I am not ignorant of a theory, which resolves instinct into sensation;* which asserts, that what appears to have a view and relation to the future, is the result only of the present disposition of the animal's body, and of pleasure or pain experienced *at the time*. Thus the incubation of eggs is accounted for by the pleasure which the bird is supposed to receive from the pressure of the smooth convex surface of the shells against the abdomen, or by the relief, which the mild temperature of the egg may afford to the heat of the lower part of the body, which is observed at this time be increased beyond its usual state. This present gratification is the only motive with the hen for sitting upon her nest: the hatching of the chickens is, with respect to her, an accidental consequence. The affection of viviparous animals for their young, is in like manner solved by the relief, and perhaps the pleasure, which they perceive from giving suck. The young animal's seeking, in so many instances, the teat of its dam, is explained from the sense of smell, which is attracted by the odour of the milk. The salmon's urging its way up the stream of fresh water rivers, is attributed to some gratification or refreshment, which, in this particular state of the fish's body, she receives from the change of element. Now of this theory it may be said,

First, that, of the cases which require solution, there are few, to which it can be applied with tolerable probability;—that there are none, to which it can be applied without strong objections, furnished by the circumstances of the case. The attention of the cow to its calf, and of the ewe to its lamb, appear to be prior to their sucking. The attraction of the calf or lamb to the teat of the dam is not explained by simply referring it to the sense of smell. What made the scent of the milk so agreeable to the lamb that it should follow it up with its nose, or seek with its mouth the place from which it proceeded? No observation, no experience, no argument could teach the new dropped animal, that the substance, from which the scent issued, was the material of its food. It had never tasted milk before its birth. None of the animals, which are not designed for that nourishment, ever offer to suck, or to seek out any such food. What is the conclusion, but that the sugescent* parts of animals are fitted for their use, and the knowledge of that use put into them?

We assert, secondly, that, even as to the cases in which the hypothesis has the fairest claim to consideration, it does not at all lessen the force of the argument for intention and design. The doctrine of instincts, is that of appetencies,* superadded to the constitution of an animal, for the effectuating of a purpose beneficial to the species. The above stated solution would derive these appetencies from organization; but then this organization is not less specifically, not less precisely, and, therefore, not less evidently adapted to the same ends, than the appetencies themselves would be upon the old hypothesis. In this way of considering the subject, sensation supplies the place of foresight: but this is the effect of contrivance on the part of the Creator. Let it be allowed, for example, that the hen is induced to brood upon her eggs by the enjoyment or relief, which, in the heated state of her abdomen, she experiences from the pressure of round smooth surfaces, or from the application of a temperate warmth. How comes this extraordinary heat or itching, or call it what you will, which you suppose to be the cause of the bird's inclination, to be felt, just at the time when the inclination itself is wanted; when it tallies so exactly with the internal constitution of the egg, and with the help which that constitution requires in order to bring it to maturity? In my opinion, this solution, if it be accepted as to the fact, ought to increase, rather than otherwise, our admiration of the contrivance. A gardener lighting up his stoves,* just when he wants to force his fruit, and when his trees require the heat, gives not a more certain evidence of design. So again; when a male and female sparrow come together, they do not meet to confer upon the expediency of perpetuating their species.* As an abstract proposition, they care not the value of a barley corn whether the species be perpetuated, or not. They follow their sensations; and all those consequences ensue, which the wisest counsels could have dictated, which the most solicitous care of futurity, which the most anxious concern for the sparrow world, could have produced. But how do these consequences ensue? The sensations, and the constitution upon which they depend, are as manifestly directed to the purpose which we see fulfilled by them; and the train of intermediate effects, as manifestly laid and planned with a view to that purpose, that is to say, design is as completely evinced by the phænomena, as it would be, even if we suppose the operations to begin, or to be carried on, from what some will allow to be alone properly called instincts, that is, from desires directed to a future end, and having no accomplishment or gratification distinct from the attainment of that end.

In a word; I should say to the patrons of this opinion, Be it so: be it, that those actions of animals which we refer to instinct, are not gone about with any view to their consequences, but that they are attended in the animal with a present gratification, and are pursued for the sake of that gratification alone; what does all this prove, but that the *prospection,** which must be somewhere, is not in the animal, but in the Creator?

In treating of the parental affection in brutes, our business lies rather with the origin of the principle, than with the effects and expressions of it. Writers recount these with pleasure and admiration. The conduct of many kinds of animals towards their young, has escaped no observer, no historian, of nature. 'How will they caress them,' says Derham,* 'with their affectionate notes; lull and quiet them with their tender parental voice; put food into their mouths; cherish, and keep them warm; teach them to pick, and eat, and gather food for themselves; and, in a word, perform the part of so many nurses, deputed by the soverign Lord and preserver of the world, to help such young and shiftless creatures?' Neither ought it, under this head, to be forgotten, how much the instinct *costs* the animal which feels it; how much a bird, for example, gives up, by sitting upon her nest; how repugnant it is to her organization, her habits, and her pleasures. An animal, formed for liberty, submits to confinement, in the very season when every thing invites her abroad: what is more; an animal delighting in motion, made for motion, all whose motions are so easy and so free, hardly a moment, at other times, at rest, is, for many hours of many days together, fixed to her nest, as close as if her limbs were tied down by pins and wires. For my part, I never see a bird in that situation, but I recognise an invisible hand,* detaining the contented prisoner from her fields and groves, for a purpose, as the event proves, the most worthy of the sacrifice, the most important, the most beneficial.

But the loss of liberty is not the whole of what the procreant bird suffers. Harvey* tells us, that he has often found the female wasted to skin and bone by sitting upon her eggs.

One observation more, and I will dismiss the subject. The *pairing* of birds, and the *non-pairing* of beasts, forms a distinction, between the two classes, which shews that the conjugal instinct is modified with a reference to utility founded in the condition of the offspring. In quadrupeds, the young animal draws its nutriment from the body of the dam. The male parent neither does, nor can, contribute any part to its sustentation. In the feathered race, the young bird is supplied by an importation of food, to procure and bring home which, in a sufficient quantity for the demand of a numerous brood, requires the industry of both parents. In this difference we see a reason, for the vagrant instinct of the quadruped, and for the faithful love of the feathered mate.

CHAPTER XIX

OF INSECTS

WE are not writing a system of natural history;* therefore, we have not attended to the classes, into which the subjects of that science are distributed. What we had to observe concerning different species of animals, fell easily, for the most part, within the divisions, which the course of our argument led us to adopt. There remain, however, some remarks upon the *insect* tribe, which could not properly be introduced under any of these heads; and which therefore we have collected into a chapter by themselves.

The structure, and the use of the parts, of insects, are less understood than that of quadrupeds and birds, not only by reason of their minuteness, or the minuteness of their parts, (for that minuteness we can, in some measure, follow with glasses)* but also, by reason of the remoteness of their manners and modes of life from those of larger animals. For instance; Insects, under all their varieties of form, are endowed with *antennæ*, which is the name given to those long feelers that rise from each side of the head; but to what common use or want of the insect kind, a provision so universal is subservient, has not yet been ascertained; and it has not been ascertained, because it admits not of a clear, or very probable, comparison, with any organs which we possess ourselves, or with the organs of animals which resemble ourselves in their functions and faculties, or with which we are better acquainted than we are with insects. We want a ground of analogy.* This difficulty stands in our way as to some particulars in the insect constitution which we might wish to be acquainted with. Nevertheless, there are many contrivances in the bodies of insects, neither dubious in their use, nor obscure in their structure, and most properly mechanical. These form parts of *our* argument.

I. The *elytra*, or scaly wings of the genus of scarabæus or beetle, furnish an instance of this kind. The true wing of the animal is a light transparent membrane, finer than the finest gauze, and not unlike it. It is also when expanded, in proportion to the size of the animal, very large. In order to protect this delicate structure, and, perhaps, also to preserve it in a due state of suppleness and humidity, a strong, hard, case is given to it, in the shape of the horny wing which we call the elytron. When the animal is at rest, the gauze wings lie folded up under this impenetrable shield. When the beetle prepares for flying, he raises the integument, and spreads out his thin membrane to the air. And it cannot be observed without admiration, what a tissue of cordage, i. e. of muscular tendons, must run, in various and complicated, but determinate directions, along this fine surface, in order to enable the animal, either to gather it up into a certain precise form, whenever it desires to place its wings under the shelter which nature hath given to them; or to expand again their folds, when wanted for action.

In some insects, the elytra cover the whole body; in others, half; in others, only a small part of it; but in all they completely hide and cover the true wings. Also,

Many or most of the beetle species lodge in holes in the earth, environed by hard, rough, substances, and have frequently to squeeze their way through narrow passages; in which situation, wings so tender, and so large, could scarcely have escaped injury, without both a firm covering to defend them, and the capacity of collecting themselves up under its protection.

II. Another contrivance, equally mechanical, and equally clear, is the *awl* or borer fixed at the tails of various species of flies; and with which they pierce, in some cases, plants; in others, wood; in others, the skin and flesh of animals; in others, the coat of the chrysalis of insects of a different species from their own; and in others, even lime, mortar, and stone. I need not add, that having pierced the substance, they deposit their eggs in the hole.* The descriptions, which naturalists give of this organ, are such as the following. It is a sharp-pointed instrument, which, in its inactive state, lies concealed in the extremity of the abdomen, and which the animal draws out at pleasure, for the purpose of making a puncture in the leaves, stem, or bark of the particular plant, which is suited to the nourishment of its young. In a sheath, which divides and opens whenever the organ is used, there is inclosed, a compact, solid, dentated stem, along which runs a *gutter* or *groove*, by which groove, after the penetration is effected, the egg, assisted, in some cases, by a peristaltic motion,* passes to its destined lodgment. In the œstrum or gadfly, the wimble *draws out* like the pieces of a spy-glass;* the last piece is armed with three hooks, and is able to bore through the hide of an ox. Can any thing more be necessary to display the mechanism, than to relate the fact?

III. The *stings* of insects, though for a different purpose, are, in their structure, not unlike the piercer. The sharpness to which the point in all of them is wrought; the temper and firmness of the substance of which it is composed; the strength of the muscles by which it is darted out, compared with the smallness and weakness of the insect, and with the soft or friable texture of the rest of the body; are properties of the sting to be noticed, and not a little to be admired. The sting of a *bee* will pierce through a goatskin glove. It penetrates the human skin more readily than the finest point of a needle. The *action* of the sting affords an example of the union of chymistry* and mechanism, such as, if it be not a proof of contrivance, nothing is. First, as to the chymistry; how highly concentrated must be the *venom*, which, in so small a quantity, can produce such powerful effects? And in the bee we may observe, that this venom is made from *honey*, the only food of the insect, but the last material from which I should have expected, that an exalted poison could, by any process or digestion whatsoever, have been prepared. In the next place, with respect to the mechanism, the sting is not a simple, but a compound instrument. The visible sting, though drawn to a point exquisitely sharp, is in strictness only a sheath; for, near to the extremity, may be perceived by the microscope two minute orifices, from which orifices, in the act of stinging, and, as it should seem, after the point of the main sting has buried itself in the flesh, are lanched out two subtile rays, which may be called the true or proper stings, as being those, through which the poison is infused into the puncture already made by the exterior sting. I have said that chymistry and mechanism are here *united*: by which observation I meant, that all this machinery would have been useless, telum imbelle,* if a supply of poison, intense in quality, in proportion to the smallness of the drop, had not been furnished to it by the chymical elaboration which was carried on in the insect's body: and that, on the other hand, the poison, the result of this process, could not have attained its effect, or reached its enemy, if, when it was collected at the extremity of the abdomen, it had not found there a machinery, fitted to conduct it to the external situations in which it was to operate, viz. an awl to bore a hole, and a syringe to inject the fluid. Yet these attributes, though combined in their action, are independent in their origin. The venom does not breed the sting; nor does the sting concoct the venom.

IV. The *proboscis*, with which many insects are endowed, comes next in order to be considered. It is a tube attached to the head of the animal. In the bee, it is composed of two pieces, connected by a joint: for, if it were constantly extended, it would be too much exposed to accidental injuries: therefore, in its indolent state, it is doubled up by means of the joint, and in that position lies secure under a scaly penthouse. In many species of the butterfly, the proboscis, when not in use, is coiled up like a watch-spring. In the same bee, the proboscis serves the office of the mouth, the insect having no other: and how much better adapted it is, than a mouth would be, for the collecting of the proper nourishment of the animal, is sufficiently evident. The food of the bee is the nectar of flowers; a drop of syrup, lodged deep in the bottom of the corollæ, in the recesses of the petals, or down the neck of a monopetalous* glove. Into these cells the bee thrusts its long narrow pump, through the cavity of which it sucks up this precious fluid, inaccessible to every other approach. The ringlets of which the proboscis of the bee is composed, the muscles by which it is extended and contracted, form so many microscopical wonders. The agility also, with which it is moved, can hardly fail to excite admiration. But it is enough for our purpose to observe in general, the suitableness of the structure to the use, of the means to the end, and especially the wisdom, by which nature has departed from its most general analogy (for animals being furnished with mouths is such) when the purpose could be better answered by the deviation.

In some insects, the proboscis, or tongue, or trunk, is shut up in a sharp-pointed sheath, which sheath, being of a much firmer texture than the proboscis itself, as well as sharpened at the point, pierces the substance which contains the food, and then *opens within the wound*, to allow the inclosed tube, through which the juice is extracted, to perform its office. Can any mechanism be plainer than this is; or surpass this?

V. The *metamorphosis* of insects from grubs* into moths and flies, is an astonishing process. A hairy caterpillar is transformed into a butterfly. Observe the change. We have four beautiful wings, where there were none before; a tubular proboscis, in the place of a mouth with jaws and teeth; six long legs, instead of fourteen feet. In another case, we see a white, smooth, soft worm, turned into a black, hard, crustaceous beetle, with gauze wings. These, as I said, are astonishing processes, and must require, as it should seem, a proportionably artificial apparatus. The hypothesis which appears to me most probable is, that, in the grub, there exist at the same time three animals, one within another, all nourished by the same digestion, and by a communicating circulation; but in different stages of maturity. The latest discoveries, made by naturalists, seem to favour this supposition. The insect already equipped with wings, is descried under the membranes both of the worm and nymph. In some species, the proboscis, the antennæ, the limbs and wings of the fly, have been observed to be folded up within the body of the caterpillar; and with such nicety, as to occupy a small space only under the two first rings. This being so, the outermost animal, which beside its own proper character, serves as an integument to the other two, being the furthest advanced, dies, as we suppose, and drops off first. The second, the pupa or chrysalis, then offers itself to observation. This also, in its turn, dies; its dead and brittle husk falls to pieces, and makes way for the appearance of the fly or moth. Now, if this be the case, or indeed whatever explication be adopted, we have a prospective contrivance of the most curious kind: we have organizations *three deep*, yet a vascular system, which supplies nutrition, growth, and life, to all of them together.

VI. Almost all insects are oviparous. Nature keeps her butterflies, moths and caterpillars, locked up during the winter in their egg state; and we have to admire the various devices, to which, if we may so speak, the same nature hath resorted, for the *security* of the egg. Many insects inclose their eggs in a silken web; others cover them with a coat of hair, torn from their own bodies; some glue them together; and others, like the moth of the silkworm, glue them to the leaves upon which they are deposited, that they may not be shaken off by the wind, or washed away by rain: some again make incisions into leaves, and hide an egg in each incision; whilst some envelope their eggs with a soft substance, which forms the first aliment of the young animal; and some again make a hole in the earth, and, having stored it with a quantity of proper food, deposit their egg in it. In all which we are to observe, that the expedient depends, not so much upon the address of the animal, as upon the physical resources of his constitution.

The art also with which the young insect is *coiled up* in the egg, presents, where it can be examined, a subject of great curiosity. The insect, furnished with all the members which it ought to have, is rolled up into a form which seems to contract it into the least possible space; by which contraction, notwithstanding the smallness of the egg, it has room enough in its apartment, and to spare. This folding of the limbs appears to me to indicate a special direction; for, if it were merely the effect of compression, the collocation of the parts would be more various than it is. In the same species, I believe, it is always the same.

These observations belong to the whole insect tribe, or to a great part of them. Other observations are limited to fewer species; but not, perhaps, less important or satisfactory.

I. The organization in the abdomen of the *silkworm* or *spider*, whereby these insects form their *thread*, is as incontestably mechanical, as a wire-drawer's mill. In the body of the silkworm are two bags, remarkable for their form, position, and use. They wind round the intestine; when drawn out they are ten inches in length, though the animal itself be only two. Within these bags, is collected a glue; and communicating with the bags, are two paps or outlets, perforated, like a grater, by a number of small holes. The glue or gum, being passed through these minute apertures, forms hairs of almost imperceptible fineness; and these hairs, when joined, compose the silk which we wind off from the cone, in which the silkworm has wrapped itself up: in the spider the web is formed of this thread. In both cases, the extremity of the thread, by means of its adhesive quality, is first attached by the animal to some external hold; and the end being now fastened to a point, the insect, by turning round its body, or by receding from that point, draws out the thread through the holes above described, by an operation, as hath been observed, exactly similar to the drawing of wire. The thread, like the wire, is formed by the hole through which it passes. In one respect there is a difference. The wire is the metal unaltered, except in figure. In the animal process, the nature of the substance is somewhat changed, as well as the form: for, as it exists within the insect, it is a soft, clammy, gum or glue. The thread acquires, it is probable, its firmness and tenacity from the action of the air upon its surface, in the moment of exposure: and a thread so fine is almost all surface. This property, however, of the paste, is part of the contrivance.

The mechanism itself consists of the bags, or reservoirs, into which the glue is collected, and of the external holes communicating with these bags: and the action of the machine is seen, in the forming of a thread, as wire is formed, by forcing the material already prepared, through holes of proper dimensions. The secretion is an act too subtle for our discernment, except as we perceive it by the produce. But one thing answers to another: the secretory glands to the quality and consistence required in the secreted substance; the bag to its reception. The outlets and orifices are constructed, not merely for relieving the reservoirs of their burthen, but for manufacturing the contents into a form and texture, of great external use, or rather indeed of future necessity, to the life and functions of the insect.

II. BEES, under one character or other, have furnished every naturalist with a set of observations. I shall, in this place, confine myself to one; and that is the *relation* which obtains between the wax and the honey. No person, who has inspected a bee-hive, can forbear remarking, how commodiously the honey is bestowed in the comb; and amongst other advantages, how effectually the fermentation of the honey is prevented by distributing it into small cells. The fact is, that when the honey is separated from the comb, and put into jars, it runs into fermentation, with a much less degree of heat than what takes place in a hive. This may be reckoned a nicety: but independently of any nicety in the matter, I would ask, what could the bee do with the honey, if it had not the wax? how, at least, could it store it up for winter? The wax, therefore, answers a purpose with respect to the honey; and the honey constitutes that purpose with respect to the wax. This is the relation between them. But the two substances, though, together, of the greatest use, and, without each other, of little, come from a different origin. The bee finds the honey, but makes the wax. The honey is lodged in the nectaria of flowers, and probably undergoes little alteration; is merely collected: whereas the wax is a ductile tenacious paste, made out of a dry powder, not simply by kneading it with a liquid, but by a digestive process in the body of the bee. What account can be rendered of facts so circumstanced, but that the animal, being intended to feed upon honey, was, by a peculiar external configuration, enabled to procure it? that, moreover, wanting the honey when it could not be procured at all, it was further endued with the no less necessary faculty of constructing repositories for its preservation? which faculty, it is evident, must depend, primarily, upon the capacity of providing suitable materials.

Two distinct functions go to make up the ability. First, the power in the bee, with respect to wax, of loading the farina of flowers upon its thighs: microscopic observers speak of the spoon-shaped appendages, with which the thighs of bees are beset for this very purpose: but inasmuch as the art and will of the bee may be supposed to be concerned in this operation, there is, secondly, that which doth not rest in art or will, a digestive faculty which converts the loose powder into a stiff substance. This is a just account of the honey and the honey comb: and this account, through every part, carries a creative intelligence along with it.

The *sting* also of the bee has this relation to the honey, that it is necessary for the protection of a treasure which invites so many robbers.

III. Our business is with mechanism. In the *panorpa** tribe of insects, there is a forceps in the tail of the male insect, with which he catches and holds the female. Are a pair of pincers more mechanical, than this provision, in their structure? or is any structure more clear and certain in its design?

IV. St Pierre* tells us,[1] that in a fly with six feet (I do not remember that he describes the species) the pair next the head, and the pair next the tail, have brushes at their extremities, with which the fly dresses, as there may be occasion, the anterior or the posterior part of its body; but that the middle pair have no such brushes, the situation of these legs not admitting of the brushes, if they were there, being converted to the same use. This is a very exact mechanical distinction.

V. If the reader, looking to our distributions of science, wish to contemplate the chymistry, as well as the mechanism of nature, the insect creation will afford him an example. I refer to the light in the tail of a *glow-worm*. Two points seem to be agreed upon by naturalists concerning it: first, that it is phosphoric;* secondly, that its use is to attract the male insect. The only thing to be enquired after, is the singularity, if any such there be, in the natural history of this animal, which should render a provision of this kind more necessary for *it*, than for other insects. That singularity seems to be the difference, which subsists between the male and the female; which difference is greater than what we find in any other species of animal whatever.

[1] Vol. i. p. 342.

The glow-worm is a female *caterpillar*; the male of which is a *fly*; lively, comparatively small, dissimilar to the female in appearance, probably also as distinguished from her in habits, pursuits, and manners, as he is unlike in form and external constitution. Here then is the adversity of the case. The caterpillar cannot meet her companion in the air. The winged rover disdains the ground. They might never therefore be brought together, did not this radiant torch direct the volatile mate to his sedentary female.

In this example we also see the resources of art anticipated. One grand operation of chymistry is the making of phosphorus; and it was thought an ingenious device, to make phosphoric matches supply the place of lighted tapers.* Now this very thing is done in the body of the glow-worm. The phosphorus is not only made, but kindled; and caused to emit a steady and genial beam, for the purpose which is here stated, and which I believe to be the true one.

VI. Nor is the last the only instance that entomology affords, in which our discoveries, or rather our projects, turn out to be imitations of nature. Some years ago, a plan was suggested, of producing propulsion by reaction in this way. By the force of a steam engine, a stream of water was to be shot out of the stern of a boat; the impulse of which stream upon the water in the river, was to push the boat itself forward: it is, in truth, the principle by which sky-rockets* ascend in the air. Of the use or the practicability of the plan I am not speaking; nor is it my concern to praise its ingenuity; but it is certainly a contrivance. Now, if naturalists are to be believed, it is exactly the device, which nature has made use of, for the motion of some species of aquatic insects. The larva of the *dragon fly*, according to Adams,* swims by ejecting water from its tail; is driven forward by the reaction of water in the pool upon the current issuing in a direction backward from its body.

VII. Again; Europe has lately been surprised by the elevation of bodies in the air by means of a balloon. The discovery consisted in finding out a manageable substance, which was, bulk for bulk, lighter than air; and the application of the discovery was, to make a body composed of this substance bear up, along with its own weight, some heavier body which was attached to it. This expedient, so new to us, proves to be no other than what the author of nature has employed in the *gossamir spider*. We frequently see this spider's thread floating in the air, and extended from hedge to hedge, across a road or brook of four or five yards width. The animal which forms the thread, has no wings wherewith to fly from one extremity to the other of this line; nor muscles to enable it to spring or dart to so great a distance. Yet its creator hath laid for it a path in the atmosphere; and after this manner. Though the animal itself be heavier than air, the thread which it spins from its bowels is specifically lighter. This is its *balloon*. The spider left to itself would drop to the ground; but, being tied to its thread, both are supported. We have here a very peculiar provision: and to a contemplative eye it is a gratifying spectacle, to see this insect wafted on her thread, sustained by a levity not her own, and traversing regions, which, if we examined only the body of the animal, might seem to have been forbidden to its nature.

I MUST now crave the reader's permission to introduce into this place, for want of a better, an observation or two upon the tribe of animals, whether belonging to land or water, which are covered by *shells*.

I. The *shells* of *snails* are a wonderful, a mechanical, and, if one might so speak concerning the works of nature, an original contrivance. Other animals have their proper retreats, their hybernacula also or winter quarters, but the snail carries these about with him. He travels with his tent; and this tent, though, as was necessary, both light and thin, is completely impervious either to moisture or air. The young snail comes out of its egg with the shell upon its back; and the gradual enlargement which the shell receives, is derived from the slime excreted by the animal's skin. Now the aptness of this excretion to the purpose, its property of hardening into a shell, and the action, whatever it be, of the animal, whereby it avails itself of its gift, and of the constitution of its glands, (to say nothing of the work being commenced before the animal is born,) are things, which can, with no probability, be referred to any other cause than to express design; and that not on the part of the animal alone, in which design, though it might build the house, could not have supplied the material. The will of the animal could not determine the quality of the excretion*. Add to which, that the shell of a snail, with its pillar and convolution, is a very artificial fabric; whilst a snail, as it should seem, is the most numb and unprovided of all artificers. In the midst of variety, there is likewise a regularity, which would hardly be expected. In the same species of snail the number of turns is, usually, if not always, the same. The sealing up of the mouth of the shell by the snail, is also well calculated for its warmth and security; but the cerate is not of the same substance with the shell.

II. Much of what has been observed of snails belongs to *shell fish* and their *shells*, particularly to those of the univalve kind; with the addition of two remarks. One of which is upon the great strength and hardness of most of these shells. I do not know, whether, the weight being given, art can produce so strong a case as are some of these shells. Which defensive strength suits well with the life of an animal, that has often to sustain the dangers of a stormy element and a rocky bottom, as well as the attacks of voracious fish. The other remark is, upon the property, in the animal excretion, not only of congealing, but of congealing or, as a builder would call it, setting in water, and into a cretaceous substance, firm and hard. This property is much more extraordinary, and, chymically speaking, more specific, than that of hardening in the air; which may be reckoned a kind of exsiccation, like the drying of clay into bricks.*

III. In the *bivalve* order of shell fish, cockles, muscles, oysters, etc. what contrivance can be so simple or so clear, as the insertion, at the back, of a tough, tendinous, substance, that becomes, at once, the ligament which binds the two shells together, and the *hinge* upon which they open and shut?

IV. The shell of a lobster's tail, in its articulations and overlappings, represents the jointed part of a coat of mail; or rather, which I believe to be the truth, a coat of mail is an imitation of a lobster's shell. The same end is to be answered by both: the same properties, therefore, are required in both, namely, hardness and flexibility, a covering which may guard the part without obstructing its motion. For this double purpose, the art of man, expressly exercised upon the subject, has not been able to devise any thing better than what nature presents to his observation. Is not this therefore mechanism, which the mechanic, having a similar purpose in view, adopts? Is the structure of a coat of mail to be referred to art? Is the same structure of the lobster, conducing to the same use, to be referred to any thing less than art?

Some, who may acknowledge the imitation, and assent to the inference which we draw from it, in the instance before us, may be disposed, possibly, to ask, why such imitations are not more frequent than they are, if it be true, as we alledge, that the same principle of intelligence, design, and mechanical contrivance, was exerted in the formation of natural bodies, as we employ in the making of the various instruments by which our purposes are served. The answers to this question are, first, that it seldom happens, that precisely the same purpose, and no other, is pursued in any work which we compare of nature and of art; secondly, that it still seldomer happens, that we *can* imitate nature, if we would. Our materials and our workmanship are equally deficient. Springs and wires, and cork and leather, produce a poor substitute for an arm or a hand. In the example which we have selected, I mean of a lobster's shell compared with a coat of mail, these difficulties stand less in the way, than in almost any other that can be assigned; and the consequence is, as we have seen, that art gladly borrows from nature her contrivance, and imitates it closely.

BUT to return to insects. I think it is in this class of animals, above all others, especially when we take in the multitude of species which the microscope discovers, that we are struck with what Cicero* has called 'the *insatiable* variety of nature.' There are said to be six thousand species of flies; seven hundred and sixty butterflies; each different from all the rest, (St Pierre). The same writer tells us from his own observation, that thirty-seven species of winged insects, with distinctions well expressed, visited a single strawberry plant in the course of three weeks.[1] *Ray** observed, within the compass of a mile or two of his own house, two hundred kinds of butterflies, nocturnal and diurnal. He likewise asserts, but, I think, without any grounds of exact computation, that the number of species of insects, reckoning all sorts of them, may not be short of ten thousand.[2] And in this vast variety of animal forms, (for the observation is not confined to insects, though more applicable perhaps to them than to any other class,) we are sometimes led to take notice of the different methods, or rather of the studiously diversified methods, by which one and the same purpose is attained. In the article of breathing, for example, which was to be provided for in some way or other, beside the ordinary varieties of lungs, gills, and breathing-holes, (for insects in general respire, not by the mouth, but through holes in the sides,) the

[1] Vol. i. p. 3.
[2] Wis. of God, p. 23.

nymphæ* of gnats have an apparatus to raise their *backs* to the top of
the water, and so take breath. The hydrocanthari* do the like by
thrusting their *tails* out of the water.[1] The maggot of the eruca labra
has a long tail, one part sheathed within another, (but which it can
draw out at pleasure,) with a starry tuft at the end, by which *tuft*,
when expanded upon the surface, the insect both supports itself in
the water, and draws in the air which is necessary. In the article of
natural clothing, we have the skins of animals invested with scales,
hair, feathers, mucus, froth; or itself turned into a shell or crust: in
the no less necessary article of offence and defence, we have teeth,
talons, beaks, horns, stings, prickles, with (the most singular expedient for the same purpose) the power of giving the electric shock, and,
as is credibly related of some animals, of driving away their pursuers
by an intolerable fœtor, or of blackening the water through which
they are pursued. The consideration of these appearances might
induce us to believe, that *variety* itself, distinct from every other
reason, was a motive in the mind of the Creator, or with the agents of
his will.

To this great variety in organized life the Deity has given, or
perhaps there arises out of it, a corresponding variety of animal
appetites. For the final cause of this we have not far to seek. Did all
animals covet the same element, retreat, or food, it is evident how
much fewer could be supplied and accommodated, than what at
present live conveniently together, and find a plentiful subsistence.
What one nature rejects, another delights in. Food, which is nauseous to one tribe of animals, becomes, by that very property which
makes it nauseous, an alluring dainty to another tribe. Carrion is a
treat to dogs, ravens, vultures, fish. The exhalations of corrupted
substances attract flies by crowds. Maggots revel in putrefaction.

---

[1] Derham,* p. 7.

CHAPTER XX

OF PLANTS

I THINK a designed and studied mechanism to be, in general, more evident in animals, than in *plants*: and it is unnecessary to dwell upon a weaker argument, where a stronger is at hand. There are, however, a few observations upon the vegetable kingdom, which lie so directly in our way, that it would be improper to pass by them without notice.

The one great intention of nature in the structure of plants, seems to be the perfecting of the *seed*; and, what is part of the same intention, the preserving of it until it be *perfected*. This intention shews itself, in the first place, by the care which appears to be taken to protect and ripen, by every advantage which can be given to them of situation in the plant, those parts which most immediately contribute to fructification, viz. the antheræ, the stamina, and the stigmata. These parts are usually lodged in the centre, the recesses, or the labyrinths of the flower; during their tender and immature state, are shut up in the stalk, or sheltered in the bud: as soon as they have acquired firmness of texture sufficient to bear exposure, and are ready to perform the important office which is assigned to them, they are disclosed to the light and air, by the bursting of the stem or the expansion of the petals: after which they have, in many cases, by the very form of the flower during its blow, the light and warmth reflected upon them from the concave side of the cup. What is called also the *sleep* of plants, is the leaves or petals disposing themselves in such a manner as to shelter the young stem, buds, or fruit. They turn up, or they fall down, according as this purpose renders either change of position requisite. In the growth of corn, whenever the plant begins to shoot, the two upper leaves of the stalk join together; embrace the ear; and protect it till the pulp has acquired a certain degree of consistency. In some water plants, the flowering and fecundation* are carried on *within* the stem, which afterwards opens to let loose the impregnated seed.[1] The *pea* or papilionaceous tribe inclose the parts of fructification within a beautiful folding of the

[1] Phil. Trans. part ii. 1796,* p. 502.

internal blossom, sometimes called from its shape the boat or keel; itself also protected under a penthouse formed by the external petals. This structure is very artificial; and, what adds to the value of it though it may diminish the curiosity, very general. It has also this further advantage (and it is an advantage strictly mechanical) that all the blossoms turn their *backs* to the wind, whenever the gale blows strong enough to endanger the delicate parts upon which the seed depends. I have observed this a hundred times in a field of peas in blossom. It is an aptitude which results from the figure of the flower, and, as we have said, is strictly mechanical; as much so, as the turning of a weather-board or tin cap upon the top of a chimney. Of the *poppy*, and of many similar species of flowers, the head, while it is growing, hangs down, a rigid curvature in the upper part of the stem giving to it that position; and in that position it is impenetrable by rain or moisture. When the head has acquired its size, and is ready to open, the stalk *erects* itself, for the purpose, as it should seem, of presenting the flower, and, with the flower, the instruments of fructification, to the genial influence of the sun's rays. This always struck me as a curious property; and specifically, as well as originally, provided for in the constitution of the plant: for, if the stem be only bent by the weight of the head, how comes it to straighten itself when the head is the heaviest? These instances shew the attention of nature to this principal object, the safety and maturation of the parts upon which the seed depends.

In *trees*, especially in those which are natives of colder climates, this point is taken up earlier. Many of these trees (observe in particular the *ash* and the *horse chesnut*) produce the embryos of the leaves and flowers in one year, and bring them to perfection the following. There is a winter therefore to be got over. Now what we are to remark is, how nature has prepared for the trials and severities of that season. These tender embryos, are, in the first place, wrapped up with a compactness, which no art can imitate: in which state, they compose what we call the bud. This is not all. The bud itself is inclosed in scales; which scales are formed from the remains of past leaves, and the rudiments of future ones. Neither is this the whole. In the coldest climates a third preservative is added, by the bud having a *coat* of gum or resin, which, being congealed, resists the strongest frosts. On the approach of warm weather this gum is softened, and ceases to be a hindrance to the expansion of the leaves and flowers. All this care is part of that system of provisions which has for its object and consummation, the production and perfecting of the seeds.

The SEEDS themselves are packed up in a *capsule*, a vessel composed of coats, which, compared with the rest of the flower, are strong and tough. From this vessel projects a tube, through which tube the farina, or some subtle fecundating effluvium that issues from it, is admitted to the seed. And here also occurs a mechanical variety, accommodated to the different circumstances under which the same purpose is to be accomplished. In flowers which are erect, the pistil is shorter than the stamina; and the pollen, shed from the antheræ* into the cup of the flower, is catched in its descent by the head of the pistil, called the stigma. But how is this managed when the flowers hang down, (as does the crown imperial, for instance,) and in which position, the farina, in its fall, would be carried from the stigma, and not towards it? The relative length of the parts is now inverted. The pistil in these flowers is usually longer, instead of shorter, than the stamina, that its protruding summit may receive the pollen as it drops to the ground. In some cases, (as in the *nigella*,) where the shafts of the pistils or styles are disproportionably long, they bend down their extremities upon the antheræ, that the necessary approximation may be effected.

But (to pursue this great work in its progress,) the impregnation, to which all this machinery relates, being completed, the other parts of the flower fade and drop off, whilst the *gravid seed-vessel*, on the contrary, proceeds to increase its bulk, always to a great, and in some species, (in the gourd, for example, and melon,) to a surprising comparative size; assuming in different plants an incalculable variety of forms, but all evidently conducing to the security of the seed. By virtue of this process, so necessary, but so diversified, we have the seed, at length, in stone fruits* and nuts, incased in a strong shell, the shell itself inclosed in a pulp or husk, by which the seed within is, or hath been, fed; or, more generally (as in grapes, oranges, and the numerous kinds of berries) plunged overhead in a glutinous syrup, contained within a skin or bladder: at other times (as in apples and pears) embedded in the heart of a firm fleshy substance; or (as in strawberries) pricked into the surface of a soft pulp.

These and many more varieties exist in what we call *fruits*.[1] In pulse, and grain, and grasses; in trees, and shrubs, and flowers; the variety of the seed-vessels is incomputable. We have the seeds (as in the pea tribe) regularly disposed in parchment pods, which, though soft and membranous, completely exclude the wet even in the heaviest rains; the pod also, not seldom (as in the bean) lined with a fine down; at other times (as in the senna*) distended like a blown bladder: or we have the seed enveloped in wool (as in the cotton plant), lodged (as in pines) between the hard and compact scales of a cone; or barricadoed (as in the artichoke and thistle) with spikes and prickles; in mushrooms, placed under a penthouse; in ferns, within slits in the back part of the leaf; or (which is the most general organization of all) we find them covered by strong, close, tunicles, and attached to the stem according to an order appropriated to each plant, as is seen in the several kinds of grain, and of grasses.

In which enumeration what we have first to notice is, unity of purpose under variety of expedients. Nothing can be more *single* than the design; more *diversified* than the means. Pellicles, shells, pulps, pods, husks, skins, scales armed with thorns, are all employed in prosecuting the same intention. Secondly; we may observe, that, in all these cases, the purpose is fulfilled within a just and *limited* degree. We can perceive, that, if the seeds of plants were more strongly guarded than they are, their greater security would interfere with other uses. Many species of animals would suffer, and many perish, if they could not obtain access to them. The plant would overrun the soil; or the seed be wasted for want of room to sow itself. It is, sometimes, as necessary to destroy particular species of plants,

---

[1] From the conformation of fruits alone, one might be led, even without experience, to suppose, that part of this provision was destined for the utilities of animals. As limited to the plant, the provision itself seems to go beyond its object. The flesh of an apple, the pulp of an orange, the meat of a plum, the 'fatness of the olive,' appear to be more than sufficient for the nourishing of the seed or kernel. The event shews, that this redundancy, if it be one, ministers to the support and gratification of animal natures: and when we observe a provision to be more than sufficient for one purpose, yet wanted for another purpose, it is not unfair to conclude that both purposes were contemplated together. It favors this view of the subject to remark, that fruits are not (which they might have been) ready all together, but that they ripen in succession throughout a great part of the year; some in summer; some in autumn; that some require the slow maturation of the winter, and supply the spring: also that the coldest fruits grow in the hottest places. Cucumbers, pine apples, melons, are the natural produce of warm climates, and contribute greatly, by their coolness, to the refreshment of the inhabitants of those countries.

as it is, at other times, to encourage their growth. Here, as in many cases, a balance is to be maintained between opposite uses. The provisions for the preservation of seeds appear to be directed, chiefly against the inconstancy of the elements, or the sweeping destruction of inclement seasons. The depredation of animals, and the injuries of accidental violence, are allowed for in the abundance of the increase. The result is, that, out of the many thousand different plants which cover the earth, not a single species, perhaps, has been lost since the creation.*

When nature has perfected her seeds, her next care is to disperse them. The seed cannot answer its purpose, while it remains confined in the capsule. After the seeds therefore are ripened, the pericarpium* opens to let them out; and the opening is not like an accidental bursting, but, for the most part, is according to a certain rule in each plant. What I have always thought very extraordinary, nuts and shells, which we can hardly crack with our teeth, divide and make way for the little tender sprout which proceeds from the kernel. Handling the nut, I could hardly conceive how the plantule* was ever to get out of it. There are cases, it is said, in which the seed-vessel by an elastic jerk, at the moment of its explosion, casts the seed to a distance. We all however know, that many seeds (those of most composite flowers, as of the thistle, dandelion, etc.) are endowed with what are not improperly called *wings*; that is, downy appendages, by which they are enabled to float in the air, and are carried oftentimes by the wind to great distances from the plant which produces them. It is the swelling also of this downy tuft within the seed-vessel, that seems to overcome the resistance of its coats, and to open a passage for the seed to escape.

But the *constitution* of seeds is still more admirable than either their preservation or their dispersion. In the body of the seed of every species of plant, or nearly of every one, provision is made for two grand purposes: first, for the safety of the *germ*;* secondly, for the temporary support of the future plant. The sprout, as folded up in the seed, is delicate and brittle, beyond any other substance. It cannot be touched without being broken. Yet, in beans, peas, grass-feeds, grain, fruits, it is so fenced on all sides, so shut up and protected, that, whilst the seed itself is rudely handled, tossed into sacks, shovelled into heaps, the miniature plant, the sacred particle, remains unhurt. It is wonderful also, how long many kinds of seed, by the help of their integuments, and perhaps of their oils, stand out against decay. A grain of mustard seed has been known to lie in the earth for a hundred years; and, as soon as it had acquired a favorable situation, to shoot as vigorously as if just gathered from the plant. Then, as to the second point, the temporary support of the future plant, the matter stands thus. In grain, and pulse, and kernels, and pippins,* the germ composes a very small part of the seed. The rest consists of a nutritious substance, from which the sprout draws its aliment for some considerable time after it is put forth; viz. until the fibres, shot out from the other end of the seed, are able to imbibe juices from the earth, in a sufficient quantity for its demand. It is owing to this constitution, that we see seeds sprout, and the sprouts make a considerable progress, without any earth at all. It is an œconomy also, in which we remark a close analogy between the seeds of plants, and the eggs of animals.* The same point is provided for, in the same manner, in both. In the egg, the residence of the living principle, the cicatrix, forms a very minute part of the contents. The white, and the white only, is expended in the formation of the chicken. The yolk, very little altered or diminished, is wrapped up in the abdomen of the young bird, when it quits the shell; and serves for its nourishment, till it have learnt to pick its own food. This perfectly resembles the first nutrition of a plant. In the plant, as well as in the animal, the structure has every character of contrivance belonging to it: in both it breaks the transition from prepared to unprepared aliment: in both it is prospective and compensatory. In animals which suck, this intermediate nourishment is supplied by a different source.

In all subjects the most common observations are the best, when it is their truth and strength which have made them common. There are, of this sort, *two* concerning plants, which it falls within our plan to notice. The *first* relates to, what has already been touched upon, their germination. When a grain of corn is cast into the ground, this is the change which takes place. From one end of the grain issues a green sprout: from the other a number of white fibrous threads. How can this be explained? Why not sprouts from both ends? Why not fibrous threads from both ends? To what is the difference to be referred, but to design; to the different uses which the parts are thereafter to serve; uses which discover themselves in the sequel of the process? The sprout, or plumule, struggles into the air; and

becomes the plant, of which, from the first, it contained the rudiments: the fibres shoot into the earth; and, thereby, both fix the plant
to the ground, and collect nourishment from the soil for its support.
Now, what is not a little remarkable, the parts issuing from the seed
take their respective directions, into whatever position the seed itself
happens to be cast. If the seed be thrown into the wrongest possible
position, that is, if the ends point in the ground the reverse of what
they ought to do, every thing, nevertheless, goes on right. The
sprout, after being pushed down a little way, makes a bend and turns
upwards; the fibres, on the contrary, after shooting at first upwards,
turn down. Of this extraordinary vegetable fact an account has lately
been attempted to be given. 'The plumule, it is said, is stimulated by
the *air* into action, and elongates itself when it is thus most excited:
the radicle is stimulated by *moisture*, and elongates itself when *it* is
thus most excited. Whence one of these grows upward in quest of its
adapted object, and the other downward.'[1] Were this account better
verified by experiment than it is, it only shifts the contrivance. It
does not disprove the contrivance; it only removes it a little further
back. Who, to use our author's own language, 'adapted the objects?'*
Who gave such a quality to these connate parts, as to be susceptible
of different 'stimulation:' as to be 'excited' each only by its own
element, and precisely by that, which the success of the vegetation
requires? I say, 'which the success of the vegetation requires,' for the
toil of the husbandman would have been in vain; his laborious and
expensive preparation of the ground in vain; if the event must, after
all, depend, upon the position in which the scattered seed was sown.
Not one seed out of a hundred would fall in a right direction.

Our *second* observation is upon a general property of climbing
plants, which is strictly mechanical. In these plants, from each knot
or joint, or, as botanists call it, axilla of the plant, issue, close to each
other, two shoots; one, bearing the flower and fruit, the other, drawn
out into a wire, a long, tapering, spiral tendril, that twists itself round
any thing which lies within its reach. Considering, that, in this class,
two purposes are to be provided for (and together) fructification and
support, the fruitage of the plant, and the sustentation of its stalk,
what means could be used more effectual, or, as I have said, more
mechanical, than what this structure presents to our eyes? Why or

[1] Darwin's Phytologia,* p. 144.

how, without a view to this double purpose, do two shoots, of such different and appropriate forms, spring from the same joint, from contiguous points of the same stalk? It never happens thus in robust plants, or in trees. 'We see not,' says Ray, 'so much as one tree, or shrub, or herb, that hath a firm and strong stem, and that is able to mount up and stand alone without assistance, *furnished with these tendrils.*' Make only so simple a comparison as that between a pea and a bean. Why does the pea put forth tendrils, the bean not; but because the stalk of the pea cannot support itself, the stalk of the bean can? We may add also, as a circumstance not to be overlooked, that, in the pea tribe, these clasps do not make their appearance, till they are wanted; till the plant has grown to a height to stand in need of support.

This word 'support,' suggests to us a reflection upon a property of grasses, of corn, and canes. The hollow stems of these classes of plants, are set, at certain intervals, with joints. These joints are not found in the trunks of trees, or in the solid stalks of plants. There may be other uses of these joints; but the fact is, and it appears to be, at least, one purpose designed by them, that they *corroborate* the stem; which, by its length and hollowness, would, otherwise, be too liable to break or bend.

Grasses are Nature's care. With these she clothes the earth: with these she sustains its inhabitants. Cattle feed upon their leaves; birds upon their smaller seeds; men upon the larger; for few readers need be told that the plants, which produce our bread corn, belong to this class. In those tribes, which are more generally considered as grasses, their extraordinary means and powers of preservation and increase, their hardiness, their almost unconquerable disposition to spread, their faculties of reviviscence, coincide with the intention of nature concerning them. They thrive under a treatment by which other plants are destroyed. The more their leaves are consumed, the more their roots increase. The more they are trampled upon, the thicker they grow. Many of the seemingly dry and dead leaves of grasses revive, and renew their verdure, in the spring. In lofty mountains, where the summer heats are not sufficient to ripen the seeds, grasses abound, which are viviparous, and consequently able to propagate themselves without seed. It is an observation likewise which has often been made, that herbivorous animals attach themselves to the leaves of grasses; and, if at liberty in their pastures to range and choose, leave untouched the straws which support the flowers.[1]

THE GENERAL properties of vegetable nature, or properties common to large portions of that kingdom, are almost all which the compass of our argument allows to bring forward. It is impossible to follow plants into their several species. We may be allowed, however, to single out three or four of these species as worthy of a particular notice, either by some singular mechanism, or by some peculiar provision, or by both.

I. In Dr Darwin's Botanic Garden,* line 395, note, is the following account of the *vallisneria*, as it has been observed in the river Rhone. 'They have roots at the bottom of the Rhone. The flowers of the *female plant* float on the surface of the water, and are furnished with an *elastic, spiral, stalk*, which extends or contracts as the water rises or falls: this rise or fall, from the torrents which flow into the river, often amounting to many feet in a few hours. The flowers of the *male plant* are produced under water; and, as soon as the fecundating farina is mature, they separate themselves from the plant; rise to the surface; and are wasted by the air, or borne by the currents, to the female flowers.' Our attention in this narrative will be directed to two particulars; first to the mechanism, the 'elastic, spiral, stalk' which lengthens or contracts itself according as the water rises or falls; secondly, to the provision which is made for bringing the male flower, which is produced *under* water, to the female flower which floats upon the surface.

II. My second example I take from Withering Arrang. vol. ii. p. 209. ed. 3. 'The *cuscuta europœa* is a parasitical plant. The feed opens, and puts forth a *little spiral body*, which does NOT seek the earth to take root; but *climbs* in a spiral direction, from right to left, up other plants, from which, by means of vessels, it draws its nourishment.' The 'little spiral body' proceeding from the seed is to be compared with the fibres which seeds send out in ordinary cases; and the comparison ought to regard both the form of the threads and the direction. They are straight; this is spiral. They shoot downwards; this points upwards. In the rule, and in the exception, we equally perceive design.

III. A better known parasitical plant is the evergreen shrub, called

[1] With.* Bot. Arr. vol. i. p. 28. ed. 2d.

the *misseltoe*. What we have to remark in it, is a singular instance of *compensation*. No art hath yet made these plants take root in the earth. Here therefore might seem to be a mortal defect in their constitution. Let us examine how this defect is made up to them. The seeds are endued with an adhesive quality so tenacious, that, if they be rubbed upon the smooth bark of almost any tree, they will stick to it. And then what follows? Roots springing from their seeds, insinuate their fibres into the woody substance of the tree; and the event is, that a misseltoe plant is produced the next winter.[1] Of no other plant do the roots refuse to shoot in the ground; of no other plant do the seeds possess this adhesive, generative, quality, when applied to the bark of trees.

IV. Another instance of the *compensatory* system is in the autumnal crocus or meadow saffron, (*cholcicum autumnale*). I have pitied this poor plant a thousand times. Its blossom rises out of the ground in the most forlorn condition possible; without a sheath, a fence, a calyx, or even a leaf to protect it; and that, not in the spring, not to be visited by summer suns, but under all the disadvantages of the declining year. When we come however to look more closely into the structure of this plant, we find that, instead of its being neglected, nature has gone out of her course to provide for its security, and to make up to it for all its defects. The seed-vessel, which in other plants is situated within the cup of the flower, or just beneath it, in this plant lies buried ten or twelve inches under ground within the bulbous root. The tube of the flower, which is seldom more than a few tenths of an inch long, in this plant extends down to the root. The styles always reach the seed-vessel; but it is in this, by an elongation unknown to any other plant. All these singularities contribute to one end. 'As this plant blossoms late in the year, and, probably, would not have time to ripen its seeds before the access of winter which would destroy them, Providence has contrived its structure such, that this important office may be performed at a depth in the earth out of reach of the usual effects of frost.'[2] That is to say, in the autumn nothing is done above ground but the business of impregnation; which is an affair between the antheræ and the stigmata. The maturation of the impregnated seed, which in other plants proceeds

---

[1] Ib. p. 203.
[2] Ib. p. 360

within a capsule, exposed together with the rest of the flower to the open air, is here carried on, and during the whole winter, within the heart, as we may say, of the earth, that is, 'out of the reach of the usual effects of frosts.' But then a new difficulty presents itself. Seeds, though perfected, are known not to vegetate at this depth in the earth. Our seeds therefore, though so safely lodged, would, after all, be lost to the purpose for which all seeds are intended. Lest this should be the case, 'a second admirable provision is made to raise them above the surface when they are perfected, and to sow them at a proper distance:' viz. the germ grows up *in the spring*, upon a fruit-stalk, accompanied with leaves. The seeds now, in common with those of other plants, have the benefit of the summer, and are sown upon the surface. The order of vegetation externally is this. The plant produces its flowers in September; its leaves and fruits in the spring following.

V. I give the account of the dionæa muscipula,* an extraordinary American plant, as some late authors have related it; but, whether we be yet enough acquainted with the plant to bring every part of this account to the test of repeated and familiar observation, I am unable to say. Its leaves are jointed, and furnished with two rows of strong prickles; their surfaces covered with a number of minute glands, which secrete a sweet liquor, that allures the approach of flies. When these parts are touched by the legs of flies, the two lobes of the leaf instantly spring up, the rows of prickles lock themselves fast together, and squeeze the unwary animal to death.[1] Here, under a new model, we recognise the ancient plan of nature; viz. the relation of parts and provisions to one another, to a common office, and to the utility of the organized body to which they belong. The attracting syrup, the rows of strong prickles, their position so as to interlock, the joints of the leaves; and, what is more than the rest, that singular irritability of their surfaces, by which they close at a touch; all bear a contributory part in producing an effect, connected either with the defence, or with the nutrition, of the plant.

[1] Smellie's Phil. of Nat. Hist.* vol. i. p. 5.

CHAPTER XXI

THE ELEMENTS

WHEN we come to the elements,* we take leave of our mechanics; because we come to those things, of the organization of which, if they be organized, we are confessedly ignorant. This ignorance is implied by their name. To say the truth, our investigations are stopped long before we arrive at this point. But then it is for our comfort to find, that a knowledge of the constitution of the elements is not necessary for us. For instance, as Addison* has well observed, 'we know *water* sufficiently, when we know how to boil, how to freeze, how to evaporate, how to make it fresh, how to make it run or spout out, in what quantity and direction we please, without knowing what water is.' The observation of this excellent writer has more propriety in it now, than it had at the time it was made: for the constitution, and the constituent parts, of water, appear in some measure to have been lately discovered;* yet it does not, I think, appear, that we can make any better or greater use of water since the discovery, than we did before it.

We can never think of the elements without reflecting upon the number of distinct uses which are *consolidated* in the same substance. The *air* supplies the lungs, supports fire, conveys sound, reflects light, diffuses smells, gives rain, wafts ships, bears up birds. Εξ ὕδατος τα παντα;* *water*, beside maintaining its own inhabitants, is the universal nourisher of plants, and through them of terrestrial animals; is the basis of their juices and fluids: dilutes their food, quenches their thirst, floats their burthens. *Fire* warms, dissolves, enlightens; is the great promoter of vegetation and life, if not necessary to the support of both.

We might enlarge, to almost any length we pleased, upon each of these uses; but it appears to me almost sufficient to state them. The few remarks, which I judge it necessary to add, are as follow.

I. AIR is essentially different from earth. There appears to be no necessity for an atmosphere's investing our globe:* (the moon has none:) yet it does invest it; and we see how many, how various, and how important are the purposes which it answers to every order of animated, not to say of organized, beings, which are placed upon the

terrestrial surface. I think that every one of these uses will be understood upon the first mention of them, except it be that of *reflecting*
light, which may be explained thus. If I had the power of seeing only
by means of rays coming directly from the sun, whenever I turned
my back upon the luminary, I should find myself in darkness. If I had
the power of seeing by reflected light, yet by means only of light
reflected from solid masses, these masses would shine, indeed, and
glisten, but it would be in the dark. The hemisphere, the sky, the
world, could only be *illuminated*, as it is illuminated, by the light of
the sun being from all sides, and in every direction, reflected to the
eye, by particles, as numerous, as thickly scattered, and as widely
diffused, as are those of the air.

Another general quality of the atmosphere is, the power of evaporating fluids. The adjustment of this quality to our use is seen in its
action upon the sea.* In the sea, water and salt are mixed together
most intimately; yet the atmosphere raises the water, and leaves the
salt. Pure and fresh as drops of rain descend, they are collected from
brine. If evaporation be solution, (which seems to be probable,) then
the air dissolves the water and not the salt. Upon whatever it be
founded, the distinction is critical; so much so, that, when we
attempt to imitate the process by art, we must regulate our distilla-
tion with great care and nicety, or, together with the water, we get the
bitterness, or, at least, the distastefulness of the marine substance:
and, after all, it is owing to this original elective power in the air,* that
we can effect the separation which we wish, by any art or means
whatever.

By evaporation water is carried up into the air; by the converse of
evaporation it falls down upon the earth. And how does it fall? Not
by the clouds being all at once reconverted into water, and descending, like a sheet; not in rushing down in columns from a spout; but in
moderate drops, as from a cullender.* Our watering-pots are made to
imitate showers of rain. Yet, a priori, I should have thought either of
the two former methods more likely to have taken place than the last.

By respiration,* flame, putrefaction, air is rendered unfit for the
support of animal life. By the constant operation of these corrupting
principles, the whole atmosphere, if there were no restoring causes,
would come at length to be deprived of its necessary degree of purity. Some of these causes seem to have been discovered, and their
efficacy ascertained by experiment. And so far as the discovery has

proceeded, it opens to us a beautiful and a wonderful œconomy. *Vegetation* proves to be one of them. A sprig of mint, corked up with a small portion of foul air placed in the light, renders it again capable of supporting life or flame. Here therefore is a constant circulation of benefits maintained between the two great provinces of organized nature. The plant purifies, what the animal had poisoned: in return, the contaminated air is more than ordinarily nutritious to the plant. *Agitation with water* turns out to be another of these restoratives. The foulest air, shaken in a bottle with water for a sufficient length of time, recovers a great degree of its purity. Here then again, allowing for the scale upon which nature works, we see the salutary effects of *storms* and *tempests*. The yesty waves, which confound the heaven and the sea, are doing the very thing which is done in the bottle. Nothing can be of greater importance to the living creation, than the salubrity of their atmosphere. It ought to reconcile us therefore to these agitations of the elements, of which we sometimes deplore the consequences, to know, that they tend powerfully to restore to the air that purity, which so many causes are constantly impairing.

II. In WATER, what ought not a little to be admired, are those negative qualities which constitute its *purity*. Had it been vinous, or oleaginous,* or acid; had the sea been filled, or the rivers flowed, with wine or milk; fish, constituted as they are, must have died; plants, constituted as they are, would have withered; the lives of animals, which feed upon plants, must have perished. Its very *insipidity*, which is one of those negative qualities, renders it the best of all menstrua. Having no taste of its own, it becomes the sincere vehicle of every other. Had there been a taste in water, be it what it might, it would have infected every thing we ate or drank, with an importunate repetition of the same flavor.

Another thing in this element, not less to be admired, is the constant *round* which it travels; and by which, without suffering either adulteration or waste, it is continually offering itself to the wants of the habitable globe. From the sea are exhaled those vapours which form the clouds. These clouds descend in showers, which, penetrating into the crevices of the hills, supply springs. Which springs flow in little streams into the valleys; and, there uniting, become rivers. Which rivers, in return, feed the ocean. So there is an incessant circulation of the same fluid; and not one drop probably more or less now, than there was at the creation. A particle of water takes its departure from the surface of the sea, in order to fulfill certain important offices to the earth; and having executed the service which was assigned to it, returns to the bosom which it left.

Some have thought that we have too much water upon the globe; the sea occupying above three quarters of its whole surface. But the expanse of ocean, immense as it is, may be no more than sufficient to fertilise the earth. Or, independently of this reason, I know not why the sea may not have as good a right to its place as the land. It may proportionably support as many inhabitants; minister to as large an aggregate of enjoyment. The land only affords a habitable surface; the sea is habitable to a great depth.

III. Of FIRE, we have said that it *dissolves*. The only idea probably which this term raised in the reader's mind was, that of fire melting metals, resins, and some other substances, fluxing ores,* running glass, and assisting us in many of our operations, chymical or culinary. Now these are only uses of an occasional kind, and give us a very imperfect notion of what fire does for us. The grand importance of this dissolving power, the great office indeed of fire in the œconomy of nature, is keeping things in a state of solution, that is to say, in a state of fluidity. Were it not for the presence of heat, or of a certain degree of it, all fluids would be frozen. The ocean itself would be a quarry of ice: universal nature stiff and dead.

We see therefore, that the elements bear, not only a strict relation to the constitution of organized bodies, but a relation to each other. Water could not perform its office to the earth without air; nor exist, as water, without fire.

IV. Of LIGHT, (whether we regard it as of the same substance with fire, or as a different substance,)* it is altogether superfluous to expatiate upon the use. No man disputes it. The observations, therefore, which I shall offer, respect that little which we seem to know of its constitution.

Light passes from the sun to the earth in eleven minutes; a distance, which it would take a cannon ball twenty-five years, in going over. Nothing more need be said to shew the velocity of light. Urged by such a velocity, with what *force* must its particles drive against, I will not say the eye, the tenderest of animal substances, but every substance, animate or inanimate, which stands in its way? It might seem to be a force sufficient to shatter to atoms the hardest bodies.*

How then is this effect, the consequence of such prodigious

velocity, guarded against? By a proportionable *minuteness* of the particles of which light is composed. It is impossible for the human
mind to imagine to itself any thing so small as a particle of light. But
this extreme exility, though difficult to conceive, it is easy to prove. A
drop of tallow, expended in the wick of a farthing candle, shall shed
forth rays sufficient to fill a hemisphere of a mile diameter; and to fill
it so full of these rays, that an aperture not larger than the pupil of an
eye, wherever it be placed within the hemisphere, shall be sure to
receive some of them. What floods of light are continually poured
from the sun we cannot estimate; but the immensity of the sphere
which is filled with its particles, even if it reached no further than the
orbit of the earth, we can in some sort compute: and we have reason
to believe, that, throughout this whole region, the particles of light
lie, in latitude at least, near to one another. The spissitude* of the
sun's rays at the earth is such, that the number which falls upon a
burning glass of an inch diameter, is sufficient, when concentrated,
to set wood on fire.

The tenuity and the velocity of particles of light, as ascertained by
separate observations, may be said to be proportioned to each other:
both surpassing our utmost stretch of comprehension; but proportioned. And it is this proportion alone, which converts a tremendous element into a welcome visitor.

It has been observed to me by a learned friend, as having often
struck his mind, that, if light had been made by a common artist, it
would have been of one uniform *colour*: whereas, by its present composition, we have that variety of colours,* which is of such infinite use
to us for the distinguishing of objects; which adds so much to the
beauty of the earth, and augments the stock of our innocent
pleasures.

With which may be joined another reflection, viz. that, considering light as compounded of rays of seven different colours, (of which
there can be no doubt, because it can be resolved into these rays by
simply passing it through a prism,) the constituent parts must be
well mixed and blended together, to produce a fluid, so clear and
colourless, as a beam of light is, when received from the sun.

CHAPTER XXII

ASTRONOMY[1]

MY opinion of Astronomy has always been, that it is *not* the best medium through which to prove the agency of an intelligent Creator*; but that, this being proved, it shews, beyond all other sciences, the magnificence of his operations. The mind which is once convinced, it raises to sublimer views of the Deity, than any other subject affords; but is not so well adapted, as some other subjects are, to the purpose of argument. We are destitute of the means of examining the constitution of the heavenly bodies. The very simplicity of their appearance is against them. We see nothing, but bright points, luminous circles, or the phases of spheres reflecting the light which falls upon them. Now we deduce design from relation, aptitude, and correspondence of *parts*. Some degree therefore of *complexity* is necessary to render a subject fit for this species of argument. But the heavenly bodies do not, except perhaps in the instance of Saturn's ring, present themselves to our observation as compounded of parts at all. This, which may be a perfection in them, is a disadvantage to us, as enquirers after their nature. They do not come within our mechanics.

And what we say of their forms, is true of their *motions*. Their motions are carried on without any sensible intermediate apparatus: whereby we are cut off from one principal ground of argumentation and analogy. We have nothing wherewith to compare them; no invention, no discovery, no operation or resource of art, which, in this respect, resembles them. Even those things which are made to imitate and represent them, such as orreries, planetaria,* cœlestial globes, etc. bear no affinity to them, in the cause and principle by which their motions are actuated. I can assign for this difference a reason of utility, viz. a reason why, though the action of *terrestrial* bodies upon each other be, in almost all cases, through the intervention of solid or fluid substances, yet central attraction does not operate in this manner. It was necessary that the intervals between the planetary orbs

---

[1] For the articles in this chapter marked with an asterisk, I am indebted to some obliging communications, received (through the hands of the Lord Bishop of Elphin*) from the Rev. J. Brinkley,* M. A. Andrew's Professor of Astronomy in the University of Dublin.

should be devoid of any *inert* matter either fluid or solid,* because such an intervening substance would, by its resistance, destroy those very motions, which attraction is employed to preserve. This may be a final cause of the difference; but still the difference destroys the analogy.

Our ignorance, moreover, of the *sensitive* natures, by which other planets are inhabited,* necessarily keeps from us the knowledge of numberless utilities, relations, and subserviencies, which we perceive upon our own globe.

After all; the real subject of admiration is, that we understand so much of astronomy as we do. That an animal confined to the surface of one of the planets; bearing a less proportion to it, than the smallest microscopic insect does to the plant it lives upon; that this little, busy, inquisitive creature, by the use of senses which were given to it for its domestic necessities, and by means of the assistance of those senses which it has had the art to procure, should have been enabled to observe the whole system of worlds to which its own belongs; the changes of place of the immense globes which compose it; and with such accuracy, as to mark out, beforehand, the situation in the heavens in which they will be found at any future point of time; and that these bodies, after sailing through regions of void and trackless space, should arrive at the place where they were expected, not within a minute, but within a few seconds of a minute, of the time prefixed and predicted: all this is wonderful, whether we refer our admiration to the constancy of the heavenly motions themselves, or to the perspicacity and precision with which they have been noticed by mankind. Nor is this the whole, nor indeed the chief part, of what astronomy teaches. By bringing reason to bear upon observation, (the acutest reasoning upon the exactest observation,) the astronomer has been able, out of the confusion (for such it is) under which the motions of the heavenly bodies present themselves to the eye of a mere gazer upon the skies, to elicit their order and their real paths.

Our knowledge therefore of astronomy is admirable though imperfect: and, amidst the confessed desiderata and desideranda, which impede our investigation of the wisdom of the Deity, in these the grandest of his works, there are to be found, in the phænomena, ascertained circumstances and laws, sufficient to indicate an intellectual agency in three of its principal operations,* viz. in chusing, in determining, in regulating; in *chusing*, out of a boundless variety of suppositions which were equally possible, that which is beneficial; in *determining*, what, left to itself, had a thousand chances against con-veniency, for one in its favour; in *regulating* subjects, as to quantity and degree, which, by their nature, were unlimited with respect to either. It will be our business to offer, under each of these heads, a few instances, such as best admit of a popular explication.

I. Amongst proofs of choice, one is, fixing the source of light and heat in the *centre* of the system. The sun is ignited and luminous; the planets, which move round him, cold and dark. There seems to be no antecedent necessity for this order. The sun might have been an opaque mass: some one, or two, or more, or any, or all, of the planets, globes of fire. There is nothing in the nature of the heavenly bodies, which requires that those which are stationary should be on fire, that those which move should be cold: for, in fact, comets are bodies on fire, yet revolve round a centre: nor does this order obtain between the primary planets and their secondaries, which are all opaque. When we consider, therefore, that the sun is one; that the planets going round it are, at least, seven;* that it is indifferent to their nature which are luminous and which are opaque; and also, in what order with respect to each other, these two kinds of bodies are disposed; we may judge of the improbability of the present arrangement taking place by chance.

If, by way of accounting for the state in which we find the solar system, it be alledged (and this is one amongst the guesses of those who reject an intelligent Creator*) that the planets themselves are only cooled or cooling masses, and were once, like the sun, many thousand times hotter than red hot iron; then it follows, that the sun also himself must be in his progress towards growing cold; which puts an end to the possibility of his having existed, as he is, from eternity. This consequence arises out of the hypothesis with still more certainty, if we make a part of it, what the philosophers who maintain it, have usually taught, that the planets were originally masses of matter struck off, in a state of fusion, from the body of the sun, by the percussion of a comet, or by a shock from some other cause with which we are not acquainted: for, if these masses, partak-ing of the nature and substance of the sun's body, have in process of time lost their heat, that body itself, in time likewise, no matter in how much longer time, must lose its heat also; and therefore be incapable of an eternal duration in the state in which we see it, either for the time to come, or the time past.

The preference of the present to any other mode of distributing luminous and opaque bodies I take to be evident. It requires more astronomy than I am able to lay before the reader, to shew, in its particulars, what would be the effect to the system, of a dark body at the centre, and of one of the planets being luminous: but I think it manifest, without either plates or calculation, first, that, supposing the necessary proportion of magnitude between the central and the revolving bodies to be preserved, the ignited planet would not be sufficient to illuminate and warm the rest of the system; secondly, that its light and heat would be imparted to the other planets, much more irregularly than light and heat are now received from the sun.

(*) II. Another thing, in which a choice appears to be exercised; and in which, amongst the possibilities out of which the choice was to be made, the number of those which were wrong, bore an infinite proportion to the number of those which were right, is in what geometricians call the *axis of rotation*. This matter I will endeavour to explain. The earth, it is well known, is not an exact globe, but an oblate spheroid, something like an orange. Now the axes of rotation, or the diameters upon which such a body may be made to turn round, are as many as can be drawn through its centre to opposite points upon its whole surface: but of these axes none are *permanent*, except either its shortest diameter, i. e. that which passes through the heart of the orange from the place where the stalk is inserted into it, and which is but one; or its longest diameters, at right angles with the former, which must all terminate in the single circumference which goes round the thickest part of the orange. This shortest diameter is that upon which in fact the earth turns; and it is, as the reader sees, what it ought to be, a permanent axis: whereas, had blind chance, had a casual impulse, had a stroke or push at random, set the earth a-spinning, the odds were infinite, but that they had sent it round upon a wrong axis. And what would have been the consequence? The difference between a permanent axis and another axis is this. When a spheroid in a state of rotatory motion gets upon a permanent axis, it keeps there; it remains steady and faithful to its position; its poles preserve their direction with respect to the plane and to the centre of its orbit: but, whilst it turns upon an axis which is *not* permanent, (and the number of those, we have seen, infinitely exceeds the number of the other,) it is always liable to shift and vacillate from one axis to another, with a corresponding change in the inclination of its poles. Therefore, if a planet once set off revolving upon any other than its shortest, or one of its longest axes, the poles on its surface would keep perpetually changing, and it never would attain a permanent axis of rotation. The effect of this unfixedness and instability would be, that the equatorial parts of the earth might become the polar, or the polar the equatorial; to the utter destruction of plants and animals, which are not capable of interchanging their situations, but are respectively adapted to their own. As to ourselves, instead of rejoicing in our temperate zone, and annually preparing for the moderate vicissitude, or rather the agreeable succession of seasons, which we experience and expect, we might come to be locked up in the ice and darkness of the arctic circle, with bodies neither inured to its rigors, nor provided with shelter or defence against them. Nor would it be much better, if the trepidation of our pole, taking an opposite course, should place us under the heats of a vertical sun. But, if it would fare so ill with the human inhabitant, who can live under greater varieties of latitude than any other animal, still more noxious would this translation of climate have proved to life in the rest of the creation; and, most perhaps of all, in plants. The habitable earth, and its beautiful variety, might have been destroyed, by a simple mischance in the axis of rotation.

(*) III. All this however proceeds upon a supposition of the earth having been formed at first an oblate spheroid. There is another supposition; and, perhaps, our limited information will not enable us to decide between them. The second supposition is, that the earth, being a mixed mass somewhat fluid, took, as it might do, its present form, by the joint action of the mutual gravitation of its parts and its rotatory motion. This, as we have said, is a point in the history of the earth, which our observations are not sufficient to determine. For a very small depth below the surface (but extremely small, less, perhaps, than an eight thousandth part, compared with the depth of the centre) we find vestiges of ancient fluidity. But this fluidity must have gone down many hundred times further than we can penetrate, to enable the earth to take its present oblate form; and, whether any traces of this kind exist to that depth, we are ignorant. Calculations were made a few years ago of the mean density of the earth,* by

comparing the force of its attraction with the force of attraction of a
rock of granite, the bulk of which could be ascertained: and the
upshot of the calculation was, that the earth upon an average,
through its whole sphere, has twice the density of granite, or about
five times that of water. Therefore it cannot be a hollow shell, as
some have formerly supposed: nor can its internal parts be occupied
by central fire, or by water. The solid parts must greatly exceed the
fluid parts: and the probability is, that it is a solid mass throughout,
composed of substances, more ponderous the deeper we go. Nevertheless, we may conceive the present face of the earth to have originated from the revolution of a sphere, covered with a surface of a
compound mixture; the fluid and solid parts separating, as the surface became quiescent. Here then comes in the *moderating* hand of
the Creator. If the water had exceeded its present proportion, even
but by a trifling quantity compared with the whole globe, all the land
would have been covered: had there been much less than there is,
there would not have been enough to fertilize the continent. Had the
exsiccation been progressive, such as we may suppose to have been
produced by an evaporating heat, how came it to stop at the point at
which we see it? Why did it not stop sooner; why at all? The mandate
of the Deity will account for this: nothing else will.

IV. OF CENTRIPETAL FORCES. By virtue of the simplest law that
can be imagined, viz. that a body *continues* in the state in which it is,
whether of motion or rest;* and, if in motion, goes on in the line in
which it was proceeding, and with the same velocity, *unless* there be
some cause for change: by virtue, I say, of this law, it comes to pass
(what may appear to be a strange consequence) that cases arise, in
which attraction, incessantly drawing a body towards a centre, never
brings, nor ever will bring, the body to that centre, but keep it in
eternal circulation round it. If it were possible to fire off a cannon
ball with a velocity of five miles in a second, and the resistance of the
air could be taken away, the cannon ball would for ever wheel round
the earth, instead of falling down upon it. This is the principle which
sustains the heavenly motions. The Deity having appointed this
law to matter, than which, as we have said before, no law could be
more simple, has turned it to a wonderful account in constructing
planetary systems.

The actuating cause in these systems, is an attraction which varies
reciprocally as the square of the distance:* that is, at double the

distance, has a quarter of the force; at half the distance, four times the strength; and so on. Now, concerning this law of variation, we have three things to observe; first, that attraction, for any thing we know about it, was just as capable of one law of variation as of another: secondly; that, out of an infinite number of possible laws, those which were admissible for the purpose of supporting the heavenly motions, lay within certain narrow limits: thirdly; that of the admissible laws, or those which come within the limits prescribed, the law that actually prevails is the most beneficial. So far as these propositions can be made out, we may be said, I think, to prove choice and *regulation*; choice, out of boundless variety; and regulation, of that which, by its own nature, was, in respect of the property regulated, indifferent and indefinite.

I. First then, attraction, for any thing we know about it, was originally indifferent to all laws of variation depending upon change of distance, i. e. just as susceptible of one law as of another. It might have been the same at all distances. It might have increased as the distance increased. Or it might have diminished with the increase of the distance, yet in ten thousand different proportions from the present. It might have followed no stated law at all. If attraction be, what Cotes* with many other Newtonians have thought it, a primordial property of matter,* not dependent upon, or traceable to, any other material cause, then, by the very nature and definition of a primordial property, it stood indifferent to all laws. If it be the agency of something immaterial, then also, for any thing we know of it, it was indifferent to all laws. If the revolution of bodies round a centre depend upon vortices, neither are these limited to one law more than another.

There is, I know, an account given of attraction, which should seem, in its very cause, to assign to it the law, which we find it to observe, and which, therefore, makes that law, a law, not of choice, but of necessity: and it is the account, which ascribes attraction to an *emanation* from the attracting body. It is probable, that the influence of such an emanation will be proportioned to the spissitude of the rays, of which it is composed: which spissitude, supposing the rays to issue in right lines on all sides from a point, will be reciprocally as the square of the distance. The mathematics of this solution we do not call in question: the question with us is, whether there be any sufficient reason to believe, that attraction is produced by an emanation. For my part, I am totally at a loss to comprehend, how particles streaming *from* a centre, should draw a body *towards* it. The impulse, if impulse it be, is all the other way. Nor shall we find less difficulty in conceiving, a conflux of particles, incessantly flowing to a centre, and carrying down all bodies along with it, that centre also itself being in a state of rapid motion through absolute space; for, by what source is the stream fed, or what becomes of the accumulation? Add to which, that it seems to imply a contrariety of properties, to suppose an æthereal fluid* to *act* but not to *resist*; powerful enough to carry down bodies with great force towards a centre, yet, inconsistently with the nature of inert matter, powerless and perfectly yielding with respect to the motions which result from the projectile impulse. By calculations drawn from ancient notices of eclipses of the moon, we can prove, that, if such a fluid exist at all, its resistance has had no sensible effect upon the moon's motion for two thousand five hundred years. The truth is, except this one circumstance of the variation of the attracting force at different distances agreeing with the variation of the spissitude, there is no reason whatever to support the hypothesis of an emanation; and, as it seems to me, almost insuperable reasons against it.

II. (*) Our second proposition is, that, whilst the possible laws of variation were infinite, the *admissible* laws, or the laws compatible with the preservation of the system, lay within narrow limits. If the attracting force had varied according to any *direct* law of the distance, let it have been what it would, great destruction and confusion would have taken place. The direct simple proportion of the distance would, it is true, have produced an ellipse; but the perturbing forces* would have acted with so much advantage, as to be continually changing the dimensions of the ellipse, in a manner inconsistent with our terrestrial creation. For instance; if the planet Saturn, so large and so remote, had attracted the earth, both in proportion to the quantity of matter contained in it, which it does; and also in any proportion to its distance, i. e. if it had pulled the harder for being the further off, (instead of the reverse of it,) it would have dragged the globe which we inhabit out of its course, and have perplexed its motions, to a degree incompatible with our security, our enjoyments, and probably our existence. Of the *inverse* laws, if the centripetal force had changed as the cube of the distance, or in any higher proportion, that is, (for I speak to the unlearned,) if, at double the distance, the attractive force had been diminished to an eighth part, or to less than that, the consequence would have been, that the planets, if they once began to approach the sun would have fallen into his body; if they once, though by ever so little, increased their distance from the centre, would for ever have receded from it. The laws therefore of attraction, by which a system of revolving bodies could be upheld in their motions, lie within narrow limits, compared with the possible laws. I much underrate the restriction, when I say, that in a scale of a mile they are confined to an inch. All direct ratios of the distance are excluded, on account of danger from perturbing forces: all reciprocal ratios, except what lie beneath the cube of the distance, by the demonstrable consequence, that every the least change of distance, would, under the operation of such laws, have been fatal to the repose and order of the system. We do not know, that is, we seldom reflect, how interested we are in this matter. Small irregularities may be endured; but, changes within these limits being allowed for, the permanency of our ellipse is a question of life and death to our whole sensitive world.

III. (⋆) That the subsisting law of attraction falls within the limits which utility requires, when these limits bear so small a proportion to the range of possibilities, upon which chance might equally have cast it, is not, with any appearance of reason, to be accounted for, by any other cause than a regulation proceeding from a designing mind. But our next proposition carries the matter somewhat further. We say, in the third place, that, out of the different laws which lie within the limits of admissible laws, the *best* is made choice of; that there are advantages in this particular law which cannot be demonstrated to belong to any other law; and, concerning some of which, it can be demonstrated that they do not belong to any other.

(⋆) 1. Whilst this law prevails between each particle of matter, the *united* attraction of a sphere, composed of that matter, observes the same law. This property of the law is necessary, to render it applicable to a system composed of spheres, but it is a property which belongs to no other law of attraction that is admissible. The law of variation of the united attraction is in no other case the same as the law of attraction of each particle, one case excepted, and that is of the attraction varying directly as the distance; the inconveniency of which law in other respects we have already noticed.

(⋆) 2. Under the subsisting law, the *apsides*, the returning points,

or points of greatest and least distance from the centre, are *quiescent*,*
and, therefore, the body moves every revolution in exactly the same
path relative to the attracting centre: which it would not do, under
any other law whatever except that of the direct ratio of the distance,
which we have seen to be objectionable. The planetary system
required that the law of attraction should be a law which gave an
orbit returning into itself. Now, out of an infinite number of laws,
admissible and inadmissible, out of a vast variety even of admissible
laws, there are few, except the actual law, which would do this. Here
then is choice.

(*) 3. All systems must be liable to *perturbations*. And therefore to
guard against these perturbations, or rather to guard against their
running to destructive lengths, is perhaps the strongest evidence of
care and foresight that can be given. Now we are able to demonstrate
of our law of attraction, what can be demonstrated of no other, and
what qualifies the dangers which arise from cross but unavoidable
influences, that the action of the parts of our system upon one
another will not cause permanently increasing irregularities, but
merely periodical ones: that is, they will come to a limit, and then go
back again. This we can demonstrate only of a system, in which the
following properties concur, viz. that the force shall be inversely as
the square of the distance; the masses of the revolving bodies small,
compared with that of the body at the centre; the orbits not much
inclined to one another; and their eccentricity* little. In such a system
the grand points are secure. The mean distances and periodic times,
upon which depend our temperature, and the regularity of our year,
are constant. The eccentricities, it is true, will still vary, but so
slowly, and to so small an extent, as to produce no inconveniency
from fluctuation of temperature and season. The same as to the
obliquity of the planes of the orbits. For instance, the inclination of
the ecliptic* to the equator will never change above two degrees, (out
of ninety,) and that will require many thousand years in performing.

It has been rightly also remarked, that, if the great planets Jupiter
and Saturn had moved in lower spheres, their influences would have
had much more effect as to disturbing the planetary motions than
they now have. While they revolve at so great distances from the rest,
they act almost equally on the Sun and on the inferior planets, which
has nearly the same consequence as not acting at all upon either.

If it be said that the planets might have been sent round the Sun in exact circles, in which case, no change of distance from the centre taking place, the law of variation of the attracting power would have never come in question; one law would have served as well as another; an answer to the scheme may be drawn from the consideration of these same perturbing forces. The system retaining in other respects its present constitution, though the planets had been at first sent round in exact circular orbits, they could not have kept them: and if the law of attraction had not been what it is, (or, at least, if the prevailing law had transgressed the limits above assigned,) every evagation would have been fatal: the planet once drawn, as drawn it necessarily must have been, out of its course, would have wandered in endless error.

(*) V. What we have seen in the law of the centripetal force, viz. a choice guided by views of utility, and a choice of one law out of thousands which might equally have taken place, we see no less in the *figures* of the planetary orbits. It was not enough to fix the law of the centripetal force, though by the wisest choice, for, even under that law, it was still competent to the planets to have moved in paths possessing so great a degree of eccentricity, as, in the course of every revolution, to be brought very near to the sun, and carried away to immense distances from him. The comets actually move in orbits of this sort: and, had the planets done so, instead of going round in orbits nearly circular, the change from one extremity of temperature to another must, in ours at least, have destroyed every animal and plant upon its surface. Now, the distance from the centre at which a planet sets off, and the absolute force of attraction at that distance, being fixed, the figure of his orbit, its being a circle, or nearer to, or further off from, a circle, viz. a rounder or a longer oval, depends upon two things, the velocity with which, and the direction in which, the planet is projected. And these, in order to produce a right result, must be both brought within certain narrow limits. One, and only one, velocity, united with one, and only one, direction, will produce a perfect circle. And the velocity must be near to this velocity, and the direction also near to this direction, to produce orbits, such as the planetary orbits are, nearly circular; that is, ellipses with small eccentricities. The velocity and the direction must *both* be right. If the velocity be wrong, no direction will cure the error; if the direction be in any considerable degree oblique, no velocity will produce the orbit required. Take for example the attraction of gravity at the surface of the earth. The force of that attraction being what it is, out of all the degrees of velocity, swift and slow, with which a ball might be shot off, none would answer the purpose of which we are speaking but what was nearly that of five miles in a second. If it were less than that, the body would not get round at all, but would come to the ground: if it were in any considerable degree more than that, the body would take one of those eccentric courses, those long ellipses, of which we have noticed the inconveniency. If the velocity reached the rate of seven miles in a second, or went beyond that, the ball would fly off from the earth, and never be heard of more. In like manner with respect to the *direction*; out of the innumerable angles in which the ball might be sent off, I mean angles formed with a line drawn to the centre, none would serve but what was nearly a right one; out of the various directions in which the cannon might be pointed, upwards and downwards, every one would fail, but what was exactly or nearly horizontal. The same thing holds true of the planets; of our own amongst the rest. We are entitled therefore to ask, and to urge the question, Why did the projectile velocity, and projectile direction of the earth happen to be nearly those which would retain it in a *circular* form? Why not one of the infinite number of velocities, one of the infinite number of directions, which would have made it approach much nearer to, or recede much further from, the sun?

The planets going round, all in the same direction, and all nearly in the same plane, afforded to Buffon* a ground for asserting, that they had all been shivered from the sun by the same stroke of a comet, and by that stroke projected into their present orbits. Now, beside that this is to attribute to chance the fortunate concurrence of velocity and direction which we have been here noticing, the hypothesis, as I apprehend, is inconsistent with the physical laws by which the heavenly motions are governed. If the planets were struck off from the surface of the sun, they would return to the surface of the sun again. Or, if, to get rid of this difficulty, we suppose, that the same violent blow, which shattered the sun's surface, and separated large fragments from it, pushed also the sun himself out of his place; a question of no less difficulty presents itself, namely, when once put into motion, what should *stop* him. The hypothesis is also contradicted by the vast difference which subsists between the *diameters* of the planetary orbits. The distance of Saturn from the sun (to say nothing of the Georgium sidus*) is nearly five-and-twenty times that of Mercury; a disparity, which it seems impossible to reconcile with Buffon's scheme. Bodies starting from the same place, with whatever difference of direction or velocity they set off, could not have been found at these different distances from the centre, still retaining their nearly circular orbits. They must have been carried to their proper distances, before they were projected.[1]

To conclude: In astronomy, the great thing is to raise the imagination to the subject, and that oftentimes in opposition to the impression made upon the senses. An illusion, for example, must be got over, arising from the distance at which we view the heavenly bodies, viz. the apparent *slowness* of their motions. The moon shall take some hours in getting half a yard from a star which it touched. A motion so deliberate, we may think easily guided. But what is the fact?* The moon, in fact, is, all this while, driving through the heavens, at the rate of considerably more than two thousand miles in an hour; which is more than double of that, with which a ball is shot off from the mouth of a cannon. Yet is this prodigious rapidity as much under government, as if the planet proceeded ever so slowly, or were conducted in its course inch by inch. It is also difficult to bring the imagination to conceive (what yet, to judge tolerably of the matter, it is necessary to conceive) how *loose*, if we may so express it, the heavenly bodies are. Enormous globes, held by nothing, confined by nothing, are turned into free and boundless space, each to seek its course by the virtue of an invisible principle; but a principle, one, common, and the same, in all; and ascertainable. To preserve such bodies from being lost, from running together in heaps, from hindering and distracting one another's motions, in a degree inconsistent with any continuing order; h. e. to cause them to form planetary

---

[1] 'If we suppose the matter of the system to be accumulated in the centre by its gravity, *no* mechanical principles, with the assistance of this power of gravity, could separate the vast mass into such parts as the sun and planets; and, after carrying them to their different distances, project them in their several directions, preserving still the equality of action and reaction, or the state of the centre of gravity of the system. Such an exquisite structure of things could only arise from the contrivance and powerful influences of an intelligent, free, and most potent agent. The same powers, therefore, which, at present, govern the material universe, and conduct its various motions, are *very different* from those, which were necessary, to have produced it from nothing, or to have disposed it in the admirable form, in which it now proceeds.' —*Maclaurin's Account of Newton's Phil.** p. 407, ed. 3.

systems, systems that, when formed, can be upheld, and, most especially, systems accommodated to the organized and sensitive natures which the planets sustain, as we know to be the case, where alone we can know what the case is, upon our earth: all this requires an intelligent interposition, because it can be demonstrated concerning it, that it requires an adjustment of force, distance, direction, and velocity, out of the reach of chance to have produced; an adjustment, in its view to utility similar to that which we see in ten thousand subjects of nature which are nearer to us, but in power, and in the extent of space through which that power is exerted, stupendous.

But many of the heavenly bodies, as the sun and fixed stars, are *stationary*. Their rest must be the effect of an absence or of an equilibrium of attractions. It proves also that a projectile impulse was originally given to some of the heavenly bodies, and not to others. But further; if attraction act at all distances, there can be only one quiescent centre of gravity in the universe: and all bodies whatever must be approaching this centre, or revolving round it. According to the first of these suppositions, if the duration of the world had been long enough to allow of it, all its parts, all the great bodies of which it is composed, must have been gathered together in a heap round this point. No changes however which have been observed, afford us the smallest reason for believing that either the one supposition or the other is true; and then it will follow, that attraction itself is controlled or suspended by a superior agent; that there is a power above the highest of the powers of material nature; a will which restrains and circumscribes the operations of the most extensive.

CHAPTER XXIII

OF THE PERSONALITY OF THE DEITY

CONTRIVANCE, if established, appears to me to prove every thing which we wish to prove. Amongst other things it proves the *personality* of the Deity,* as distinguished from what is sometimes called nature, sometimes called a principle: which terms, in the mouths of those who use them philosophically, seem to be intended, to admit and to express an efficacy, but to exclude and to deny a personal agent. Now that which can contrive, which can design, must be a person.* These capacities constitute personality, for they imply consciousness, and thought. They require that which can perceive an end or purpose; as well as the power of providing means, and of directing them to their end.[1] They require a centre in which perceptions unite, and from which volitions flow; which is mind. The acts of a mind prove the existence of a mind: and in whatever a mind resides is a person. The seat of intellect is a person. We have no authority to limit the properties of mind to any particular corporeal form, or to any particular circumscription of space. These properties subsist, in created nature, under a great variety of sensible forms. Also every animated being has its *sensorium*, that is, a certain portion of space, within which perception and volition are exerted. This sphere may be enlarged to an indefinite extent; may comprehend the universe:* and, being so imagined, may serve to furnish us with as good a notion, as we are capable of forming, of the *immensity* of the divine nature, i. e. of a Being, infinite, as well in essence, as in power; yet nevertheless a person.

'No man hath seen God at any time.'* And this, I believe, makes the great difficulty. Now it is a difficulty, which chiefly arises from our not duly estimating the state of our faculties. The Deity, it is true, is the object of none of our senses: but reflect what limited capacities animal senses are. Many animals seem to have but one sense, or perhaps two at the most, touch and taste. Ought such an animal to conclude against the existence of smells, sounds, and colours? To another species is given the sense of smelling. This is an

---

[1] Priestley's Letters to a Philosophical Unbeliever,* p. 153, ed. 2.

advance in the knowledge of the powers and properties of nature: but, if this favored animal should infer from its superiority over the class last described, that it perceived every thing which was perceptible in nature, it is known to us, though perhaps not suspected by the animal itself, that it proceeded upon a false and presumptuous estimate of its faculties. To another is added the sense of hearing; which lets in a class of sensations entirely unconceived by the animal before spoken of; not only distinct, but remote from any which *it* had ever experienced, and greatly superior to them. Yet this last animal has no more ground for believing, that its senses comprehend all things, and all properties of things, which exist, than might have been claimed by the tribes of animals beneath it: for we know, that it is still possible to possess another sense, that of sight, which shall disclose to the percipient a new world. This fifth sense makes the animal what the human animal is: but to infer that possibility stops here; that either this fifth sense is the last sense, or that the five comprehend all existence, is just as unwarrantable a conclusion, as that which might have been made by any of the different species which possessed fewer, or even by that, if such there be, which possessed only one. The conclusion of the one sense animal, and the conclusion of the five sense animal, stand upon the same authority. There may be more and other senses than those which we have. There may be senses suited to the perception of the powers, properties, and substance of spirits.* These may belong to higher orders of rational agents; for there is not the smallest reason for supposing that we are the highest, or that the scale of creation stops with us.

The great *energies* of nature are known to us only by their effects. The substances which produce them, are as much concealed from our senses as the divine essence itself. *Gravitation*, though constantly present, though constantly exerting its influence, though every where around us, near us, and within us; though diffused throughout all space, and penetrating the texture of all bodies with which we are acquainted, depends, if upon a fluid, upon a fluid, which, though both powerful and universal in its operation, is no object of sense to us; if upon any other kind of substance or action, upon a substance and action from which *we* receive no distinguishable impressions. Is it then to be wondered at, that it should, in some measure, be the same with the divine nature?

Of this however we are certain, that, whatever the Deity be, neither the *universe*, nor any part of it which we see, can be he. The universe itself is merely a collective name: its parts are all which are real; or which are *things*. Now inert matter is out of the question; and organized substances include marks of contrivance. But whatever includes marks of contrivance, whatever, in its constitution, testifies design, necessarily carries us to something beyond itself, to some other being, to a designer prior to, and out of, itself. No animal, for instance, can have contrived its own limbs and senses; can have been the author to itself of the design with which they were constructed. That supposition involves all the absurdity of self-creation, i. e. of acting without existing. Nothing can be God, which is ordered by a wisdom and a will, which itself is void of; which is indebted for any of its properties to contrivance *ab extra*.* The *not* having that in his nature which requires the exertion of another prior being, (which property is sometimes called self-sufficiency, and sometimes self-comprehension,) appertains to the Deity, as his essential distinction, and removes his nature from that of all things which we see. Which consideration contains the answer to a question that has sometimes been asked, namely, Why, since something or other must have existed from eternity, may not the present universe be that something? The contrivance, perceived in it, proves that to be impossible. Nothing contrived, can, in a strict and proper sense, be eternal, forasmuch as the contriver must have existed before the contrivance.

Wherever we see marks of contrivance, we are led for its cause to an *intelligent* author. And this transition of the understanding is founded upon uniform experience. We see intelligence constantly contriving, that is, we see intelligence constantly producing effects, marked and distinguished by certain properties; not certain particular properties, but by a kind and class of properties, such as relation to an end, relation of parts to one another, and to a common purpose. We see, wherever we are witnesses to the actual formation of things, nothing except intelligence producing effects so marked and distinguished. Furnished with this experience, we view the productions of nature. We observe *them* also marked and distinguished in the same manner. We wish to account for their origin. Our experience suggests a cause perfectly adequate to this account. No experience, no single instance or example, can be offered in favor of any other. In this cause therefore we ought to rest: in this cause the common sense of mankind* has in fact rested, because it agrees with that, which, in all cases, is the foundation of knowledge, the undeviating course of their experience. The reasoning is the same, as that, by which we conclude any ancient appearances to have been the effects of volcanos or inundations,* namely, because they resemble the effects which fire and water produce before our eyes; and because we have never known these effects to result from any other operation. And this resemblance may subsist in so many circumstances, as not to leave us under the smallest doubt in forming our opinion. Men are not deceived by this reasoning; for whenever it happens, as it sometimes does happen, that the truth comes to be known by direct information, it turns out to be what was expected. In like manner, and upon the same foundation, (which in truth is that of experience,) we conclude that the works of nature proceed from intelligence and design, because, in the properties of relation to a purpose, subserviency to an use, they resemble what intelligence and design are constantly producing, and what nothing except intelligence and design ever produce at all. Of every argument, which would raise a question as to the safety of this reasoning, it may be observed, that, if such argument be listened to, it leads to the inference, not only that the present order of nature is insufficient to prove the existence of an intelligent Creator, but that no imaginable order would be sufficient to prove it; that *no* contrivance, were it ever so mechanical, ever so precise, ever so clear, ever so perfectly like those which we ourselves employ, would support this conclusion. A doctrine, to which, I conceive, no sound mind can assent.

The force however of the reasoning is sometimes sunk by our taking up with mere names. We have already noticed,[1] and we must here notice again, the misapplication of the term 'law,'* and the mistake concerning the idea which that term expresses in physics, whenever such idea is made to take the place of power, and still more of an intelligent power, and, as such, to be assigned for the cause of any thing, or of any property of any thing, that exists. This is what we are secretly apt to do when we speak of organized bodies (plants, for instance, or animals) owing their production, their form, their growth, their qualities, their beauty, their use, to any law or laws of nature: and when we are contented to sit down with that answer to our enquiries concerning them. I say once more, that it is a perversion

[1] Ch. I. s. vii.

of language to assign any law, as the efficient, operative, cause of any thing. A law presupposes an agent; for it is only the mode according to which an agent proceeds: it implies a power, for it is the order according to which that power acts. Without this agent, without this power, which are both distinct from itself, the 'law' does nothing; is nothing.

What has been said concerning 'law,' holds true of *mechanism*. Mechanism is not itself power. Mechanism, without power, can do nothing. Let a watch be contrived and constructed ever so ingeniously; be its parts ever so many, ever so complicated, ever so finely wrought or artificially put together, it cannot *go* without a weight or spring, i. e. without a force independent of, and ulterior to, its mechanism. The spring acting at the centre, will produce different motions and different results, according to the variety of the intermediate mechanism. One and the self-same spring, acting in one and the same manner, viz. by simply expanding itself, may be the cause of a hundred different, and all useful movements, if a hundred different and well-devised sets of wheels be placed between it and the final effect, e. g. may point out the hour of the day, the day of the month, the age of the moon, the position of the planets, the cycle of the years, and many other serviceable notices; and these movements may fulfill their purposes with more or less perfection, according as the mechanism is better or worse contrived, or better or worse executed, or in a better or worse state of repair: *but, in all cases, it is necessary that the spring act at the centre*. The course of our reasoning upon such a subject would be this. By inspecting the watch, even when standing still, we get a proof of contrivance, and of a contriving mind, having been employed about it. In the form and obvious relation of its parts we see enough to convince us of this. If we pull the works in pieces, for the purpose of a closer examination, we are still more fully convinced. But, when we see the watch *going*, we see proof of another point, viz. that there is a power somewhere and somehow or other, applied to it; a power in action; that there is more in the subject than the mere wheels of the machine; that there is a secret spring or a gravitating plummet; in a word, that there is force and energy, as well as mechanism.

So then, the watch in motion establishes to the observer two conclusions: one; that thought, contrivance, and design, have been employed in the forming, proportioning, and arranging of its parts;

and that, whoever or wherever he be, or were, such a contriver there is, or was: the other; that force or power, distinct from mechanism, is, at this present time, acting upon it. If I saw a hand-mill even at rest, I should see contrivance; but, if I saw it grinding, I should be assured that a hand was at the windlass, though in another room. It is the same in nature. In the works of nature we trace mechanism; and this alone proves contrivance: but living, active, moving, productive nature, proves also the exertion of a power at the centre; for, wherever the power resides, may be denominated the centre.

The intervention and disposition of what are called 'second causes'* fall under the same observation. This disposition is or is not mechanism, according as we can or cannot trace it by our senses, and means of examination. That is all the difference there is; and it is a difference which respects our faculties, not the things themselves. Now where the order of second causes is mechanical, what is here said of mechanism strictly applies to it. But it would be always mechanism (natural chymistry, for instance, would be mechanism) if our senses were acute enough to descry it. Neither mechanism, therefore in the works of nature, nor the intervention of what are called second causes, (for I think that they are the same thing,) excuse the necessity of an agent distinct from both.

If, in tracing these causes, it be said, that we find certain general properties of matter, which have nothing in them that bespeaks intelligence, I answer, that, still, the *managing* of these properties, the pointing and directing them to the uses which we see made of them, demands intelligence in the highest degree. For example, suppose animal secretions to be elective attractions,* and that such and such attractions universally belong to such and such substances; in all which there is no intellect concerned; still the choice and collocation of these substances, the fixing upon right substances and disposing them in right places, must be an act of intelligence. What mischief would follow, were there a single transposition of the secretary organs; a single mistake in arranging the glands which compose them?

There may be many second causes, and many courses of second causes, one behind another, between what we observe of nature, and the Deity; but there must be intelligence somewhere; there must be more in nature than what we see; and, amongst the things unseen, there must be an intelligent, designing, author. The philosopher beholds with astonishment the production of things around him. Unconscious particles of matter take their stations, and severally range themselves in an order, so as to become collectively plants or animals, i. e. organized bodies, with parts bearing strict and evident relation to one another, and to the utility of the whole: and it should seem that these particles could not move in any other way than as they do, for they testify not the smallest sign of choice, or liberty, or direction. There may be plastic natures, particular intelligent beings, guiding these motions in each case: or they may be the result of trains of mechanical dispositions, fixed beforehand by an intelligent appointment,* and kept in action by a power at the centre. But in either case, there must be intelligence.

The minds of most men are fond of what they call a *principle*, and of the appearance of simplicity, in accounting for phænomena. Yet this principle, this simplicity, is sometimes nothing more than in the *name*; which name, comprises, perhaps, under it a diversified, multifarious, or progressive operation, distinguishable into parts. The power, in organized bodies, of producing bodies like themselves, is one of these principles. Give a philosopher this, and he can get on. But he does not reflect, what this principle, (if such he chuse to call it,) what this mode of production requires; how much it presupposes; what an apparatus of instruments, some of which are strictly mechanical, is necessary to its success; what a train it includes of operations and changes, one succeeding another, one related to another, one ministring to another; all advancing, by intermediate, and, frequently, by sensible steps, to their ultimate result. Yet, because the whole of this complicated action is wrapped up in a single term, *generation*, we are to set it down as an elementary principle; and to suppose, that, when we have resolved the things which we see into this principle, we have sufficiently accounted for their origin, without the necessity of a designing, intelligent Creator. The truth is, generation is not a principle, but a *process*. We might as well call the casting of metals a principle: we might, so far as appears to me, as well call spinning and weaving principles: and then, referring the texture of cloths, the fabric of muslins and callicoes, the patterns of diapers and damasks, to these as principles, pretend to dispense with intention, thought, and contrivance, on the part of the artist; or to dispense, indeed, with the necessity of any artist at all, either in the manufactory of the article,

or in the fabrication of the machinery by which the manufactory
was carried on.

And, after all, how, or in what sense, is it true, that animals produce their *like*? A butterfly, with a proboscis instead of a mouth, with
four wings and six legs, produces a hairy caterpillar, with jaws and
teeth, and fourteen feet. A frog produces a tadpole. A black beetle,
with gauze wings and a crusty covering, produces a white, smooth,
soft worm: an ephemeron fly, a cod-bait maggot. These, by a progress through different stages of life, and action, and enjoyment,
(and, in each state, provided with implements and organs appropriated to the temporary nature which they bear,) arrive at last at
the form and fashion of the parent animal. But all this is process, not
principle; and proves, moreover, that the property of animated
bodies of producing their like, belongs to them, not as a primordial
property, not by any blind necessity in the nature of things, but
as the effect of œconomy, wisdom, and design; because the property
itself, assumes diversities, and submits to deviations, dictated
by intelligible utilities, and serving distinct purposes of animal
happiness.

The opinion, which would consider 'generation' as a *principle* in
nature; and which would assign this principle as the cause, or
endeavour to satisfy our minds with such a cause, of the existence of
organized bodies, is confuted, in my judgment, not only by every
mark of contrivance discoverable in those bodies, for which it gives
us no contriver, offers no account, whatever; but also by the further
consideration, that things generated possess a clear relation to things
not generated. If it were merely one part of a generated body bearing
a relation to another part of the same body, as the mouth of an animal
to the throat, the throat to the stomach, the stomach to the intestines, those to the recruiting of the blood, and, by means of the
blood, to the nourishment of the whole frame: or if it were only one
generated body bearing a relation to another generated body, as the
sexes of the same species to each other, animals of prey to their prey,
herbivorous and graminivorous animals to the plants or seeds upon
which they feed, it might be contended, that the whole of this correspondency was attributable to generation, the common origin from
which these substances proceeded. But what shall we say to agree-
ments which exist between things generated and things *not generated*? Can it be doubted, was it ever doubted, but that the *lungs* of

animals bear a relation to the *air*, as a permanently elastic fluid? They act in it and by it: they cannot act without it. Now, if generation produced the animal, it did not produce the air; yet their properties correspond. The *eye* is made for *light*, and light for the eye. The eye would be of no use without light, and light perhaps of little without eyes: yet one is produced by generation; the other not. The *ear* depends upon *undulations* of air. Here are two sets of motions; first, of the pulses of the air; secondly, of the drum, bones, and nerves of the ear; sets of motions bearing an evident reference to each other: yet the one, and the apparatus for the one, produced by the intervention of generation; the other altogether independent of it.

If it be said, that the air, the light, the elements, the world itself, is *generated*, I answer, that I do not comprehend the proposition. If the term mean any thing, similar to what it means, when applied to plants or animals, the proposition is certainly without proof; and, I think, draws as near to absurdity, as any proposition can do, which does not include a contradiction in its terms. I am at a loss to conceive, how the formation of the world can be compared to the generation of an animal. If the term generation signify something quite different from what it signifies upon ordinary occasions, it may, by the same latitude, signify any thing. In which case a word or phrase taken from the language of Otaheite,* would convey as much theory concerning the origin of the universe, as it does to talk of its being generated.

We know a cause (intelligence) adequate to the appearances, which we wish to account for: we have this cause continually producing similar appearances: yet, rejecting this cause, the sufficiency of which we know, and the action of which is constantly before our eyes, we are invited to resort to suppositions, destitute of a single fact for their support, and confirmed by no analogy with which we are acquainted. Were it necessary to enquire into the *motives* of men's opinions, I mean their motives separate from their arguments, I should almost suspect, that, because the proof of a Deity drawn from the constitution of nature is not only popular but vulgar, (which may arise from the cogency of the proof, and be indeed its highest recommendation,) and because it is a species almost of puerility to take up with it, for these reasons, minds, which are habitually in search of invention and originality, feel a resistless inclination to strike off into

other solutions and other expositions. The truth is, that many minds
are not so indisposed to any thing which can be offered to them, as
they are to the *flatness* of being content with common reasons; and,
what is most to be lamented, minds conscious of superiority are the
most liable to this repugnancy.

The 'suppositions' here alluded to all agree in one character. They
all endeavour to dispense with the necessity in nature of a particular,
personal, intelligence; that is to say, with the exertion of an intending, contriving mind, in the structure and formation of the organized constitutions which the world contains. They would resolve all
productions into *unconscious* energies, of a like kind, in that respect,
with attraction, magnetism, electricity, etc.; without any thing
further.

In this the old systems of atheism and the new agree.* And I much
doubt, whether the new schemes have advanced anything upon the
old, or done more than changed the terms of the nomenclature. For
instance, I could never see the difference between the antiquated
system of atoms,* and Buffon's organic molecules. This philosopher,
having made a planet by knocking off from the sun a piece of melted
glass, in consequence of the stroke of a comet; and having set it in
motion, by the same stroke, both round its own axis and the sun,
finds his next difficulty to be, how to bring plants and animals upon
it. In order to solve this difficulty, we are to suppose the universe
replenished with particles, endowed with life, but without organization or senses of their own; and endowed also with a tendency to
marshal themselves into organized forms. The concourse of these
particles, by virtue of this tendency, but without intelligence, will, or
direction, (for I do not find that any of these qualities are ascribed to
them,) has produced the living forms which we now see.

Very few of the conjectures, which philosophers hazard upon
these subjects, have more of pretension in them, than the challenging
you to shew the direct impossibility of the hypothesis. In the present
example, there seemed to be a positive objection to the whole scheme
upon the very face of it; which was, that, if the case were as here
represented, *new* combinations ought to be perpetually taking place;
new plants and animals, or organized bodies which were neither,
ought to be starting up before our eyes every day. For this, however,
our philosopher has an answer. Whilst so many forms of plants
and animals are already in existence, and, consequently, so many

'internal molds,* as he calls them, are prepared and at hand, the organic particles run into these molds, and are employed in supplying an accession of substance to them, as well for their growth, as for their propagation. By which means things keep their antient course. But, says the same philosopher, should any general loss or destruction of the present constitution of organized bodies take place, the particles, for want of 'molds' into which they might enter, would run into different combinations, and replenish the waste with new species of organized substances.

Is there any history to countenance this notion? Is it known, that any destruction has been so repaired? any desart thus repeopled?

So far as I remember, the only natural appearance mentioned by our author, by way of fact whereon to build his hypothesis, the only support on which it rests, is the formation of *worms* in the intestines of animals, which is here ascribed to the coalition of superabundant organic particles, floating about in the first passages; and which have combined themselves into these simple animal forms, for want of internal molds, or of vacancies in those molds, into which they might be received. The thing referred to is rather a species of facts, than a single fact; as some other cases may, with equal reason, be included under it. But to make it a fact at all, or, in any sort, applicable to the question, we must begin with asserting an *equivocal* generation contrary to analogy, and without necessity: contrary to an analogy, which accompanies us to the very limits of our knowledge or enquiries, for wherever, either in plants or animals, we are able to examine the subject, we find procreation from a parent form; without necessity, for I apprehend that it is seldom difficult to suggest methods, by which the eggs, or spawn, or yet invisible rudiments of these vermin, may have obtained a passage into the cavities in which they are found.[1] Add to this, that their *constancy to their species*, which, I believe, is as regular in these as in the other vermes,* decides the question against our philosopher, if, in truth, any question remained upon the subject.

Lastly; these wonder-working instruments, these 'internal molds,' what are they after all? what, when examined, but a name without

---

[1] I trust I may be excused, for not citing, as another fact which is to confirm the hypothesis, a grave assertion of this writer, that the branches of trees upon which the stag feeds, break out again in his horns. Such *facts* merit no discussion.

signification; unintelligible, if not self-contradictory; at the best, differing nothing from the 'essential forms'* of the Greek philosophy? One short sentence of Buffon's work exhibits his scheme as follows. 'When this nutritious and prolific matter, which is diffused throughout all nature, passes through the *internal mold* of an animal or vegetable, and finds a proper matrix or receptacle, it gives rise to an animal or vegetable of the same species.' Does any reader annex a meaning to the expression, 'internal mold,' in this sentence? Ought it then to be said, that, though we have little notion of an internal mold, we have not much more of a designing mind? The very contrary of this assertion is the truth. When we speak of an artificer or an architect, we talk of what is comprehensible to our understanding, and familiar to our experience. We use no other terms, than what refer us for their meaning to our consciousness and observation; what express the constant objects of both: whereas names, like that we have mentioned, refer us to nothing; excite no idea; convey a sound to the ear, but I think do no more.

ANOTHER system, which has lately been brought forward, and with much ingenuity, is that of *appetencies.** The principle, and the short account, of the theory, is this. Pieces of soft, ductile, matter, being endued with propensities or appetencies for particular actions, would, by continual endeavours, carried on through a long series of generations, work themselves gradually into suitable forms; and, at length, acquire, though perhaps by obscure and almost imperceptible improvements, an organization fitted to the action which their respective propensities led them to exert. A piece of animated matter, for example, that was endued with a propensity to *fly*, though ever so shapeless, though no other we will suppose than a round ball to begin with, would, in a course of ages, if not in a million of years, perhaps in a hundred millions of years,* (for our theorists, having eternity to dispose of, are never sparing in time,) acquire *wings*. The same tendency to loco-motion in an aquatic animal, or rather in an animated lump which might happen to be surrounded by water, would end in the production of *fins*: in a living substance, confined to the solid earth, would put out *legs* and *feet*; or, if it took a different turn, would break the body into ringlets, and conclude by *crawling* upon the ground.

Although I have introduced the mention of this theory into this place, I am unwilling to give to it the name of an *atheistic* scheme, for two reasons; first, because, so far as I am able to understand it, the original propensities and the numberless varieties of them (so different, in this respect, from the laws of mechanical nature, which are few and simple) are, in the plan itself, attributed to the ordination and appointment of an intelligent and designing Creator: secondly, because, likewise, that large postulatum, which is all along assumed and presupposed, the faculty in living bodies of producing other bodies organized like themselves, seems to be referred to the same cause at least is not attempted to be accounted for by any other. In one important respect, however, the theory before us coincides with atheistic systems, viz. in that, in the formation of plants and animals, in the structure and use of their parts, it does away final causes. Instead of the parts of a plant or animal, or the particular structure of the parts, having been intended for the action or the use to which we see them applied, according to this theory they have themselves grown out of that action, sprung from that use. The theory therefore dispenses with that which we insist upon, the necessity, in each particular case, of an intelligent, designing, mind, for the contriving and determining of the forms which organized bodies bear. Give our philosopher these appetencies; give him a portion of living irritable matter (a nerve, or the clipping of a nerve,) to work upon; give also to his incipient or progressive forms, the power, in every stage of their alteration, of propagating their like; and, if he is to be believed, he could replenish the world with all the vegetable and animal productions which we at present see in it.

The scheme under consideration is open to the same objection with other conjectures of a similar tendency, viz. a total defect of evidence. No changes, like those which the theory requires, have ever been observed. All the changes in Ovid's Metamorphoses* might have been effected by these appetencies, if the theory were true; yet not an example, nor the pretence of an example, is offered, of a single change being known to have taken place. Nor is the order of generation obedient to the principle upon which this theory is built. The mammæ[1]* of the male have not vanished by inusitation;* *nec curtorum, per multa sæcula, Judæorum propagini deest præputium.** It is easy to

---

[1] I confess myself totally at a loss to guess at the reason, either final or efficient, for this part of the animal frame, unless there be some foundation for an opinion, of which I draw the hint from a paper of Mr Everard Home's,* (Phil. Transac. 1799, p. 2.) viz. that the mammæ of the fœtus may be formed before the sex is determined.

say, and it has been said, that the alterative process is too slow to be perceived; that it has been carried on through tracts of immeasurable time; and that the present order of things is the result of a gradation, of which no human record can trace the steps. It is easy to say this; and yet it is still true, that the hypothesis remains destitute of evidence.

The *analogies* which have been alledged are of the following kind. The *bunch* of a camel, is said to be no other than the effect of carrying burthens; a service in which the species has been employed from the most antient times of the world. The first race, by the daily loading of the back, would probably find a small grumous* tumour to be formed in the flesh of that part. The next progeny would bring this tumour into the world with them. The life, to which they were destined, would increase it. The cause, which first generated the tubercle, being continued, it would go on, through every succession, to augment its size, till it attained the form and the bulk under which it now appears. This may serve for one instance: another, and that also of the passive sort, is taken from certain species of birds. Birds of the *crane* kind, as the crane itself, the heron, bittern, stork, have, in general, their thighs bare of feathers. This privation is accounted for from the habit of wading in water, and from the effect of that element to check the growth of feathers upon these parts: in consequence of which, the health and vegetation of the feathers declined through each generation of the animal: the tender down, exposed to cold and wetness, became weak, and thin, and rare, till the deterioration ended in the result which we see, of absolute nakedness. I will mention a third instance because it is drawn from an active habit, as the two last were from passive habits;* and that is the *pouch* of the pelican. The description, which naturalists give of this organ is as follows: 'From the lower edges of the under chap, hangs a bag, reaching from the whole length of the bill to the neck, which is said to be capable of containing fifteen quarts of water. This bag the bird has a power of wrinkling up into the hollow of the under chap. When the bag is empty it is not seen: but when the bird has fished with success, it is incredible to what an extent it is often dilated. The first thing the pelican does in fishing, is to fill the bag; and then it returns to digest its burthen at leisure. The bird preys upon the large fishes, and hides them by dozens in its pouch. When the bill is opened to its widest extent, a person may run his head into the bird's mouth; and conceal

it in this monstrous pouch, thus adapted for very singular purposes.'[1]
Now this extraordinary conformation, is nothing more, say our
philosophers, than the result of habit; not of the habit or effort of a
single pelican, or of a single race of pelicans, but of a habit perpetuated through a long series of generations. The pelican soon found
the conveniency, of reserving in its mouth, when its appetite was
glutted, the remainder of its prey, which is fish. The fullness produced by this attempt, of course stretched the skin which lies
between the under chaps, as being the most yielding part of the
mouth. Every distension increased the cavity. The original bird, and
many generations which succeeded him, might find difficulty
enough in making the pouch answer this purpose: but future pelicans, entering upon life with a pouch derived from their pro-
genitors, of considerable capacity, would more readily accelerate its
advance to perfection, by frequently pressing down the sac with the
weight of fish which it might now be made to contain.

These, or of this kind, are the analogies relied upon. Now in the
first place, the instances themselves are unauthenticated by testimony;* and, in theory, to say the least of them, open to great objections. Who ever read of camels without bunches, or with bunches
less than those with which they are at present usually formed? A
bunch, not unlike the camel's, is found between the shoulders of the
buffalo; of the origin of which it is impossible to give the account
which is here given. In the second example. Why should the application of water, which appears to promote and thicken the growth of
feathers upon the bodies and breasts of geese and swans and other
water fowls, have divested of this covering the thighs of cranes? The
third instance, which appears to me as plausible as any that can be
produced, has this against it, that it is a singularity restricted to the
species; whereas, if it had its commencement in the cause and manner which have been assigned, the like conformation might be
expected to take place in other birds, which fed upon fish. How
comes it to pass, that the pelican alone was the inventress, and her
descendants the only inheritors, of this curious resource?

But it is the less necessary to contravert the instances themselves,
as it is a straining of analogy beyond all limits of reason and cred-
ibility, to assert that birds, and beasts, and fish, with all their variety

---

[1] Goldsmith, vol. vi. p. 52.

and complexity of organization, have been brought into their forms, and distinguished into their several kinds and natures, by the same process (even if that process could be demonstrated, or had ever been actually noticed) as might seem to serve for the gradual generation of a camel's bunch, or a pelican's pouch.

The solution, when applied to the works of nature *generally*, is contradicted by many of the phænomena, and totally inadequate to others. The *ligaments* or strictures, by which the tendons are tied down at the angles of the joints, could, by no possibility, be formed by the motion or exercise of the tendons themselves; by any appetency exciting these parts into action; or by any tendency arising therefrom. The tendency is all the other way: the conatus in constant opposition to them. Length of time does not help the case at all, but the reverse. The *valves* also in the blood-vessels, could never be formed in the manner, which our theorist proposes. The blood, in its right and natural course, has no tendency to form them. When obstructed or refluent, it has the contrary. These parts could not grow out of their use, though they had eternity to grow in.

The *senses* of animals appear to me altogether incapable of receiving the explanation of their origin which this theory affords. Including under the word 'sense' the organ and the perception, we have no account of either. How will our philosopher get at *vision*, or make an eye? How should the blind animal affect sight, of which blind animals, we know, have neither conception nor desire? Affecting it, by what operation of its will, by what endeavour to see, could it so determine the fluids of its body, as to inchoate the formation of an eye? or, suppose the eye formed, would the perception follow? The same of the other senses. And this objection holds its force, ascribe what you will to the hand of time, to the power of habit, to changes, too slow to be observed by man, or brought within any comparison which he is able to make of past things with the present: concede what you please to these arbitrary and unattested suppositions, how will they help you? Here is no inception. No laws, no course, no powers of nature which prevail at present, nor any analogous to these, could give commencement to a new sense. And it is in vain to enquire, how that might proceed, which could never *begin*.

I think the senses, to be the most inconsistent with the hypothesis before us, of any part of the animal frame. But other parts are sufficiently so. The solution does not apply to the parts of animals, which have little in them of motion. If we could suppose joints and muscles to be gradually formed by action and exercise, what action or exercise could form a skull, or fill it with brains? No effort of the animal could determine the clothing of its skin. What conatus* could give prickles to the porcupine or hedgehog, or to the sheep its fleece?

In the last place; what do these appetencies mean when applied to plants? I am not able to give a signification to the term, which can be transferred from animals to plants; or which is common to both. Yet a no less successful organization is found in plants, than what obtains in animals. A solution is wanted for one, as well as the other.

Upon the whole; after all the struggles of a reluctant philosophy the necessary resort is to a Deity. The marks of *design* are too strong to be got over. Design must have had a designer. That designer must have been a person. That person is God.

CHAPTER XXIV

OF THE NATURAL ATTRIBUTES OF THE DEITY

IT is an immense conclusion, that there is a God; a perceiving, intelligent, designing Being; at the head of creation, and from whose will it proceeded. The *attributes* of such a Being,* suppose his reality to be proved, must be adequate to the magnitude, extent, and multiplicity of his operations: which are not only vast beyond comparison with those performed by any other power, but, so far as respects our conceptions of them, infinite, because they are unlimited on all sides.

Yet the contemplation of a nature so exalted, however surely we arrive at the proof of its existence, overwhelms our faculties. The mind feels its powers sink under the subject. One consequence of which is, that from painful abstraction the thoughts seek relief in sensible images. From whence may be deduced the ancient, and almost universal, propensity to idolatrous substitutions. They are the resources of a labouring imagination. False religions usually fall in with the natural propensity: true religions, or such as have derived themselves from the true, resist it.

It is one of the advantages of the revelations which we acknowledge,* that, whilst they reject idolatry with its many pernicious accompaniments, they introduce the Deity to human apprehension, under an idea more personal, more determinate, more within its compass, than the theology of nature can do. And this they do by representing him exclusively under the relation in which he stands to ourselves; and, for the most part, under some precise character, resulting from that relation, or from the history of his providences. Which method suits the span of our intellects much better, than the universality which enters into the idea of God, as deduced from the views of nature. When, therefore, these representations are well founded in point of authority, (for all depends upon that,) they afford a condescension to the state of our faculties, of which, those, who have reflected most upon the subject, will be the first to acknowledge the want and the value.

Nevertheless, if we be careful to imitate the documents of our religion, by confining our explanations to what concerns ourselves, and do not affect more precision in our ideas than the subject allows of, the several terms, which are employed to denote the attributes of the Deity, may be made, even in natural religion, to bear a sense, consistent with truth and reason, and not surpassing our comprehension.

These terms are, omnipotence, omniscience, omnipresence,* eternity, self-existence, necessary existence, spirituality.

'Omnipotence,' 'omniscience;' infinite power, infinite knowledge, are *superlatives*; expressing our conception of these attributes in the strongest, and most elevated, terms, which language supplies. We ascribe power to the Deity under the name of 'omnipotence,' the strict and correct conclusion being, that a power, which could create such a world as this is, must be, beyond all comparison, greater, than any which we experience in ourselves, than any which we observe in other visible agents; greater, also, than any which we can want, for our individual protection and preservation, in the Being upon whom we depend. It is a power likewise, to which we are not authorised by our observation or knowledge, to assign any limits of space or duration.

Very much of the same sort of remark is applicable to the term 'omniscience,' infinite knowledge, or infinite wisdom. In strictness of language, there is a difference between knowledge and wisdom; wisdom always supposing action, and action directed by it. With respect to the first, viz. knowledge, the Creator must *know*, intimately, the constitution and properties of the things which he created; which seems also to imply a foreknowledge of their action upon one another, and of their changes;* at least, so far as the same result from trains of physical and necessary causes. His omniscience also, as far as respects things present, is deducible from his nature, as an intelligent being, joined with the extent, or rather the universality, of his operations. Where he acts, he is; and, where he is, he perceives. The *wisdom* of the Deity, as testified in the works of creation, surpasses all idea we have of wisdom, drawn from the highest intellectual operations of the highest class of intelligent Beings with whom we are acquainted; and, which is of the chief importance to us, whatever be its compass or extent, which it is evidently impossible that we should be able to determine, it must be adequate to the conduct of that order of things under which we live. And this is enough. It is of very inferior consequence, by what terms we express our notion, or rather our admiration, of this attribute. The terms, which the piety and the usage of language have rendered habitual to us, may be as proper as any other. We can trace this attribute much beyond what is necessary for any conclusion to which we have occasion to apply it. The degree of knowledge and power, requisite for the formation of created nature, cannot, with respect to us, be distinguished from infinite.

The divine 'omnipresence' stands, in natural theology, upon this foundation. In every part and place of the universe, with which we are acquainted, we perceive the exertion of a power, which we believe, mediately or immediately, to proceed from the Deity. For instance; in what part or point of space, that has ever been explored, do we not discover attraction? In what regions, do we not find light? In what accessible portion of our globe, do not we meet with gravity, magnetism, electricity; together with the properties also and powers of organized substances, of vegetable or of animated nature? Nay, further, we may ask, what kingdom is there of nature, what corner of space, in which there is any thing that can be examined by us, where we do not fall upon contrivance and design? The only reflection perhaps which arises in our minds from this view of the world around us is, that the laws of nature every where prevail; that they are uniform, and universal. But what do we mean by the laws of nature, or by any law? Effects are produced by power, not by laws. A law cannot execute itself. A law refers us to an agent. Now an agency so general, as that we cannot discover its absence, or assign the place in which some effect of its continued energy is not found, may, in popular language at least, and, perhaps, without much deviation from philosophical strictness, be called universal: and, with not quite the same, but with no inconsiderable propriety, the person or Being, in whom that power resides, or from whom it is derived, may be taken to be *omnipresent*. He who upholds all things by his power, may be said to be every where present.

This is called a virtual presence. There is also what metaphysicians denominate an essential ubiquity: and which idea the language of scripture seems to favour: but the former, I think, goes as far as natural theology carries us.

'Eternity,' is a negative idea, clothed with a positive name. It supposes, in that to which it is applied, a present existence; and is the negation of a beginning, or an end of that existence. As applied to the Deity, it has not been contraverted by those who acknowledged a Deity at all. Most assuredly, there never was a time in which nothing existed, because that condition must have continued. The universal blank must have remained; nothing could rise up out of it; nothing could ever have existed since; nothing could exist now. In strictness, however, we have no concern with duration prior to that of the visible world. Upon this article therefore of theology, it is sufficient to know, that the contriver necessarily existed before the contrivance.

'Self-existence,' is another negative idea, viz. the negation of a preceding cause, as of a progenitor, a maker, an author, a creator.

'Necessary existence' means demonstrable existence.

'Spirituality' expresses an idea, made up of a negative part, and of a positive part. The negative part, consists in the exclusion of some of the known properties of matter, especially of solidity, of the vis inertiæ, and of gravitation. The positive part, comprises perception, thought, will, power, action, by which last term is meant, the origination of motion; the quality, perhaps, in which resides the essential superiority of spirit over matter, 'which cannot move, unless it be moved; and cannot but move, when impelled by another.'[1] I apprehend that there can be no difficulty in applying to the Deity both parts of this idea.

---

[1] Bishop Wilkins's Principles of Nat. Rel.* p. 106.

CHAPTER XXV

THE UNITY OF THE DEITY

OF the 'unity of the Deity' the proof is, the *uniformity* of plan observable in the universe. The universe itself is a system; each part either depending upon other parts, or being connected with other parts by some common law of motion, or by the presence of some common substance. One principle of gravitation causes a stone to drop towards the earth, and the moon to wheel round it. One law of attraction carries all the different planets about the sun. This philosophers demonstrate. There are also other points of agreement amongst them, which may be considered as marks of the identity of their origin, and of their intelligent author. In all are found the conveniency and stability derived from gravitation. They all experience vicissitudes of days and nights, and changes of season. They all, at least Jupiter, Mars, and Venus, have the same advantages from their atmospheres as we have. In all the planets the axes of rotation are permanent. Nothing is more probable, than that the same attracting influence, acting according to the same rule, reaches to the fixed stars: but, if this be only probable, another thing is certain, viz. that the same element of light does.* The light from a fixed star affects our eyes in the same manner, is refracted and reflected according to the same laws, as the light of a candle. The velocity of the light of the fixed stars, is also the same as the velocity of the light of the sun, reflected from the satellites of Jupiter. The heat of the sun, in kind, differs nothing from the heat of a coal fire.

In our own globe the case is clearer. New countries are continually discovered, but the old laws of nature are always found in them: new plants perhaps or animals, but always in company with plants and animals, which we already know; and always possessing many of the same general properties. We never get amongst such original, or totally different, modes of existence, as to indicate, that we are come into the province of a different Creator, or under the direction of a different will. In truth, the same order of things attends us, wherever we go. The elements act upon one another, electricity operates, the tides rise and fall, the magnetic needle elects its position, in one region of the earth and sea, as well as in another. One atmosphere invests all parts of the globe, and connects all: one sun illuminates; one moon exerts its specific attraction upon all parts. If there be a variety in natural effects, as, e.g. in the tides of different seas, that very variety is the result of the same cause, acting under different circumstances. In many cases this is proved; in all is probable.

The inspection and comparison of *living* forms, add to this argument examples without number. Of all large terrestrial animals the structure is very much alike. Their senses nearly the same. Their natural functions and passions nearly the same. Their viscera nearly the same, both in substance, shape, and office. Digestion, nutrition, circulation, secretion, go on, in a similar manner, in all. The great circulating fluid is the same: for, I think, no difference has been discovered in the properties of *blood*, from whatever animal it be drawn. The experiment of transfusion proves, that the blood of one animal will serve for another. The *skeletons* also of the larger terrestrial animals, shew particular varieties, but still under a great general affinity. The resemblance is somewhat less, yet sufficiently evident, between quadrupeds and birds. They are alike in five respects, for one in which they differ.

In *fish*, which belong to another department, as it were, of nature, the points of comparison become fewer. But we never lose sight of our analogy, e. g. we still meet with a stomach, a liver, a spine; with bile and blood; with teeth; with eyes, which eyes are only slightly varied from our own, and which variation, in truth, demonstrates, not an interruption, but a continuance, of the same exquisite plan; for it is the adaptation of the organ to the element, viz. to the different refraction of light passing into the eye out of a denser medium. The provinces, also, themselves of water and earth, are connected by the species of animals which inhabit both; and also by a large tribe of aquatic animals, which closely resemble the terrestrial in their internal structure: I mean the cetaceous tribe,* which have hot blood, respiring lungs, bowels, and other essential parts, like those of land animals. This similitude, surely, bespeaks the same creation and the same Creator.

Insects and *shell fish* appear to me to differ from other classes of animals the most widely of any. Yet even here, beside many points of particular resemblance, there exists a general relation of a peculiar kind. It is the relation of inversion: the law of contrariety: namely, that, whereas, in other animals, the bones, to which the muscles are attached, lie *within* the body, in insects and shell fish they lie on the *outside* of it. The shell of a lobster performs to the animal the office of a *bone*, by furnishing to the tendons that fixed basis or immoveable fulcrum, without which mechanically they could not act. The crust of an insect is its shell, and answers the like purpose. The shell also of an oyster stands in the place of a *bone*; the bases of the muscles being fixed to it, in the same manner, as, in other animals, they are fixed to the bones. All which (under wonderful varieties, indeed, and adaptations of form) confesses an imitation, a remembrance, a carrying on, of the same plan.

The observations, here made, are equally applicable to plants; but I think unnecessary to be pursued. It is a very striking circumstance, and alone sufficient to prove all which we contend for, that, in this part likewise of organized nature, we perceive a continuation of the *sexual* system.

Certain however it is, that the whole argument for the divine unity, goes no further than to an unity of counsel.

It may likewise be acknowledged, that no arguments which we are in possession of, exclude the ministry of subordinate agents. If such there be, they act under a presiding, a controlling, will; because they act according to certain general restrictions, by certain common rules, and, as it should seem, upon a general plan: but still such agents, and different ranks, and classes, and degrees of them, may be employed.

CHAPTER XXVI

THE GOODNESS OF THE DEITY

THE proof of the *divine goodness** rests upon two propositions, each, as we contend, capable of being made out by observations drawn from the appearances of nature.

The first is, 'that, in a vast plurality of instances in which contrivance is perceived, the design of the contrivance is *beneficial*.'*

The second, 'that the Deity has superadded *pleasure* to animal sensations, beyond what was necessary for any other purpose, or when the purpose, so far as it was necessary, might have been effected by the operation of pain.'

First, 'in a vast plurality of instances, in which contrivance is perceived, the design of the contrivance is *beneficial*.'

No productions of nature display contrivance so manifestly as the parts of animals: and the parts of animals have all of them, I believe, a real, and, with very few exceptions, all of them a known and intelligible, subserviency to the use of the animal. Now, when the multitude of animals is considered, the number of parts in each, their figure and fitness, the faculties depending upon them, the variety of species, the complexity of structure, the success, in so many cases, and felicity of the result, we can never reflect, without the profoundest adoration, upon the character of that Being from whom all these things have proceeded: we cannot help acknowledging, what an exertion of benevolence creation was; of a benevolence, how minute in its care, how vast in its comprehension.

When we appeal to the parts and faculties of animals, and to the limbs and senses of animals in particular, we state, I conceive, the proper medium of proof for the conclusion which we wish to establish. I will not say, that the insensible parts of nature are made solely for the sensitive parts; but this I say, that, when we consider the benevolence of the Deity, we can only consider it in relation to sensitive Being. Without this reference, or referred to any thing else, the attribute has no object; the term has no meaning. Dead matter is nothing. The parts, therefore, especially the limbs and senses, of animals, although they constitute, in mass and quantity, a small portion of the material creation, yet, since they alone are instruments of perception, they compose what may be called the whole of visible nature, estimated with a view to the disposition of its author. Consequently, it is in these that we are to seek his character. It is by these that we are to prove, that the world was made with a benevolent design.

Nor is the design abortive. It is a happy world after all.* The air, the earth, the water, teem with delighted existence. In a spring noon, or a summer evening, on whichever side I turn my eyes, myriads of happy beings crowd upon my view. 'The insect youth are on the wing.' Swarms of new-born *flies* are trying their pinions in the air. Their sportive motions, their wanton mazes, their gratuitous activity, their continual change of place without use or purpose, testify their joy, and the exultation which they feel in their lately discovered faculties. A *bee* amongst the flowers in spring, is one of the cheerfullest objects that can be looked upon. Its life appears to be all enjoyment: so busy, and so pleased: yet it is only a specimen of insect life, with which, by reason of the animal being half domesticated, we happen to be better acquainted than we are with that of others. The *whole* winged insect tribe, it is probable, are equally intent upon their proper employments, and, under every variety of constitution, gratified, and perhaps equally gratified, by the offices which the author of their nature has assigned to them. But the atmosphere is not the only scene of enjoyment for the insect race. Plants are covered with aphides, greedily sucking their juices, and constantly, as it should seem, in the act of sucking. It cannot be doubted but that this is a state of intense gratification. What else should fix them so close to their operation, and so long? Other species are *running about* with an alacrity in their motions which carries with it every mark of pleasure. Large patches of ground are sometimes half covered with these brisk and sprightly natures. If we look to what the *waters* produce, shoals of the fry of fish frequent the margins of rivers, of lakes, and of the sea itself. These are so happy, that they know not what to do with themselves. Their attitudes, their vivacity; their leaps out of the water, their frolics in it, (which I have noticed a thousand times with equal attention and amusement,) all conduce to shew their excess of spirits, and are simply the effects of that excess. Walking by the sea side, in a calm evening, upon a sandy shore, and with an ebbing tide, I have frequently remarked the appearance of a dark cloud, or, rather, very thick mist, hanging over the edge of the water, to the height, perhaps, of half a yard, and of the breadth of two or three yards, stretching along the coast as far as the eye could reach, and always retiring with the water. When this cloud came to be examined, it proved to be nothing else than so much space, filled with young *shrimps*, in the act of bounding into the air from the shallow margin of the water, or from the wet sand. If any motion of a mute animal could express delight, it was this: if they had meant to make signs of their happiness, they could not have done it more intelligibly. Suppose then, what I have no doubt of, each individual of this number to be in a state of positive enjoyment, what a sum, collectively, of gratification and pleasure have we here before our view?

The *young* of all animals appear to me to receive pleasure simply from the exercise of their limbs and bodily faculties, without reference to any end to be attained, or any use to be answered by the exertion. A child, without knowing any thing of the use of language, is, in a high degree, delighted with being able to speak. Its incessant repetition of the few articulate sounds, or, perhaps, of the single word, which it has learnt to pronounce, proves this point clearly. Nor is it less pleased with its first successful endeavours to walk, or rather to run, (which precedes walking) although entirely ignorant of the importance of the attainment to its future life: and even without applying it to any present purpose. A child is delighted with speaking, without having any thing to say; and with walking, without knowing where to go. And, prior to both these, I am disposed to believe, that the waking hours of infancy are agreeably taken up with the exercise of vision, or perhaps, more properly speaking, with learning to see.

But it is not for youth alone, that the great Parent of creation hath provided. Happiness is found with the purring cat, no less than with the playful kitten; in the arm-chair of dozing age, as well as in either the sprightliness of the dance, or the animation of the chace. To novelty, to acuteness of sensation, to hope, to ardor of pursuit, succeeds, what is, in no inconsiderable degree, an equivalent for them all, 'perception of ease.' Herein is the exact difference between the young and the old. The young are not happy, but when enjoying pleasure; the old are happy, when free from pain. And this constitution suits with the degrees of animal power which they respectively possess. The vigor of youth was to be stimulated to action by impatience of rest; whilst, to the imbecility of age, quietness and repose become positive gratifications. In one important respect the advantage is with the old. A state of ease is, generally speaking, more attainable than a state of pleasure. A constitution, therefore, which can enjoy ease, is preferable to that which can taste only pleasure. This same perception of care oftentimes renders old age a condition of great comfort; especially when riding at its anchor, after a busy or tempestuous life. It is well described by Rousseau,* to be the interval of repose and enjoyment, between the hurry and the end of life. How far the same cause extends to other animal natures cannot be judged of with certainty. The appearance of satisfaction, with which most animals, as their activity subsides, seek and enjoy rest, affords reason to believe, that this source of gratification is appointed to advanced life, under all, or most, of its various forms. In the species with which we are best acquainted, namely our own, I am far, even as an observer of human life, from thinking, that youth is its happiest season, much less the only happy one: as a Christian, I am willing to believe that there is a great deal of truth in the following representation given by a very pious writer, as well as excellent man.[1] 'To the intelligent and virtuous, old age presents a scene of tranquil enjoyments, of obedient appetites, of well regulated affections, of maturity in knowledge, and of calm preparation for immortality. In this serene and dignified state, placed, as it were, on the confines of two worlds, the mind of a good man, reviews what is past with the complacency of an approving conscience, and looks forward, with humble confidence in the mercy of God, and with devout aspirations towards his eternal and ever increasing favor.'

What is seen in different stages of the same life, is still more exemplified in the lives of different animals. Animal enjoyments are infinitely *diversified*. The modes of life, to which the organization of different animals respectively determines them, are not only of various, but of opposite kinds. Yet each is happy in its own. For instance; animals of prey, live much alone; animals of a milder constitution, in society. Yet the herring, which lives in shoals, and the sheep, which lives in flocks, are not more happy in a crowd, or more contented amongst their companions, than is the pike, or the lion, with the deep solitudes of the pool, or the forest.

---

[1] Father's Instructions, by Dr Percival of Manchester,* p. 317.

But it will be said, that the instances which we have here brought forward, whether of vivacity or repose, or of apparent enjoyment derived from either, are picked and favorable instances. We answer that they are instances, nevertheless, which comprise large provinces of sensitive existence; that every case which we have described, is the case of millions.* At this moment, in every given moment of time, how many myriads of animals are eating their food, gratifying their appetites, ruminating in their holes, accomplishing their wishes, pursuing their pleasures, taking their pastimes? In each individual how many things must go right for it to be at ease; yet how large a proportion out of every species, are so in every assignable instant? Secondly, we contend, in the terms of our original proposition, that throughout the whole of life, as it is diffused in nature, and as far as we are acquainted with it, looking to the average of sensations, the plurality and the preponderancy is in favor of happiness by a vast excess. In our own species, in which perhaps the assertion may be more questionable than in any other, the prepollency* of good over evil, of health, for example, and ease, over pain and distress, is evinced by the very notice which calamities excite. What enquiries does the sickness of our friends produce? What conversation their misfortunes? This shews that the common course of things is in favor of happiness; that happiness is the rule; misery, the exception. Were the order reversed, our attention would be called to examples of health and competency, instead of disease and want.

One great cause of our insensibility to the goodness of the Creator is the very *extensiveness* of his bounty. We prize but little, what we share only in common with the rest, or with the generality, of our species. When we hear of blessings, we think forthwith of successes, of prosperous fortunes, of honors, riches, preferments, i. e. of those advantages and superiorities over others, which we happen either to possess, or to be in pursuit of, or to covet. The common benefits of our nature entirely escape us. Yet these are the great things. These constitute, what most properly ought to be accounted blessings of Providence; what alone, if we might so speak, are worthy of its care. Nightly rest and daily bread, the ordinary use of our limbs, and senses, and understandings, are gifts which admit of no comparison with any other. Yet, because almost every man we meet with possesses these, we leave them out of our enumeration. They raise no sentiment: they move no gratitude. Now, herein, is our judgment perverted by our selfishness. A blessing ought in truth to be the *more* satisfactory, the bounty at least of the donor is rendered more conspicuous, by its very diffusion, its commonness, its cheapness; by its falling to the lot, and forming the happiness, of the great bulk and body of our species, as well as of ourselves. Nay even when we do not possess it, it ought to be matter of thankfulness that others do. But we have a different way of thinking. We court distinction. That I don't quarrel with: but we can *see* nothing but what has distinction to recommend it. This necessarily contracts our view of the Creator's beneficence within a narrow compass; and most unjustly. It is in those things which are so common as to be no distinction, that the amplitude of the divine benignity is perceived.

But pain, no doubt, and privations, exist, in numerous instances, and to a degree, which, collectively, would be very great, if they were compared with any other thing than with the mass of animal fruition. For the application, therefore, of our proposition to that *mixed* state of things which these exceptions induce, two rules are necessary, and both, I think, just and fair rules. One is, that we regard those effects alone which are accompanied with proofs of intention: The other, that, when we cannot resolve all appearances into benevolence of design, we make the few give place to the many; the little to the great; that we take our judgment from a large and decided preponderancy, if there be one.

I crave leave to transcribe into this place, what I have said upon this subject in my Moral Philosophy.* 'When God created the human species, either he wished their happiness, or he wished their misery, or he was indifferent and unconcerned about either.

'If he had wished our misery, he might have made sure of his purpose, by forming our senses to be so many sores and pains to us, as they are now instruments of gratification and enjoyment; or by placing us amidst objects, so ill suited to our perceptions, as to have continually offended us, instead of ministering to our refreshment and delight. He might have made, for example, every thing we tasted bitter; every thing we saw loathsome; every thing we touched a sting; every smell a stench; and every sound a discord.

'If he had been indifferent about our happiness or misery, we must impute to our good fortune (as all design by this supposition is excluded) both the capacity of our senses to receive pleasure, and the supply of external objects fitted to produce it.

'But either of these, and still more both of them, being too much to be attributed to accident, nothing remains but the first supposition, that God, when he created the human species, wished their happiness; and made for them the provision which he has made, with that view, and for that purpose.

'The same argument may be proposed in different terms, thus: Contrivance proves design; and the predominant tendency of the contrivance indicates the disposition of the designer. The world abounds with contrivances; and all the contrivances which we are acquainted with, are directed to beneficial purposes. Evil no doubt exists; but is never, that we can perceive, the object of contrivance. Teeth are contrived to eat, not to ache; their aching now and then is incidental to the contrivance, perhaps inseparable from it: or even, if you will, let it be called a defect in the contrivance; but it is not the object of it. This is a distinction which well deserves to be attended to. In describing implements of husbandry, you would hardly say of the sickle, that it is made to cut the reaper's hand, though, from the construction of the instrument, and the manner of using it, this mischief often follows. But if you had occasion to describe instruments of torture or execution, This engine, you would say, is to extend the sinews; this to dislocate the joints; this to break the bones; this to scorch the soles of the feet. Here pain and misery are the very objects of the contrivance. Now, nothing of this sort is to be found in the works of nature. We never discover a train of contrivance to bring about an evil purpose. No anatomist ever discovered a system of organization, calculated to produce pain and disease; or, in explaining the parts of the human body, ever said, This is to irritate; this to inflame; this duct is to convey the gravel to the kidneys; this gland to secrete the humour which forms the gout: if by chance he come at a part of which he knows not the use, the most he can say is, that it is useless; no one ever suspects that it is put there to incommode, to annoy, or to torment.'

The two cases which appear to me to have the most of difficulty in them, as forming the most of the appearance of exception to the representation here given, are those of *venomous* animals, and of animals *preying* upon one another. These properties of animals, wherever they are found, must, I think, be referred to design; because there is, in all cases of the first, and in most cases of the second, an express and distinct organization provided for the producing of them. Under the first head, the fangs of vipers, the stings of wasps and scorpions, are as clearly intended for their purpose, as any animal structure is for any purpose the most incontestably beneficial. And the same thing must, under the second head, be acknowledged of the talons and beaks of birds, of the tusks, teeth, and claws of beasts of prey, of the shark's mouth, of the spider's web, and of numberless weapons of offence belonging to different tribes of voracious insects. We cannot, therefore, avoid the difficulty by saying, that the effect was not intended. The only question open to us is, whether it be ultimately evil. From the confessed and felt imperfection of our knowledge, we ought to presume, that there may be consequences of this œconomy which are hidden from us: from the benevolence which pervades the general designs of nature, we ought also to presume, that these consequences, if they could enter into our calculation, would turn the balance on the favorable side. Both these I contend to be reasonable presumptions. Not reasonable presumptions, if these two cases were the only cases which nature presented to our observation; but reasonable presumptions under the reflection, that the cases in question are combined with a multitude of intentions, all proceeding from the same author, and all, except these, directed to ends of undisputed utility. Of the vindications, however, of this œconomy, which we are able to assign, such as most extenuate the difficulty are the following.

With respect to *venomous* bites and stings, it may be observed,

1. That, the animal itself being regarded, the faculty complained of is *good*; being conducive, in all cases, to the defence of the animal; in some cases, to the subduing of its prey; and, in some probably, to the killing of it, when caught, by a mortal wound inflicted in the passage to the stomach, which may be no less merciful to the victim, than salutary to the devourer. In the viper, for instance, the poisonous fang may do that, which, in other animals of prey, is done by the crush of the teeth. Frogs and mice might be swallowed alive without it.

2. But it will be said, that this provision, when it comes to the case of bites, deadly even to human bodies and to those of large quadrupeds, is greatly *overdone*; that it might have fulfilled its use, and yet have been much less deleterious than it is. Now I believe the case of bites, which produce death in large animals, (of stings I think there are none,) to be very few. The experiments of the Abbé Fontana,*

which were numerous, go strongly to the proof of this point. He found that it required the action of five exasperated vipers to kill a dog of a moderate size; but that, to the killing of a mouse or a frog, a single bite was sufficient; which agrees with the use which we assign to the faculty. The Abbé seemed to be of opinion, that the bite even of the rattlesnake would not usually be mortal; allowing, however, that in certain particularly unfortunate cases, as when the puncture had touched some very tender part, pricked a principal nerve for instance, or, as it is said, some more considerable lymphatic vessel, death might speedily ensue.

3. It has been, I think, very justly remarked concerning serpents, that, whilst only a few species possess the venomous property, that property guards the whole tribe. The most innocuous snake is avoided with as much care as a viper. Now the terror, with which large animals regard this class of reptiles, is its protection; and this terror is founded in the formidable revenge, which a few of the number, compared with the whole, are capable of taking. The species of serpents, described by Linnæus, amount to two hundred and eighteen, of which thirty-two only are poisonous.

4. It seems to me, that animal constitutions are provided, not only for each element, but for each state of the elements, i. e. for every climate, and for every temperature; and that part of the mischief complained of, arises from animals (the human animal most especially) occupying situations upon the earth which do not belong to them, nor were ever intended for their habitation. The folly and wickedness of mankind, and necessities proceeding from these causes, have driven multitudes of the species to seek a refuge amongst burning sands, whilst countries blessed with hospitable skies, and with the most fertile soils, remain almost without a human tenant. We invade the territories of wild beasts and venomous reptiles, and then complain that we are infested by their bites and stings. Some accounts of Africa place this observation in a strong point of view. 'The desarts,' says Adanson,* 'are entirely barren, except where they are found to produce serpents; and in such quantities, that some extensive plains are almost entirely covered with them.' These are the natures appropriated to the situation. Let them enjoy their existence: let them have their country.* Surface enough will be left to man, though his numbers were increased an hundred fold, and left to him, where he might live, exempt from these annoyances.

The *second* case, viz. that of animals *devouring* one another, furnishes a consideration of much larger extent. To judge, whether, as a general provision, this can be deemed an *evil*, even so far as we understand its consequences, which, probably, is a partial understanding, the following reflections are fit to be attended to.

1. Immortality upon this earth is out of the question. Without death there could be no generation, no sexes, no parental relation, i. e. as things are constituted, no animal happiness. The particular duration of life, assigned to different animals, can form no part of the objection; because, whatever that duration was, whilst it remained finite and limited, it might always be asked, why it was no longer. The natural age of different animals varies from a single day to a century of years. No account can be given of this; nor could any be given, whatever other proportion of life had obtained amongst them.

The term then of life in different animals being the same as it is, the question is, what mode of taking it away is the best even for the animal itself.

Now, according to the established order of nature, (which we must suppose to prevail, or we cannot reason at all upon the subject,) the three methods by which life is usually put an end to, are acute diseases, decay, and violence. The simple and natural life of *brutes*, is not often visited by acute distempers; nor could it be deemed an improvement of their lot, if they were. Let it be considered therefore, in what a condition of suffering and misery a brute animal is placed, which is left to perish by *decay*. In human sickness or infirmity, there is the assistance of man's rational fellow creatures, if not to alleviate his pains, at least to minister to his necessities, and to supply the place of his own activity. A brute, in his wild and natural state, does every thing for himself. When his strength therefore, or his speed, or his limbs, or his senses fail him, he is delivered over, either to absolute famine, or to the protracted wretchedness of a life slowly wasted by scarcity of food. Is it then to see the world filled with drooping, superannuated, half starved, helpless and unhelped animals, that you would alter the present system of pursuit and prey?

2. Which system is also to them the spring of motion and activity on both sides. The pursuit of its prey, forms the employment, and appears to constitute the pleasure, of a considerable part of the animal creation. The using of the means of defence, or flight, or

precaution, forms also the business of another part. And even of this latter tribe, we have no reason to suppose, that their happiness is much molested by their fears. Their danger exists continually; and in some cases they seem to be so far sensible of it, as to provide, in the best manner they can, against it; but it is only when the attack is actually made upon them, that they appear to suffer from it. To contemplate the insecurity of their condition with anxiety and dread, requires a degree of reflection, which (happily for themselves) they do not possess. A *hare*, notwithstanding the number of its dangers and its enemies, is as playful an animal as any other.

3. But, to do justice to the question, the system of animal *destruction* ought always to be considered in strict connection with another property of animal nature, viz. *superfecundity*. They are countervailing qualities. One subsists by the correction of the other. In treating, therefore, of the subject under this view, (which is, I believe, the true one,) our business will be, first, to point out the advantages which are gained by the powers in nature of a superabundant multiplication; and, then, to shew, that these advantages are so many reasons for appointing that system of animal hostilities, which we are endeavouring to account for.

In almost all cases nature produces her supplies with profusion. A single cod fish spawns, in one season, a greater number of eggs, than all the inhabitants of England amount to. A thousand other instances of prolific generation might be stated, which, though not equal to this, would carry on the increase of the species with a rapidity which outruns calculation, and to an immeasurable extent. The advantages of such a constitution are two: first, that it tends to keep the world always full; whilst, secondly, it allows the proportion between the several species of animals to be differently modified, as different purposes require, or as different situations may afford for them room and food. Where this vast fecundity meets with a vacancy fitted to receive the species, there it operates with its whole effect; there it pours in its numbers, and replenishes the waste. We complain of what we call the exorbitant multiplication of some troublesome insects, not reflecting that large portions of nature might be left void without it. If the accounts of travellers may be depended upon, immense tracts of forest in North America would be nearly lost to sensitive existence if it were not for *gnats*. 'In the thinly inhabited regions of America, in which the waters stagnate, and the climate is warm, the whole air is filled with crowds of these insects.' Thus it is, that, where we looked for solitude and deathlike silence, we meet with animation, activity, enjoyment; with a busy, a happy, and a peopled world. Again; hosts of *mice* are reckoned amongst the plagues of the north-east part of Europe; whereas vast plains in Siberia, as we learn from good authority, would be lifeless without them. The Caspian desarts are converted by their presence into crowded warrens. Between the Volga and the Yaik, and in the country of Hyrcania, the ground, says Pallas,* is in many places *covered* with little hills, raised by the earth cast out in forming the burrows. Do we then so envy these blissful abodes, as to pronounce the fecundity by which they are supplied with inhabitants, to be an evil; a subject of complaint, and not of praise? Further; by virtue of this same superfecundity, what we term destruction, becomes almost instantly the parent of life. What we call blights, are, oftentimes, legions of animated beings claiming their portion in the bounty of nature. What corrupts the produce of the earth to us, prepares it for them. And it is by means of their rapid multiplication, that they take possession of their pasture: a slow propagation would not meet the opportunity.

But in conjunction with the occasional use of this fruitfulness, we observe, also, that it allows the proportion between the several species of animals to be differently modified, as different purposes of utility may require. When the forests of America come to be cleared, and the swamps drained, our gnats will give place to other inhabitants. If the population of Europe should spread to the north and the east, the mice will retire before the husbandman and the shepherd, and yield their station to herds and flocks. In what concerns the human species, it may be a part of the scheme of Providence that the earth should be inhabited by a shifting, or perhaps a circulating population. In this œconomy it is possible that there may be the following advantages. When old countries are become exceedingly corrupt, simpler modes of life, purer morals, and better institutions may rise up in new ones, whilst fresh soils reward the cultivator with more plentiful returns. Thus the different portions of the globe come into use in succession as the residence of man; and, in his absence, entertain other guests, which, by their rapid multiplication soon fill the chasm. In domesticated animals we find the effect of their fecundity to be, that we can always command *numbers*: we can always have as many of any particular species as we please, or as we can support. Nor do we complain of its excess; it being much more easy to regulate abundance, than to supply scarcity.

But then this *superfecundity*, though of great occasional use and importance, exceeds the ordinary capacity of nature to receive or support its progeny. All superabundance supposes destruction, or must destroy itself.* Perhaps there is no species of terrestrial animals whatever, which would not overrun the earth, if it were permitted to multiply in perfect safety; or of fish, which would not fill the ocean: at least, if any single species were left to their natural increase without disturbance or restraint, the food of other species would be exhausted by their maintenance. It is necessary, therefore, that the effects of such prolific faculties be curtailed. In conjunction with other checks and limits, all subservient to the same purpose, are the thinnings which take place among animals, by their action upon one another. In some instances we ourselves experience, very directly, the use of these hostilities. One species of insects rids us of another species; or reduces their ranks. A third species perhaps keeps the second within bounds: and birds or lizards are a fence against the inordinate increase by which even these last might infest us. In other, more numerous, and possibly more important instances, this disposition of things, although less necessary or useful to us, and of course less observed by us, may be necessary and useful to certain other species; or even for the preventing of the loss of certain species* from the universe: a misfortune which seems to be studiously guarded against. Though there may be the appearance of failure in some of the details of Nature's works, in her great purposes there never are. Her species never fail. The provision which was originally made for continuing the replenishment of the world has proved itself to be effectual through a long succession of ages.

What further shews, that the system of destruction amongst animals holds an express relation to the system of fecundity; that they are parts indeed of one compensatory scheme; is, that, in each species, the fecundity bears a proportion to the smallness of the animal, to the weakness, to the shortness of its natural term of life, and to the dangers and enemies by which it is surrounded. An elephant produces but one calf: a butterfly lays six hundred eggs. Birds of prey seldom produce more than two eggs: the sparrow tribe, and the duck tribe, frequently sit upon a dozen. In the rivers, we meet with a thousand minnows for one pike; in the sea, a million of herrings for a single shark. Compensation obtains throughout. Defencelessness and devastation are repaired by fecundity.

We have dwelt the longer upon these considerations, because the subject to which they apply, namely, that of animals *devouring* one another, forms the chief, if not the only instance, in the works of the Deity, of an œconomy, stamped by marks of design, in which the character of utility can be called in question. The case of *venomous* animals is of much inferior consequence to the case of prey, and, in some degree, is also included under it. To both cases it is probable that many more reasons belong, than those of which we are in possession.

Our *first* proposition, and that which we have hitherto been defending, was, 'that in a vast plurality of instances, in which *contrivance* is perceived, the design of the contrivance is beneficial.'

Our *second* proposition is, 'that the Deity has added *pleasure** to animal sensations, beyond what was necessary for any other purpose, or when the purpose, so far as it was necessary, might have been effected by the operation of pain.'

This proposition may be thus explained. The capacities, which, according to the established course of nature, are *necessary* to the support or preservation of an animal, however manifestly they may be the result of an organization contrived for the purpose, can only be deemed an act or a part of the same will, as that which decreed the existence of the animal itself; because, whether the creation proceeded from a benevolent or a malevolent being, these capacities must have been given, if the animal existed at all. Animal properties therefore, which fall under this description, do not strictly prove the goodness of God. They may prove the existence of the Deity: they may prove a high degree of power and intelligence: but they do not prove his goodness; forasmuch as they must have been found in any creation which was capable of continuance, although it is possible to suppose, that such a creation might have been produced by a being whose views rested upon misery.

But there is a class of properties, which may be said to be super-added from an intention expressly directed to happiness; an intention to give a happy existence distinct from the general intention of providing the means of existence; and that is, of capacities for pleasure, in cases, wherein, so far as the conservation of the individual or of the species is concerned, they were not wanted, or wherein the purpose might have been secured by the operation of pain. The provision which is made of a variety of objects, not necessary to life, and ministring only to our pleasures; and the properties given to the necessaries of life themselves, by which they contribute to pleasure as well as preservation; shew a further design, than that of giving existence.[1]

A single instance will make all this clear. Assuming the necessity of food for the support of animal life, it is requisite, that the animal be provided with organs, fitted for the procuring, receiving, and digesting of its food. It may be also necessary, that the animal be impelled by its sensations to exert its organs. But the pain of hunger would do all this. Why add pleasure to the act of eating; sweetness and relish to food? Why a new and appropriate sense for the perception of the pleasure? Why should the juice of a peach applied to the palate, affect the part so differently from what it does when rubbed upon the palm of the hand? This is a constitution, which, so far as appears to me, can be resolved into nothing but the pure benevolence of the Creator.* Eating is necessary; but the pleasure attending it is not necessary: and that this pleasure, depends not only upon our being in possession of the sense of taste, which is different from every other, but upon a particular state of the organ in which it resides, a felicitous adaptation of the organ to the object, will be confessed by any one, who may happen to have experienced that vitiation of taste which frequently occurs in fevers, when every taste is irregular, and every one bad.

In mentioning the gratifications of the palate, it may be said that we have made choice of a trifling example. I am not of that opinion. They afford a share of enjoyment to man; but to brutes, I believe, that they are of very great importance. A horse at liberty passes a great part of his waking hours in eating. To the ox, the sheep, the deer, and other ruminating animals, the pleasure is doubled. Their whole time almost is divided between browsing upon their pasture and chewing their cud. Whatever the pleasure be, it is spread over a large portion of their existence. If there be animals, such as the

---

[1] See this topic considered in Dr Balguy's treatise upon the Divine Benevolence.* This excellent author first, I think, proposed it; and nearly in the terms in which it is here stated. Some other observations also under this head are taken from that treatise.

lupous fish, which swallow their prey whole, and at once, without any time, as it should seem, for either drawing out, or relishing, the taste in the mouth, is it an improbable conjecture that the seat of taste with them is in the stomach; or, at least, that a sense of pleasure, whether it be taste or not, accompanies the dissolution of the food in that receptacle, which dissolution in general is carried on very slowly? If this opinion be right, they are more than repaid for their defect of palate. The feast lasts as long as the digestion.

In seeking for argument we need not stay to insist upon the comparative importance of our example, for the observation holds equally of all, or of three at least, of the other senses. The necessary purposes of hearing might have been answered without harmony; of smell, without fragrance; of vision, without beauty. Now 'If the Deity had been indifferent about our happiness or misery, we must impute to our good fortune (as all design by this supposition is excluded) both the capacity of our senses to receive pleasure, and the supply of external objects fitted to excite it.' I alledge these as *two* felicities, for they are different things yet both necessary: the sense being formed, the objects, which were applied to it, might not have suited it; the objects being fixed, the sense might not have agreed with them. A coincidence is here required which no accident can account for. There are three possible suppositions upon the subject, and no more. The first; that the sense, by its original constitution, suited the object: the second; that the object, by its original constitution, suited the sense: the third; that the sense is so constituted, as to be able, either universally, or within certain limits, by habit and familiarity, to render every object pleasant. Whichever of these suppositions we adopt, the effect evinces, on the part of the Author of nature, a studious benevolence. If the pleasures which we derive from any of our senses, depend upon an original congruity between the sense and the properties perceived by it, we know by experience, that the adjustment demanded, with respect to the qualities which were conferred upon the objects that surround us, not only choice and selection, out of a boundless variety of possible qualities with which these objects might have been endued, but a *proportioning also of degree*, because an excess or defect of intensity spoils the perception, as much almost as an error in the kind and nature of the quality. Likewise the degree of dullness or acuteness in the sense itself, is no arbitrary thing, but, in order to preserve the congruity here spoken

of, requires to be in an exact or near correspondency with the
strength of the impression. The dullness of the senses forms the
complaint of old age. Persons in fevers, and, I believe, in most maniacal cases,* experience great torment from their preternatural acuteness. An increased, no less than an impaired sensibility, induces a
state of disease and suffering.

The doctrine of a specific congruity between animal senses and
their objects, is strongly savored by what is observed of insects in the
election of their food. Some of these will feed upon one kind of plant
or animal, and upon no other: some caterpillars upon the cabbage
alone; some upon the black currant alone. The species of caterpillar,
which eats the vine, will starve upon the elder; nor will that which
we find upon fennel, touch the rose bush. Some insects confine
themselves to two or three kinds of plants or animals. Some again
shew so strong a preference, as to afford reason to believe that,
though they may be driven by hunger to others, they are led by the
pleasure of taste to a few particular plants alone; and all this, as it
should seem, independently of habit* or imitation.

But should we accept the third hypothesis, and even carry it so far,
as to ascribe every thing, which concerns the question, to habit, (as
in certain species, the human species most particularly, there is reason to attribute something) we have then before us an animal capacity, not less perhaps to be admired, than the native congruities
which the other scheme adopts. It cannot be shewn to result from
any fixed necessity in nature, that what is frequently applied to the
senses should of course become agreeable to them. It is, so far as it
subsists, a power of accommodation considered and provided by the
author of their structure, and forms a part of their perfection.

In whichever way we consider the senses, they appear to be specific gifts, ministring, not only to preservation, but to pleasure. But
what we usually call the *senses* are probably themselves far from being
the only vehicles of enjoyment, or the whole of our constitution,
which is calculated for the same purpose. We have many internal
sensations of the most agreeable kind, hardly referable to any of the
five senses. Some physiologists have held, that all secretion is pleasurable; and that the complacency which in health, without any
external, assignable, object to excite it, we derive from life itself, is
the effect of our secretions going on well within us. All this may be
true: but, if true, what reason can be assigned for it, except the

will of the Creator? It may reasonably be asked, why is any thing a pleasure? and I know no answer which can be returned to the question, but that which refers it to appointment. We can give no account whatever of our pleasures in the simple and original perception;* and, even when physical sensations are assumed, we can seldom account for them in the secondary and complicated shapes, in which they take the name of diversions. I never yet met with a sportsman, who could tell me in what the sport consisted; who could resolve it into its principle, and state that principle. I have been a great follower of fishing myself,* and in its chearful solitude have passed some of the happiest hours of a sufficiently happy life; but, to this moment, I could never trace out the source of the pleasure which it afforded me.

The 'quantum in rebus inane,'* whether applied to our amusements, or to our graver pursuits, (to which, in truth, it sometimes equally belongs,) is always an unjust complaint. If trifles engage, and if trifles make us happy, the true reflection suggested by the experiment, is upon the tendency of nature to gratification and enjoyment; which is, in other words, the goodness of its author towards his sensitive creation.

Rational natures also, as such, exhibit qualities which help to confirm the truth of our position. The degree of understanding found in mankind, is usually much greater than what is necessary for preservation. The pleasure of chusing for themselves, and of prosecuting the object of their choice, should seem to be an original source of enjoyment. The pleasures received from things, great, beautiful, or new, from imitation, or from the liberal arts, are, in some measure, not only superadded, but unmixed gratifications, having no pains to balance them.[1]

I do not know whether our attachment to *property* be not something more than the mere dictate of reason, or even than the mere effect of association. Property communicates a charm to whatever is the object of it. It is the first of our abstract ideas; it cleaves to us the closest and the longest. It endears to the child its plaything, to the peasant his cottage, to the landholder his estate. It supplies the place of prospect and scenery. Instead of coveting the beauty of distant situations, it teaches every man to find it in his own. It gives boldness and grandeur to plains and fens,* tinge and colouring to clays and fallows.

---

[1] Balguy on the Divine Benevolence.

All these considerations come in aid of our *second* proposition. The reader will now bear in mind what our *two* propositions were. They were, first; that, in a vast plurality of instances, in which contrivance is perceived, the design of the contrivance is beneficial: secondly; that the Diety has added pleasure to animal sensations beyond what was necessary for any other purpose; or when the purpose, so far as it was necessary, might have been effected by the operation of pain.

Whilst these propositions can be maintained, we are authorized to ascribe to the Deity the character of benevolence:* and what is benevolence at all, must in him be *infinite* benevolence, by reason of the infinite, that is to say, the incalculably great, number of objects, upon which it is exercised.

Of the ORIGIN OF EVIL no universal solution has been discovered:* I mean no solution which reaches to all cases of complaint. The most comprehensive is that which arises from the consideration of *general rules*. We may, I think, without much difficulty, be brought to admit the four following points; first, that important advantages may accrue to the universe from the order of nature proceeding according to general laws: secondly; that general laws, however well set and constituted, often thwart and cross one another: thirdly; that from these thwartings and crossings frequent particular inconveniences will arise: and fourthly; that it agrees with our observation to suppose, that some degree of these inconveniences takes place in the works of nature. These points may be allowed: and it may also be asserted that the general laws with which we are acquainted, are directed to beneficial ends. On the other hand, with many of these laws we are not acquainted at all, or we are totally unable to trace them in their branches and in their operation: the effect of which ignorance is, that they cannot be of importance to us as measures by which to regulate our conduct. The conservation of them may be of importance in other respects, or to other beings, but we are uninformed of their value or use: consequently when, and how far, they may or may not be suspended, or their effects turned aside, by a presiding and benevolent will, without incurring greater evils than those which would be avoided. The consideration, therefore, of general laws, although it may concern the question of the origin of evil very nearly, (which I think it does,) rests in views disproportionate to our faculties, and in a knowledge which we do not possess. It serves rather to account for the obscurity of the subject, than to supply us with distinct answers to our difficulties. However, whilst we assent to the above stated propositions as principles, whatever uncertainty we may find in the application, we lay a ground for believing, that cases, of apparent evil, for which *we* can suggest no particular reason, are governed by reasons, which are more general, which lie deeper in the order of second causes, and on that account are removed to a greater distance from us.

The doctrine of *imperfections*,* or, as it is called, of evils of imperfection, furnishes an account, founded like the former, in views of universal nature. The doctrine is briefly this. It is probable that creation may be better replenished, by sensitive beings of different sorts, than by sensitive beings all of one sort. It is likewise probable, that it may be better replenished, by different orders of being rising one above another in gradation, than by beings possessed of equal degrees of perfection. Now a gradation of such beings implies a gradation of imperfections. No class can justly complain of the imperfections which belong to its place in the scale, unless it were allowable for it to complain, that a scale of being was appointed in nature: for which appointment there appear to be reasons of wisdom and goodness.

In like manner, *finiteness*, or what is resolvable into finiteness, in inanimate subjects, can never be a just subject of complaint, because, if it were ever so, it would be always so: we mean, that we can never reasonably demand that things should be larger or more, when the same demand might be made, whatever the quantity or number was.

And to me it seems, that the sense of mankind has so far acquiesced in these reasons, as that we seldom complain of evils of this class, when we clearly perceive them to be such. What I have to add therefore is, that we ought not to complain of some other evils, which stand upon the same foot of vindication as evils of confessed imperfection. We never complain that the globe of our earth is too small: nor should we complain, if it were even much smaller. But where is the difference to us, between a less globe, and part of the present being uninhabitable? The inhabitants of an island, may be apt enough to murmur at the sterility of some parts of it, against its rocks, or sands, or swamps; but no one thinks himself authorised to murmur, simply because the island is not larger than it is. Yet these are the same griefs.

The above are the two metaphysical answers which have been given to this great question. They are not the worse for being metaphysical, provided they be founded (which, I think, they are) in right reasoning; but they are of a nature too wide to be brought under our survey, and it is often difficult to apply them in the detail. Our speculations, therefore, are perhaps better employed when they confine themselves within a narrower circle.

The observations which follow are of this more limited, but more determinate kind.

Of *bodily pain* the principal observation, no doubt, is, that which we have already made, and already dwelt upon, viz. 'that it is seldom the object of contrivance; that, when it is so, the contrivance rests ultimately in good.'

To which however may be added, that the annexing of pain to the means of destruction is a salutary provision: inasmuch as it teaches vigilance and caution; both gives notice of danger, and excites those endeavours which may be necessary to preservation. The evil consequence, which sometimes arises from the want of that timely intimation of danger which pain gives, is known to the inhabitants of cold countries by the example of frost-bitten limbs. I have conversed with patients who have lost toes and fingers by this cause. They have in general told me, that they were totally unconscious of any local uneasiness at the time. Some I have heard declare, that, whilst they were about their employment, neither their situation, nor the state of the air, was unpleasant. They felt no pain: they suspected no mischief: till, by the application of warmth, they discovered, too late, the fatal injury which some of their extremities had suffered. I say that this shews the use of pain, and that we stand in need of such a monitor. I believe also that the use extends further than we suppose, or can now trace; that to disagreeable sensations, we, and all animals, owe, or have owed, many habits of action which are salutary, but which are become so familiar as not easily to be referred to their origin.

Pain also itself is not without its *alleviations*. It may be violent and frequent; but it is seldom both violent and long continued: and its pauses and intermissions become positive pleasures. It has the power of shedding a satisfaction over intervals of ease, which, I believe, few enjoyments exceed. A man resting from a fit of the stone or gout,* is, for the time, in possession of feelings which undisturbed health cannot impart. They may be dearly bought, but still they are to be set against the price. And, indeed, it depends upon the duration and urgency of the pain, whether they be dearly bought or not. I am far from being sure, that a man is not a gainer by suffering a moderate interruption of bodily ease for a couple of hours out of the four-and-twenty. Two very common observations favor this opinion: one is, that remissions of pain call forth, from those who experience them, stronger expressions of satisfaction and of gratitude towards both the author and the instruments of their relief, than are excited by advantages of any other kind: the second is, that the spirits of sick men do not sink in proportion to the acuteness of their sufferings; but rather appear to be roused and supported, not by pain, but by the high degree of comfort which they derive from its cessation, whenever that occurs: and which they taste with a relish, that diffuses some portion of mental complacency over the whole of that mixed state of sensations in which disease has placed them.

In connection with bodily pain may be considered bodily *disease*, whether painful or not. Few diseases are fatal. I have before me the account of a dispensary* in the neighbourhood which states six years experience as follows: 'admitted 6,420—*cured* 5,476—dead 234.' And this I suppose nearly to agree with what other similar institutions exhibit. Now, in all these cases, some disorder must have been felt, or the patients would not have applied for a remedy; yet we see how large a proportion of the maladies which were brought forward, have either yielded to proper treatment, or, what is more probable, ceased of their own accord. We owe these frequent recoveries, and, where recovery does not take place, this patience of the human constitution under many of the distempers by which it is visited, to two benefactions of our nature. One is, that she works within certain limits; allows of a certain latitude, within which health may be preserved, and within the confines of which it only suffers a graduated diminution. Different quantities of food, different degrees of exercise, different portions of sleep, different states of the atmosphere,* are compatible with the possession of health. So likewise is it with the secretions and excretions, with many internal functions of the body, and with the state probably of most of its internal organs. They may vary considerably, not only without destroying life, but without occasioning any high degree of inconveniency. The other property of our nature to which we are still more beholden, is its constant endeavour to restore itself, when disordered, to its regular course. The fluids of the body appear to possess a power of separating and expelling any noxious substance which may have mixed itself with them. This they do, in eruptive fevers, by a kind of despumation,* as Sydenham* calls it, analogous in some measure to the intestine action by which fermenting liquors work the yest to the surface. The solids, on their part, when their action is obstructed, not only resume that action, as soon as the obstruction is removed, but they struggle with the impediment: they take an action as near to the true one, as the difficulty and the disorganization, with which they have to contend, will allow of.

Of *mortal* diseases the great use is to reconcile us to death. The horror of death proves the value of life. But it is in the power of disease to abate, or even extinguish, this horror; which it does in a wonderful manner, and, oftentimes, by a mild and imperceptible gradation. Every man who has been placed in a situation to observe it, is surprised with the change which has been wrought in himself, when he compares the view which he entertains of death upon a sick bed, with the heart-sinking dismay with which he should some time ago have met it in health. There is no similitude between the sensations of a man led to execution, and the calm expiring of a patient at the close of his disease. Death to him is only the last of a long train of changes: in his progress through which, it is possible that he may experience no shocks or sudden transitions.

Death itself, as a mode of removal and of succession, is so connected with the whole order of our animal world, that almost every thing in that world must be changed, to be able to do without it. It may seem likewise impossible to separate the fear of death from the enjoyment of life, or the perception of that fear from rational natures. Brutes are in a great measure delivered from all anxiety on this account by the inferiority of their faculties; or rather they seem to be armed with the apprehension of death just sufficiently to put them upon the means of preservation, and no further. But would a human being wish to purchase this immunity by the loss of those mental powers which enable him to look forward to the future?

Death implies *separation*: and the loss of those whom we love must necessarily be accompanied with pain. To the brute creation, nature seems to have stepped in with some secret provision for their relief, under the rupture of their attachments. In their instincts* towards their offspring, and of their offspring to them, I have often been surprised to observe, how ardently they love, and how soon they forget. The pertinacity of human sorrow (upon which time also, at length, lays its softening hand) is probably, therefore, in some manner connected with the qualities of our rational or moral nature. One thing however is clear, viz. that it is better that we should possess affections, the sources of so many virtues and so many joys, although they be exposed to the incidents of life, as well as the interruptions of mortality, than, by the want of them, be reduced to a state of selfishness, apathy, and quietism.

Of other external evils (still confining ourselves to what are called physical or natural evils*) a considerable part come within the scope of the following observation. The great principle of human satisfaction is *engagement*. It is a most just distinction, which the late Mr Tucker* has dwelt upon so largely in his works, between pleasures in which we are passive, and pleasures in which we are active. And, I believe, every attentive observer of human life will assent to his position, that, however grateful the sensations may occasionally be in which we are passive, it is not these, but the latter class of our pleasures, which constitute satisfaction; which supply that regular stream of moderate and miscellaneous enjoyments, in which happiness, as distinguished from voluptuousness, consists. Now for rational occupation, which is, in other words, for the very material of contented existence, there would be no place left, if either the things with which we had to do were absolutely impracticable to our endeavours, or if they were too obedient to our uses. A world furnished with advantages on one side, and beset with difficulties, wants, and inconveniences on the other, is the proper abode of free, rational, and active natures, being the fittest to stimulate and exercise their faculties. The very *refractoriness* of the objects they have to deal with contributes to this purpose. A world in which nothing depended upon ourselves, however it might have suited an imaginary race of beings, would not have suited mankind. Their skill, prudence, industry; their various arts, and their best attainments, from the application of which they draw, if not their highest, their most permanent gratifications, would be insignificant, if things could be either molded by our volitions, or, of their own accord, conformed themselves to our views and wishes. Now it is in this refractoriness that we discern the seed and principle of *physical* evil, as far as it arises from that which is external to us.

Civil evils, or the evils of civil life,* are much more easily disposed of than physical evils; because they are, in truth, of much less magnitude, and also because they result by a kind of necessity, not only from the constitution of our nature, but from a part of that constitution which no one would wish to see altered. The case is this. Mankind will in every country *breed up* to a certain point of distress. That point may be different in different countries or ages according to the established usages of life in each. It will also shift upon the scale, so as to admit of a greater or less number of inhabitants, as the quantity of provision which is either produced in the country or supplied to it from others may happen to vary. But there must always be such a point, and the species will always breed up to it. The order of generation proceeds by something like a geometrical progression. The increase of provision, under circumstances even the most advantageous, can only assume the form of an arithmetic series. Whence it follows, that the population will always overtake the provision, will pass beyond the line of plenty, and will continue to increase till checked by the difficulty of procuring subsistence.[1] Such difficulty therefore, along with its attendant circumstances, *must* be found in every old country; and these circumstances constitute what we call poverty, which, necessarily, imposes labour, servitude, restraint.

It seems impossible to people a country with inhabitants who shall be all in easy circumstances. For suppose the thing to be done, there would be such marrying and giving in marriage amongst them, as would in a few years change the face of affairs entirely; i. e. as would increase the consumption of those articles, which supplied the natural or habitual wants of the country, to such a degree of scarcity, as must leave the greatest part of the inhabitants unable to procure them without toilsome endeavours, or, out of the different kinds of these articles, to procure any kind except that which was most easily produced. And this, in fact, describes the condition of the mass of the community in all countries; a condition, unavoidably, as it should seem, resulting from the provision which is made in the human,

[1] See this subject stated in a late treatise upon population.*

in common with all animal constitutions, for the perpetuity and multiplication of the species.

It need not however dishearten any endeavours for the public service, to know that population naturally treads upon the heels of improvement. If the condition of a people be meliorated, the consequence will be, either that the *mean* happiness will be increased, or a greater number partake of it; or, which is most likely to happen, that both effects will take place together. There may be limits fixed by nature to both, but they are limits not yet attained, nor even approached, in any country of the world.

And when we speak of limits at all, we have respect only to provisions for animal wants. There are sources, and means, and auxiliaries, and augmentations of human happiness, communicable without restriction of numbers; as capable of being possessed by a thousand persons, as by one. Such are those, which flow from a mild, contrasted with a tyrannic government, whether civil or domestic; those which spring from religion; those which grow out of a sense of security; those which depend upon habits of virtue, sobriety, moderation, order; those, lastly, which are founded in the possession of well directed tastes and desires, compared with the dominion of tormenting, pernicious, contradictory, unsatisfied, and unsatisfiable passions.

The *distinctions* of civil life are apt enough to be regarded as evils, by those who sit under them: but, in my opinion, with very little reason.

In the first place the advantages which the higher conditions of life are supposed to confer, bear no proportion in value to the advantages which are bestowed by nature. The gifts of nature always surpass the gifts of fortune. How much, for example, is activity better than attendance; beauty, than dress; appetite, digestion, and tranquil bowels, than the artifices of cookery, or than forced, costly, and farfetched dainties?

Nature has a strong tendency to equalization. Habit, the instrument of nature, is a great leveller; the familiarity which it induces, taking off the edge both of our pleasures and our sufferings. Indulgences which are habitual keep us in ease, and cannot be carried much further. So that, with respect to the gratifications of which the senses are capable, the difference is by no means proportionable to the apparatus. Nay, so far as superfluity generates fastidiousness, the difference is on the wrong side.

It is not necessary to contend, that the advantages derived from wealth are none, (under due regulations they are certainly considerable) but that they are not greater than they ought to be. *Money* is the sweetener of human toil; the substitute for coercion; the reconciler of labour with liberty. It is, moreover, the stimulant of enterprise in all projects and undertakings, as well as of diligence in the most beneficial arts and employments. Now did affluence, when possessed, contribute nothing to happiness, or nothing beyond the mere supply of necessaries; and the secret should come to be discovered; we might be in danger of losing great part of the uses, which are, at present, derived to us through this important medium. Not only would the tranquillity of social life be put in peril by the want of a motive to attach men to their private concerns; but the satisfaction which all men receive from success in their respective occupations, which collectively constitutes the great mass of human comfort, would be done away in its very principle.

With respect to *station*, as it is distinguished from riches, whether it confer authority over others, or be invested with honors which apply solely to sentiment and imagination, the truth is, that what is gained by rising through the ranks of life, is not more than sufficient to draw forth the exertions of those who are engaged in the pursuits which lead to advancement, and which, in general, are such as ought to be encouraged. Distinctions of this sort are subjects much more of competition than of enjoyment: and in that competition their use consists. It is not, as hath been rightly observed, by what the *Lord Mayor* feels in his coach, but by what the *apprentice* feels who gazes at him, that the public is served.

As we approach the summits of human greatness, the comparison of good and evil, with respect to personal comfort, becomes still more problematical; even allowing to ambition all its pleasures. The poet asks, 'What is grandeur, what is power?' The philosopher answers, 'Constraint and plague; et in maximâ quâque fortunâ minimum licere.'* One very common error misleads the opinion of mankind upon this head, viz. that, universally, authority is pleasant, submission painful. In the general course of human affairs, the very reverse of this is nearer to the truth. Command is anxiety, obedience ease.

Artificial distinctions sometimes promote real equality. Whether they be hereditary, or be the homage paid to office, or the respect attached by public opinion to particular professions, they serve to *confront* that grand and unavoidable distinction which arises from property, and which is most overbearing where there is no other. It is of the nature of property, not only to be irregularly distributed, but to run into large masses. Public laws should be so constructed as to favor its diffusion as much as they can. But all that can be done by laws, consistently with that degree of government over his property which ought to be left to the subject, will not be sufficient to counteract this tendency. There must always therefore be the difference between rich and poor;* and this difference will be the more grinding, when no pretension is allowed to be set up against it.

So that the evils, if evils they must be called, which spring either from the necessary subordinations of civil life, or from the distinctions which have, naturally though not necessarily, grown up in most societies, so long as they are unaccompanied by privileges injurious or oppressive to the rest of the community, are such, as may, even by the most depressed ranks, be endured, with very little prejudice to their comfort.

The mischiefs of which mankind are the occasion to one another, by their private wickednesses and cruelties; by tyrannical exercises of power, by rebellions against just authority, by wars, by national jealousies and competitions operating to the destruction of third countries, or by other instances of misconduct either in individuals or societies, are all to be resolved into the character of man, as a *free agent*. Free agency in its very essence contains liability to abuse. Yet, if you deprive man of his free agency, you subvert his nature. You may have order from him and regularity, as you may from the tides or the trade winds, but you put an end to his moral character, to virtue, to merit, to accountableness, to the use indeed of reason. To which must be added the observation, that even the bad qualities of mankind have an origin in their good ones. The case is this. Human passions are either necessary to human welfare, or capable of being made, and, in a great majority of instances, in fact made, conducive to its happiness. These passions are strong and general; and, perhaps, would not answer their purpose unless they were so. But strength and generality, when it is expedient that particular circumstances should be respected, become, if left to themselves, excess and misdirection. From which excess and misdirection the vices of mankind (the causes, no doubt, of much misery) appear to spring. This account, whilst it shews us the principle of vice, shews us, at the same time, the province of reason and of self-government; the want also of every support which can be procured to either from the aids of religion; and that, without having recourse to any native gratuitous malignity in the human constitution. Mr Hume in his posthumous dialogues,* asserts, indeed, of *idleness* or aversion to labour, (which he states to lie at the root of a considerable part of the evils which mankind suffer,) that it is simply and merely bad. But how does he distinguish idleness from the love of ease? or is he sure, that the love of ease in individuals is not the chief foundation of social tranquillity? It will be found, I believe, to be true, that in every community there is a large class of its members, whose idleness is the best quality about them, being the corrective of other bad ones. If it were possible, in every instance, to give a right determination to industry, we could never have too much of it. But this is not possible, if men are to be free. And without this, nothing would be so dangerous, as an incessant, universal, indefatigable activity. In the civil world as well as in the material, it is the *vis inertiæ** which keeps things in their order.

NATURAL THEOLOGY has ever been pressed with this question, Why, under the regency of a supreme and benevolent Will, should there be, in the world, so much, as there is, of the appearance of *chance*?*

The question in its whole compass lies beyond our reach, but there are not wanting, as in the origin of evil, answers which seem to have considerable weight in particular cases, and also to embrace a considerable number of cases.

I. There must be *chance* in the midst of design: by which we mean, that events which are not designed, necessarily arise from the pursuit of events which are designed. One man travelling to York meets another man travelling to London. Their meeting is by chance, is accidental, and so would be called and reckoned, though the journeys which produced the meeting, were, both of them, undertaken with design and from deliberation. The meeting, though accidental, was nevertheless hypothetically necessary, (which is the only sort of necessity that is intelligible); for, if the two journeys were commenced at the time, pursued in the direction, and with the speed, in which and with which they were in fact begun and performed, the meeting could not be avoided. There was not, there-fore, the less necessity in it for its being by chance. Again, the meet-ing might be most unfortunate, though the errands, upon which each party set out upon his journey, were the most innocent or the most laudable. The bye effect may be unfavourable, without impeachment of the proper purpose, for the sake of which, the train, from the operation of which these consequences ensued, was put in motion. Although no cause act without a good purpose, accidental consequences, like these, may be either good or bad.

II. The *appearance of chance* will always bear a proportion to the ignorance of the observer. The cast of a die, as regularly follows the laws of motion, as the going of a watch; yet, because we can trace the operation of those laws through the works and movements of the watch, and cannot trace them in the shaking and throwing of the die, (though the laws be the same, and prevail equally in both cases,) we call the turning up of the number of the die chance, the pointing of the index of the watch, machinery, order, or by some name which excludes chance. It is the same in those events which depend upon the will of a free and rational agent. The verdict of a jury, the sentence of a judge, the resolution of an assembly, the issue of a contested election, will have more or less of the appearance of chance, might be more or less the subject of a wager, according as we were less or more acquainted with the reasons which influenced the deliberation. The difference resides in the information of the obser-ver, and not in the thing itself; which, in all the cases proposed, proceeds from intelligence, from mind, from counsel, from design.

Now when this one cause of the appearance of chance, viz. the ignorance of the observer, comes to be applied to the operations of the Deity, it is easy to foresee how fruitful it must prove of difficul-ties, and of seeming confusion. It is only to think of the Deity to perceive, what variety of objects, what distance of time, what extent of space and action, his counsels may, or rather must, comprehend. Can it be wondered at, that, of the purposes which dwell in such a mind as this, so small a part should be known to us? It is only necessary therefore to bear in our thought, that, in proportion to the inadequateness of our information, will be the quantity, in the world, of apparent chance.

III. In a great variety of cases, and of cases comprehending numerous subdivisions, it appears, for many reasons, to be better, that events rise up by *chance*, or, more properly speaking, with the appearance of chance, than according to any observable rule whatever. This is not seldom the case even in human arrangements. Each person's place and precedency in a public meeting may be determined by *lot*. Work and labour may be *allotted*. Tasks and burthens may be *allotted*.

——Operumque laborem
Partibus æquabat justis, aut *sorte* trahebat.*

Military service and station may be *allotted*. The distribution of provision may be made by *lot*, as it is in a sailor's mess; in some cases also, the distribution of favors may be made by *lot*. In all these cases it seems to be acknowledged, that there are advantages in permitting events to chance, superior to those, which would or could arise from regulation. In all these cases also, though events rise up in the way of chance, it is by appointment that they do so.

In other events, and such as are independent of human will, the reasons for this preference of uncertainty to rule appear to be still stronger. For example, it seems to be expedient, that the period of human life should be *uncertain*. Did mortality follow any fixed rule, it would produce a security in those that were at a distance from it, which would lead to the greatest disorders, and a horror in those who approached it, similar to that which a condemned prisoner feels on the night before his execution. But, that death be uncertain, the young must sometimes die, as well as the old. Also were deaths never *sudden*, they, who are in health, would be too confident of life. The strong and the active, who want most to be warned and checked, would live without apprehension or restraint. On the other hand; were sudden deaths very frequent, the sense of constant jeopardy would interfere too much with the degree of ease and enjoyment intended for us; and human life be too precarious for the business and interests which belong to it. There could not be dependance either upon our own lives, or the lives of those with whom we were connected, sufficient to carry on the regular offices of human society. The manner, therefore, in which death is made to occur, conduces to the purposes of admonition, without overthrowing the necessary stability of human affairs.

Disease being the forerunner of death, there is the same reason for its attacks coming upon us under the appearance of chance, as there is for uncertainty in the time of death itself.

The *seasons* are a mixture of regularity and chance. They are regular enough to authorize expectation, whilst their being, in a considerable degree, irregular, induces, on the part of the cultivators of the soil, a necessity for personal attendance, for activity, vigilance, precaution. It is this necessity which creates farmers; which divides the profit of the soil between the owner and the occupier; which, by requiring expedients, by increasing employment, and by rewarding expenditure, promotes agricultural arts and agricultural life, of all modes of life the best. I believe it to be found in fact, that where the soil is the most fruitful and the seasons the most constant, there the condition of the cultivators of the earth is the most depressed. Uncertainty, therefore, has its use even to those who sometimes complain of it the most. Seasons of scarcity themselves are not without their advantages; the most conducive to health, to virtue, to enjoyment. They call forth new exertions; they set contrivance and ingenuity at work; they give birth to improvements in agriculture and œconomy; they promote the investigation and management of public resources.

Again; there are strong intelligible reasons, why there should exist in human society great disparity of wealth and station. Not only as these things are acquired in different degrees, but at the first setting out of life. In order, for instance, to answer the various demands of civil life, there ought to be amongst the members of every civil society a diversity of education, which can only belong to an original diversity of circumstances. As this sort of disparity, which ought to take place from the beginning of life, must, ex hypothesi,* be previous to the merit or demerit of the persons upon whom it falls, can it be better disposed of than by chance? *Parentage* is that sort of chance: yet it is the commanding circumstance, which in general fixes each man's place in civil life, along with every thing which appertains to its distinctions. It may be the result of a beneficial rule, that the fortunes or honors of the father devolve upon the son; and, as it should seem, of a still more necessary rule, that the low or laborious condition of the parent be communicated to his family; but, with respect to the successor himself, it is the drawing of a ticket in a lottery. Inequalities therefore of fortune, at least the greatest part of them, viz. those which attend us from our birth, and depend upon our birth, may be left, as they are left, to *chance*, without any just cause for questioning the regency of a supreme Disposer of events.

But not only the donation, when by the necessity of the case they must be gifts, but even the *acquirability* of civil advantages, ought, perhaps, in a considerable degree, to lie at the mercy of chance. Some would have all the virtuous rich, or, at least, removed from the evils of poverty, without perceiving, I suppose, the consequence, that all the poor must be wicked. And how such a society could be kept in subjection to government has not been shewn, for the poor, that is, they who seek their subsistence by constant manual labour, must still form the mass of the community; otherwise the necessary labour of life could not be carried on; the work would not be done, which the wants of mankind in a state of civilization, and still more in a state of refinement, require to be done.

It appears to be also true, that the exigencies of social life call not only for an original diversity of *external* circumstances, but for a mixture of different faculties, tastes, and tempers. Activity and contemplation, restlessness and quiet, courage and timidity, ambition and contentedness, not to say even indolence and dullness, are all wanted in the world, all conduce to the well going on of human affairs, just as the rudder, the sails, and the ballast, of a ship, all perform their part in the navigation. Now since these characters require for their foundation, different original talents, different dispositions, perhaps also different bodily constitutions; and since, likewise, it is apparently expedient, that they be promiscuously scattered amongst the different classes of society, can the distribution of talents, dispositions, and the constitutions upon which they depend, be better made than by *chance*?

The opposites of apparent chance, are constancy and sensible interposition; every degree of secret direction being consistent with it. Now of *constancy*, or of fixed and known rules, we have seen in some cases the inapplicability: and inconveniences, which we do not see, might attend their application in other cases.

Of *sensible* interposition we may be permitted to remark, that a Providence, always and certainly distinguishable, would be neither more nor less than miracles rendered frequent and common. It is difficult to judge of the state into which this would throw us. It is enough to say, that it would cast us upon a quite different dispensation from that under which we live. It would be a total and radical change. And the change would deeply affect, or perhaps subvert, the whole conduct of human affairs. I can readily believe, that, other circumstances being adapted to it, such a state might be better than our present state. It may be the state of other beings: it may be ours hereafter. But the question with which we are now concerned is, how far it would be consistent with our condition, supposing it in other respects to remain as it is? And in this question there seem to be reasons of great moment on the negative side. For instance, so long as bodily labour continues, on so many accounts, to be necessary for the bulk of mankind, any dependency upon supernatural aid, by unfixing those motives which promote exertion, or by relaxing those habits which engender patient industry, might introduce negligence, inactivity, and disorder, into the most useful occupations of human life; and thereby deteriorate the condition of human life itself.

As moral agents we should experience a still greater alteration, of which more will be said under the next article.*

Although therefore the Deity, who possesses the power of winding and turning, as he pleases, the course of causes which issue from himself, do in fact interpose to alter or intercept effects, which without such interposition would have taken place, yet is it by no means incredible, that his Providence, which always rests upon final good, may have made a *reserve* with respect to the manifestation of his interference, a part of the very plan which he has appointed for our terrestrial existence, and a part conformable with, or, in some sort, required by, other parts of the same plan. It is at any rate evident, that a large and ample province remains for the exercise of Providence, without its being naturally perceptible by us; because obscurity, when applied to the interruption of laws, bears a necessary proportion to the imperfection of our knowledge when applied to the laws themselves, or rather to the effects, which these laws, under their various and incalculable combinations, would of their own accord produce. And if it be said, that the doctrine of divine Providence,* by reason of the ambiguity under which its exertions present themselves, can be attended with no *practical* influence upon our conduct; that, although we believe ever so firmly that there is a Providence; we must prepare, and provide, and act, as if there were none; I answer, that this is admitted: and that we further alledge, that so to prepare, and so to provide, is consistent with the most perfect assurance of the reality of a Providence; and not only so, but that it is, probably, one advantage of the present state of our information, that our provisions and preparations are not disturbed by it. Or if it be still asked, of what use at all then is the doctrine, if it neither alter our measures nor regulate our conduct, I answer again, that it is of the greatest use, but that, it is a doctrine of sentiment and piety, not (immediately at least) of action or conduct; that it applies to the consolation of men's minds, to their devotions, to the excitement of gratitude, the support of patience, the keeping alive and the strengthening of every motive for endeavouring to please our Maker; and that these are great uses.

Of all views under which human life has ever been considered, the most reasonable in my judgment is that, which regards it as a state of *probation*. If the course of the world were separated from the contrivances of nature, I do not know that it would be necessary to look for any other account of it, than what, if it may be called an account, is contained in the answer, that events rise up by chance. But since the contrivances of nature decidedly evince *intention*; and since the course of the world and the contrivances of nature have the same author; we are, by the force of this connection, led to believe, that the appearance, under which events take place, is reconcileable with the supposition of design on the part of the Deity. It is enough that they be reconcileable with this supposition (and it is undoubtedly true, that they may be reconcileable, though we cannot reconcile them): the mind, however, which contemplates the works of nature, and, in those works, sees so much of means directed to ends, of beneficial effects brought about by wise expedients, of concerted trains of causes terminating in the happiest results; so much, in a word, of counsel, intention, and benevolence: a mind, I say, drawn into the habit of thought which these observations excite, can hardly turn its view to the condition of our own species, without endeavouring to suggest to itself some purpose, some design, for which the state in which we are placed is fitted, and which it is made to serve. Now we assert the most probable supposition to be, that it is a state of moral probation; and that many things in it suit with this hypothesis, which suit with no other. It is not a state of unmixed happiness, or of happiness simply: it is not a state of designed misery, or of misery simply: it is not a state of retribution: it is not a state of punishment. It suits with none of these suppositions. It accords much better with the idea of its being a condition calculated for the production, exercise, and improvement, of moral qualities, with a view to a future state, in which, these qualities, after being so produced, exercised, and improved, may, by a new and more favoring constitution of things, receive their reward, or become their own. If it be said, that this is to enter upon a religious rather than a philosophical consideration, I answer that the name of religion ought to form no objection, if it shall turn out to be the case, that the more religious our views are, the more probable they become. The degree of beneficence, of benevolent intention, and of power, exercised in the construction of sensitive beings, goes strongly in favor, not only of a creative, but of a continuing care, that is, of a ruling Providence. The degree of chance which appears to prevail in the world requires to be reconciled with this hypothesis. Now it is one thing to maintain the doctrine of Providence along with that of a future state, and another thing without it. In my opinion the two doctrines must stand or fall together. For although more of this apparent chance, may perhaps, upon other principles, be accounted for, than is generally supposed, yet a future state alone rectifies all disorders; and if it can be shewn that the appearance of disorder, is consistent with the uses of life, as a preparatory state, or that in some respects it promotes these uses, then, so far as this hypothesis may be accepted, the ground of the difficulty is done away.

In the wide scale of human condition, there is not perhaps one of its manifold diversities, which does not bear upon the design here suggested. Virtue is infinitely various. There is no situation in which a rational being is placed, from that of the best instructed Christian, down to the condition of the rudest barbarian, which affords not room for moral agency; for the acquisition, exercise, and display of voluntary qualities, good and bad. Health and sickness, enjoyment and suffering, riches and poverty, knowledge and ignorance, power and subjection, liberty and bondage, civilization and barbarity, have all their offices and duties, all serve for the *formation* of character: for, when we speak of a state of trial, it must be remembered, that characters are not only tried, or proved, or detected, but that they are generated also, and *formed*, by circumstances. The *best* dispositions may subsist under the most depressed, the most afflicted fortunes. A West Indian slave, who, amidst his wrongs, retains his benevolence, I, for my part, look upon, as amongst the foremost of human candidates for the rewards of virtue. The kind master of such a slave, that is, he, who, in the exercise of an inordinate authority, postpones, in any degree, his own interest to his slave's comfort, is likewise a meritorious character; but still he is inferior to his slave.* All however which I contend for, is, that these destinies, opposite as they may be in every other view, are both trials; and equally such. The observation may be applied to every other condition; to the whole range of the scale, not excepting even its lowest extremity. *Savages* appear to us all alike, but it is owing to the distance at which we view savage life, that we perceive in it no discrimination of character. I make no doubt, but that moral qualities, both good and bad, are called into action as much, and that they subsist in as great variety, in these inartificial societies, as they are, or do, in polished life. Certain at least it is, that the good and ill treatment, which each individual meets with, depends more upon the choice and voluntary conduct of those about him, than it does, or ought to do, under regular civil institutions, and the coercion of public laws. So again, to turn our eyes to the other end of the scale, namely, that part of it, which is occupied by mankind, enjoying the benefits of learning together with the lights of revelation,* there also, the advantage is all along *probationary*. Christianity itself, I mean the revelation of Christianity, is not only a blessing but a trial. It is one of the diversified means by which the character is exercised; and they who require of Christianity, that the revelation of it should be universal, may possibly be found to require, that one species of probation should be adopted, if not to the exclusion of others, at least to the narrowing of that variety which the wisdom of the Deity hath appointed to this part of his moral œconomy.[1]

Now if this supposition be well founded; that is, if it be true, that our ultimate, or our most permanent happiness, will depend, not upon the temporary condition into which we are cast, but upon our behaviour in it; then is it a much more fit subject of *chance* than we usually allow or apprehend it to be, in what manner, the variety of external circumstances, which subsist in the human world, is

---

[1] The reader will observe, that I speak of the revelation of Christianity as distinct from Christianity itself. The *dispensation* may already be universal. That part of mankind which never heard of Christ's name, may nevertheless be redeemed, that is, be placed in a better condition with respect to their future state, by his intervention; be the objects of his benignity and intercession, as well as of the propitiatory virtue of his passion. But this is not 'natural Theology,' therefore I will not dwell longer upon it.

distributed amongst the individuals of the species. 'This life being a state of probation, it is immaterial,' says Rousseau, 'what kind of trials we experience in it, provided they produce their effects.' Of two agents, who stand indifferent to the moral Governor of the universe, one may be exercised by riches, the other by poverty. The treatment of these two shall appear to be very opposite, whilst in truth it is the same: for, though in many respects, there be great disparity between the conditions assigned, in one main article there may be none, viz. in that they are alike trials; have both their duties and temptations, not less arduous or less dangerous, in one case than the other: so that, if the final award follow the character, the original distribution of the circumstances under which that character is formed, may be defended upon principles not only of justice but equality. What hinders, therefore, but that mankind may draw lots for their condition? They take their portion of faculties and opportunities, as any unknown cause, or concourse of causes, or as causes acting for other purposes, may happen to set them out, but the event is governed by that which depends upon themselves, the application of what they have received. In dividing the talents,* no rule was observed; none was necessary: in rewarding the use of them, that of the most correct justice. The chief difference at last appears to be, that the right use of more talents, i. e. of a greater trust, will be more highly rewarded, than the right use of fewer talents, i. e. of a less trust. And since, for other purposes, it is expedient, that there be an inequality of concredited talents here, as well, probably, as an inequality of conditions hereafter, though all remuneratory, can any rule, adapted to that inequality, be more agreeable even to our apprehensions of distributive justice, than this is?

We have said, that the appearance of *casualty*, which attends the occurrences and events of life, not only does not interfere with its uses, as a state of probation, but that it promotes these uses.

Passive virtues, of all others the severest and the most sublime; of all others, perhaps, the most acceptable to the Deity; would, it is evident, be excluded from a constitution, in which happiness and misery regularly followed virtue and vice. Patience and composure under distress, affliction, and pain; a steadfast keeping up of our confidence in God, and of our reliance upon his final goodness, at the time when every thing present is adverse and discouraging; and (what is no less difficult to retain) a cordial desire for the happiness of others, even when we are deprived of our own; these dispositions, which constitute, perhaps, the perfection of our moral nature, would not have found their proper office and object in a state of avowed retribution; and in which, consequently, endurance of evil would be only submission to punishment.

Again; one man's sufferings may be another man's trial. The family of a sick parent is a school of filial piety. The charities of domestic life, and not only these, but all the social virtues, are called out by distress. But then, misery, to be the proper object of mitigation, or of that benevolence which endeavours to relieve, must be really or apparently casual. It is upon such sufferings alone that benevolence can operate. For were there no evils in the world, but what were punishments, properly and intelligibly such, benevolence would only stand in the way of justice. Such evils, consistently with the administration of moral government, could not be prevented or alleviated, that is to say, could not be remitted in whole or in part, except by the authority which inflicted them, or by an appellate or superior authority. This consideration, which is founded in our most acknowledged apprehensions of the nature of penal justice, may possess its weight in the Divine councils. Virtue perhaps is the greatest of all ends. In human beings relative virtues form a large part of the whole. Now relative virtue presupposes, not only the existence of evil, without which it could have no object, no material to work upon, but that evils be, apparently at least, *misfortunes*; that is, the effects of apparent chance. It may be in pursuance, therefore, and in furtherance of the same scheme of probation, that the evils of life are made so to present themselves.

I have already observed that, when we let in religious considerations, we often let in light upon the difficulties of nature. So in the fact now to be accounted for, the *degree* of happiness, which we usually enjoy in this life, may be better suited to a state of trial and probation, than a greater degree would be. The truth is, we are rather too much delighted with the world, than too little. Imperfect, broken, and precarious as our pleasures are, they are more than sufficient to attach us to the eager pursuit of them. A regard to a *future* state can hardly keep its place as it is. If we were designed therefore to be influenced by that regard, might not a more indulgent system, a higher, or more uninterrupted state of gratification, have interfered with the design? At least it seems expedient, that mankind should be susceptible of this influence, when presented to them; that the condition of the world should not be such, as to exclude its operation, or even to weaken it more than it does. In a religious view (however we may complain of them in every other) privation, disappointment, and satiety, are not without the most salutary tendencies.

CHAPTER XXVII

CONCLUSION

In all cases, wherein the mind feels itself in danger of being confounded by variety, it is sure to rest upon a few strong points, or perhaps upon a single instance. Amongst a multitude of proofs, it is one that does the business. If we observe in any argument, that hardly two minds fix upon the same instance, the diversity of choice shews the strength of the argument, because it shews the number and competition of the examples. There is no subject in which the tendency to dwell upon select or single topics is so usual, because there is no subject, of which, in its full extent, the latitude is so great, as that of natural history applied to the proof of an intelligent Creator. For my part, I take my stand in human anatomy: and the examples of mechanism I should be apt to draw out from the copious catalogue which it supplies, are the pivot upon which the head turns, the ligament within the socket of the hip joint, the pulley or trochlear muscle of the eye, the epiglottis, the bandages which tie down the tendons of the wrist and instep, the slit or perforated muscles at the hands and feet, the knitting of the intestines to the mesentery, the course of the chyle into the blood, and the constitution of the sexes as extended throughout the whole of the animal creation. To these instances, the reader's memory will go back, as they are severally set forth in their places: there is not one of the number which I do not think decisive; not one which is not strictly mechanical: nor have I read or heard of any solution of these appearances, which, in the smallest degree, shakes the conclusion that we build upon them.

But, of the greatest part of those, who, either in this book or any other, read arguments to prove the existence of a God, it will be said, that they leave off only where they began; that they were never ignorant of this great truth, never doubted of it; that it does not therefore appear, what is gained by researches from which no new opinion is learnt, and upon the subject of which no proofs were wanted. Now I answer, that, by *investigation*, the following points are always gained, in favor of doctrines even the most generally acknowledged, (supposing them to be true,) viz. stability and impression. Occasions will arise to try the firmness of our most habitual opinions. And, upon these occasions, it is a matter of incalculable use to feel our foundation; to find a support in argument for what we had taken up upon authority. In the present case, the arguments upon which the conclusion rests, are exactly such, as a truth of universal concern ought to rest upon. 'They are sufficiently open to the views and capacities of the unlearned, at the same time that they acquire new strength and lustre from the discoveries of the learned.' If they had been altogether abstruse and recondite, they would not have found their way to the understandings of the mass of mankind; if they had been merely popular; they might have wanted solidity.

But, secondly, what is gained by research in the stability of our conclusion, is also gained from it in *impression*. Physicians tell us, that there is a great deal of difference between taking a medicine, and the medicine getting into the constitution. A difference not unlike which, obtains with respect to those great moral propositions, which ought to form the directing principles of human conduct. It is one thing to assent to a proposition of this sort; another, and a very different thing, to have properly imbibed its influence. I take the case to be this. Perhaps almost every man living has a particular train of thought, into which his mind falls, when at leisure from the impressions and ideas that occasionally excite it: perhaps also, the train of thought here spoken of, more than any other thing, determines the character. It is of the utmost consequence, therefore, that this property of our constitution be well regulated. Now it is by frequent or continued meditation upon a subject, by placing a subject in different points of view, by induction of particulars, by variety of examples, by applying principles to the solution of phænomena, by dwelling upon proofs and consequences, that mental exercise is drawn into any particular channel. It is by these means, at least, that we have any power over it. The train of spontaneous thought, and the choice of that train, may be directed to different ends, and may appear to be more or less judiciously fixed, according to the purpose, in respect of which we consider it: but, in a *moral view*, I shall not, I believe, be contradicted when I say, that, if one train of thinking be more desirable than another, it is that which regards the phænomena of nature with a constant reference to a supreme intelligent Author. To have made this the ruling, the habitual sentiment of our minds, is to have laid the foundation of every thing which is religious. The world from thenceforth becomes a temple, and life itself one continued act of adoration. The change is no less than this, that, whereas formerly God was seldom in our thoughts, we can now scarcely look upon any thing without perceiving its relation to him. Every organized natural body, in the provisions which it contains for its sustentation and propagation, testifies a care on the part of the Creator expressly directed to these purposes. We are on all sides surrounded by such bodies; examined in their parts, wonderfully curious; compared with one another, no less wonderfully diversified. So that the mind, as well as the eye, may either expatiate* in variety and multitude, or fix itself down to the investigation of particular divisions of the science. And in either case it will rise up from its occupation, possessed by the subject, in a very different manner, and with a very different degree of influence, from what a mere assent to any verbal proposition which can be formed concerning the existence of the Deity, at least that merely complying assent with which those about us are satisfied, and with which we are too apt to satisfy ourselves, can or will produce upon the thoughts. More especially may this difference be perceived, in the degree of admiration and of awe, with which the Divinity is regarded, when represented to the understanding by its own remarks, its own reflections, and its own reasonings, compared with what is excited by any language that can be used by others. The works of nature want only to be contemplated.* When contemplated, they have every thing in them which can astonish by their greatness: for, of the vast scale of operation, through which our discoveries carry us, at one end we see an intelligent Power arranging planetary systems, fixing, for instance, the trajectory of *Saturn*, or constructing a ring of a hundred thousand miles diameter, to surround his body, and be suspended like a magnificent arch over the heads of his inhabitants; and, at the other, bending a hooked tooth, concerting and providing an appropriate mechanism, for the clasping and reclasping of the filaments of the feather of a humming bird. We have proof, not only of both these works proceeding from an intelligent agent, but of their proceeding from the same agent: for, in the first place, we can trace an identity of plan, a connection of system, from Saturn to our own globe; and when arrived upon our globe, we can, in the second place, pursue the connection through all the organized, especially the animated, bodies, which it supports. We can observe marks of a common relation, as well to one another, as to the elements of which their habitation is composed. Therefore one mind hath planned, or at least hath prescribed a general plan for, all these productions. One Being has been concerned in all.

Under this stupendous Being we live. Our happiness, our existence, is in his hands. All we expect must come from him. Nor ought we to feel our situation insecure. In every nature, and in every portion of nature, which we can descry, we find attention bestowed upon even the minutest parts. The hinges in the wings of an *earwig*, and the joints of its antennæ, are as highly wrought, as if the Creator had had nothing else to finish. We see no signs of diminution of care by multiplicity of objects, or of distraction of thought by variety. We have no reason to fear, therefore, our being forgotten, or overlooked, or neglected.

The existence and character of the Deity, is, in every view, the most interesting of all human speculations. In none, however, is it more so, than as it facilitates the belief of the fundamental articles of *Revelation*. It is a step to have it proved, that there must be something in the world more than what we see. It is a further step to know, that, amongst the invisible things of nature, there must be an intelligent mind, concerned in its production, order, and support. These points being assured to us by Natural Theology, we may well leave to Revelation the disclosure of many particulars, which our researches cannot reach, respecting either the nature of this Being as the original cause of all things, or his character and designs as a moral governor; and not only so, but the more full confirmation of other particulars, of which, though they do not lie altogether beyond our reasonings and our probabilities, the certainty is by no means equal to the importance. The true Theist will be the first to listen to *any* credible communication of divine knowledge. Nothing which he has learnt from Natural Theology, will diminish his desire of further instruction, or his disposition to receive it with humility and thankfulness. He wishes for light: he rejoices in light. His inward veneration of this great Being, will incline him to attend with the utmost seriousness, not only to all that can be discovered concerning him by researches into nature, but to all that is taught by a revelation, which gives reasonable proof of having proceeded from him.

But, above every other article of revealed religion, does the anterior belief of a Deity, bear with the strongest force, upon that grand point, which gives indeed interest and importance to all the rest—the resurrection of the human dead. The thing might appear hopeless, did we not see a power at work adequate to the effect, a power under the guidance of an intelligent will, and a power penetrating the inmost recesses of all substance. I am far from justifying the opinion of those, who 'thought it a thing incredible that God should raise the dead;' but I admit that it is first necessary to be persuaded, that there *is* a God to do so. This being thoroughly settled in our minds, there seems to be nothing in this process (concealed and mysterious as we confess it to be,) which need to shock our belief. They who have taken up the opinion,* that the acts of the human mind depend upon *organization,* that the mind itself indeed consists in organization, are supposed to find a greater difficulty than others do, in admitting a transition by death to a new state of sentient existence, because the old organization is apparently dissolved. But I do not see that any impracticability need be apprehended even by these; or that the change, even upon their hypothesis, is far removed from the analogy of some other operations, which we know with certainty that the Deity is carrying on. In the ordinary derivation of plants and animals from one another, a particle, in many cases, minuter than all assignable, all conceivable dimension; an aura, an effluvium, an infinitesimal; determines the organization of a future body: does no less than fix, whether that which is about to be produced, shall be a vegetable, a merely sentient, or a rational being; an oak, a frog, or a philosopher; makes all these differences; gives to the future body its qualities, and nature, and species. And this particle, from which springs, and by which is determined a whole future nature, itself proceeds from, and owes its constitution to, a prior body: nevertheless, which is seen in plants most decisively, the incepted organization, though formed within, and through, and by a preceding organization, is not corrupted by its corruption, or destroyed by its dissolution; but, on the contrary, is sometimes extricated and developed by those very causes; survives and comes into action, when the purpose, for which it was prepared, requires its use. Now an œconomy which nature has adopted, when the purpose was to transfer an organization from one individual to another, may have something analogous to it, when the purpose is to transmit an organization from one state of being to another state: and they who found thought in organization, may see something in this analogy applicable to their difficulties; for, whatever can transmit a similarity of organization will answer their purpose, because, according even to their own theory, it may be the vehicle of consciousness, and because consciousness, without doubt, carries identity and individuality* along with it through all changes of form or of visible qualities. In the most general case, that, as we have said, of the derivation of plants and animals from one another, the latent organization is either itself similar to the old organization, or has the power of communicating to new matter the old organic form. But it is not restricted to this rule. There are other cases, especially in the progress of insect life, in which the dormant organization does not much resemble that which incloses it, and still less suits with the situation in which the inclosing body is placed, but suits with a different situation to which it is destined. In the larva of the libellula,* which lives constantly, and has still long to live, under water, are descried the wings of a fly, which two years afterwards is to mount into the air. Is there nothing in this analogy? It serves at least to shew, that, even in the observable course of nature, organizations are formed one beneath another; and, amongst a thousand other instances, it shews completely, that the Deity can mold and fashion the parts of material nature, so as to fulfill any purpose whatever which he is pleased to appoint.

They who refer the operations of mind to a substance totally and essentially different from matter, as, most certainly, these operations, though affected by material causes, hold very little affinity to any properties of matter with which we are acquainted, adopt, perhaps, a juster reasoning and a better philosophy; and by these the consider-ations above suggested are not wanted, at least in the same degree. But to such as find, which some persons do find, an insuperable difficulty in shaking off an adherence to those analogies, which the corporeal world is continually suggesting to their thoughts; to such, I say, every consideration will be a relief, which manifests the extent of that intelligent power which is acting in nature, the fruitfulness of its resources, the variety, and aptness, and success of its means; most especially every consideration, which tends to shew, that, in the translation of a conscious existence, there is not, even in their own way of regarding it, any thing greatly beyond, or totally unlike, what takes place in such parts (probably small parts) of the order of nature, as are accessible to our observation.

Again; if there be those who think, that the contractedness and debility of the human faculties in our present state, seem ill to accord with the high destinies which the expectations of religion point out to us, I would only ask them, whether any one, who saw a child two hours after its birth, could suppose that it would ever come to understand *fluxions*;[1] or who then shall say, what further amplification of intellectual powers, what accession of knowledge, what advance and improvement, the rational faculty, be its constitution what it will, may not admit of, when placed amidst new objects, and endowed with a sensorium, adapted, as it undoubtedly will be, and as our present senses are, to the perception of those substances, and of those properties of things, with which our concern may lie.

Upon the whole; in every thing which respects this awful, but, as we trust, glorious change, we have a wise and powerful Being, (the author, in nature, of infinitely various expedients for infinitely various ends,) upon whom to rely for the choice and appointment of means, adequate to the execution of any plan which his goodness or his justice may have formed, for the moral and accountable part of his terrestrial creation. That great office rests with him: be it ours to hope and to prepare; under a firm and settled persuasion, that, living and dying, we are his; that life is passed in his constant presence, that death resigns us to his merciful disposal.

FINIS

[1] See Search's Light of Nature,* passim.

APPENDIX
FURTHER READING

The following works address specific aspects of the authors and ideas mentioned by Paley within their historical context. Although some of these sources are readily available in most research libraries, a good number of them are indexed in speciality databases or do not have a title which reflects their relevancy to Paley's work.

Natural Theology

Ashworth, W. B., 'Christianity and the Mechanistic Universe', in D. C. Lindberg and R. L. Numbers (eds.), *When Science and Christianity Meet* (Chicago, 2003).

Barr, J., *Biblical Faith and Natural Theology: The Gifford Lectures for 1991* (Oxford, 1993).

Brooke, J. H., *Science and Religion: Some Historical Perspectives* (Cambridge, 1991).

—— and Cantor, G., *Reconstructing Nature: The Engagement of Science and Religion* (Oxford, 1998).

Chadwick, H., *The Early Church* (Harmondsworth, 1993).

Flew, A., *God and Philosophy* (London, 1966).

Gascoigne, J., 'Rise and Fall of British Newtonian Natural Theology', *Science in Context*, 2 (1988), 219–56.

Gillespie, N. C., 'Divine Design and the Industrial Revolution: William Paley's Abortive Reform of *Natural Theology*', *Isis*, 81 (1990), 214–29.

Heimann, P. M., 'Voluntarism and Immanence: Conceptions of Nature in Eighteenth Century Thought', in J. W. Yolton (ed.), *Philosophy, Religion and Science in the Seventeenth and Eighteenth Centuries* (Rochester, NY, 1990).

Hoffman, J., and Rosenkrantz, G. S., *The Divine Attributes* (Oxford, 2002).

Hooykaas, R., *Religion and the Rise of Modern Science* (Edinburgh, 1972).

Kaiser, C., *Creation and the History of Science* (Grand Rapids, Mich., 1991).

Knight, D. M., *Science and Spirituality: The Volatile Connection* (London, 2004).

Lewis, R., 'The Publication of John Wilkins's Essay (1668): Some Contextual Considerations', *Notes and Records of the Royal Society of London*, 56 (2002), 133–46.

Livingstone, D. N., 'The Historical Roots of Our Ecological Crisis: A Reassessment', *Fides et Historia*, 26 (1994), 38–55.

Louth, A., *The Origins of the Christian Mystical Tradition: From Plato to Denys* (Oxford, 1981).

McAdoo, H. R., *The Spirit of Anglicanism: A Survey of Anglican Theological Method in the Seventeenth Century* (London, 1965).

McGrath, A. E., *A Scientific Theology*, vol. i: *Nature* (Edinburgh, 2001).

Mason, S. F., 'Analogies of Thought-Style in the Protestant Reformation and Early Modern Science', *Notes and Records of the Royal Society of London*, 46 (1992), 1–21.

Paley, E., *An Account of the Life and Writings of William Paley* (Farnborough, 1970; originally the first volume of *The Works of William Paley*, London, 1825).

Pelikan, J., *Christianity and Classical Culture: The Metamorphosis of Natural Theology in the Christian Encounter with Hellenism* (New Haven, 1993).

Plantinga, A., *God, Freedom, and Evil* (Grand Rapids, Mich., 1974).

Raven, C., *Natural Religion and Christian Theology* (Cambridge, 1953).

Sharpe, A. B., 'Evil', in C. G. Herbermann, E. A. Pace, C. B. Pallen, T. J. Shahan, and J. J. Wynne (eds.), *The Catholic Encyclopedia: An International Work of Reference on the Constitution, Doctrine, Discipline, and History of the Catholic Church*, vol. v (New York, 1907).

Webb, C. C. J., *Studies in Natural Theology* (Oxford, 1915).

Rhetoric and Literature

Addison, Joseph, *Addison's Works*, ed. R. Hurd (London, 1854).

Benjamin, A. E., Cantor, G. N., and Christie, J. R. R. (eds.), *The Figural and the Literal: Problems of Language in the History of Science and Philosophy, 1630–1800* (Manchester: 1987).

Bettestin, M. C., 'Tom Jones: The Argument of Design', in H. K. Miller, E. Rothstein, and G. S. Rousseau (eds.), *The Augustan Milieu* (Oxford, 1970).

Bloom, E., and Bloom, L., *Joseph Addison's Sociable Animal: In the Market Place, On the Hustings, In the Pulpit* (Providence, RI, 1971).

—— —— and E. Leites, *Educating the Audience: Addison, Steele, & Eighteenth Century Culture* (Los Angeles, 1984).

Dear, P. (ed.), *The Literary Structure of Scientific Argument* (Philadelphia, 1991).

Fahnestock, J., *Rhetorical Figures in Science* (Oxford, 1999).

Golinski, J. V., 'Language, Discourse and Science', in R. C. Olby, G. N. Cantor, J. R. R. Christie, and M. J. S. Hodge (eds.), *Companion to the History of Modern Science* (London, 1996), 110–23.

Grafton, A., *Defenders of the Text: The Traditions of Scholarship in the Age of Science 1450–1800* (Cambridge, Mass., 1991).

Hilson, J. C., 'Hume: The Historian as Man of Feeling', in J. C. Hilson, M. M. B. Jones, and J. R. Watson (eds.), *Augustan Worlds* (Bristol, 1978), 205–22.

Howell, W. S., *Poetics, Rhetoric, and Logic: Studies in the Basic Discipline of Criticism* (London, 1975).

Jones, W. P., *The Rhetoric of Science: A Study of Scientific Ideas and Imagery in Eighteenth-Century English Poetry* (London, 1966).

Keller, E., 'Embryonic Individuals: The Rhetoric of Seventeenth-Century Embryology and the Construction of Early-Modern Identity', *Eighteenth-Century Studies*, 33 (2000), 321–48.

Knight, D. M., 'Science Fiction of the Seventeenth Century', *Seventeenth Century*, 1 (1986), 69–79.

Potkay, A., *The Fate of Eloquence in the Age of Hume* (Ithaca, NY, 1994).

Skinner, Q., *Reason and Rhetoric in the Philosophy of Hobbes* (Cambridge, 1996).

Smithers, P., *The Life of Joseph Addison* (Oxford, 1954).

Struever, N. S., 'The Conversable World: Eighteenth-Century Transformations of the Relation of Rhetoric and Truth', in D. H. Bialostosky and L. D. Needham (eds.), *Rhetorical Traditions and British Romantic Literature* (Bloomington, Ind., 1995), 233–49.

Vickers, B., 'Territorial Disputes: Philosophy *versus* Rhetoric', in B. Vickers (ed.), *Rhetoric Revalued: Papers from the International Society for the History of Rhetoric* (Binghamton, NY, 1982), 247–66.

Natural Philosophy

Alston, W. P., 'Thomas Reid on Epistemic Principles', *History of Philosophy Quarterly*, 2 (1985), 435–52.

Beeson, D., *Maupertuis: An Intellectual Biography* (Oxford, 1992).

Cantor, G. N., and Hodge, M. J. S. (eds.), *Conceptions of Ether: Studies in the History of Ether Theories, 1740–1900* (Cambridge, 1981).

Cummins, P. D., 'Reid's Realism', *Journal of the History of Philosophy*, 12 (1974), 317–40.

Daston, L., *Classical Probability in the Enlightenment* (Princeton, 1988).

Dick, S. J., *Plurality of Worlds: The Origins of the Extraterrestrial Life Debate from Democritus to Kant* (Cambridge, 1982).

Ducharme, H., 'Personal Identity in Samuel Clarke', *Journal of the History of Philosophy*, 24 (1986), 359–83.

Force, J. E., 'Samuel Clarke's Four Categories of Deism, Isaac Newton, and the Bible', in R. Popkin (ed.), *Scepticism in the History of Philosophy* (London, 1996), 53–74.

Garrett, A. V. (ed.), *Animal Rights and Souls in the Eighteenth Century* (6 vols., Bristol, 2000).

Gay, J. H., 'Matter and Freedom in the Thought of Samuel Clarke', *Journal of the History of Ideas*, 24 (1963), 85–105.

Grabiner, J. V., 'A Mathematician Among the Molasses Barrels: Maclaurin's Unpublished Memoir on Volumes', *Proceedings of the Edinburgh Mathematical Society*, 39 (1996), 193–240.

Hacking, I., *The Taming of Chance* (Cambridge, 1990).

Harman, P. M., *Energy, Force, and Matter: The Conceptual Development of Nineteenth-Century Physics* (Cambridge, 1982).

—— 'Dynamics and Intelligibility: Bernoulli and Maclaurin', in R. S. Woolhouse (ed.), *Metaphysics and Philosophy of Science in the Seventeenth and Eighteenth Centuries* (London, 1988), 213–26.

Knight, D. M., *Atoms and Elements: A Study of Theories of Matter in England in the Nineteenth Century* (London, 1967).

Lowe, E. J., *Locke on Human Understanding* (London, 1995)

Murdoch, P., 'An Account of the Life and Writings of the Author', in Colin Maclaurin, *An Account of Sir Isaac Newton's Philosophical Discoveries* (Edinburgh, 1748).

O'Connor, T., 'Thomas Reid on Free Agency', *Journal of the History of Philosophy*, 32 (1994), 605–22.

Poovey, M., *A History of the Modern Fact: Problems of Knowledge in the Sciences of Wealth of Society* (Chicago, 1998).

Richards, R. J., *The Romantic Conception of Life: Science and Philosophy in the Age of Goethe* (Chicago, 2002).

Rossi, P., *Logic and the Art of Memory*, trans. Stephen Clucas (Chicago, 2000), 145–75.

Scarre, G. F., *Utilitarianism* (London, 1996).

Shapin, S., *A Society History of Truth: Civility and Science in Seventeenth-Century England* (Chicago, 1994).

Shapiro, A. E., *Fits, Passions, and Paroxysms: Physics, Method, and Chemistry and Newton's Theories of Colored Bodies and Fits of Easy Reflection* (Cambridge, 1993).

Stewart, L., 'Samuel Clarke, Newtonianism and the Factions of Post-Revolutionary England', *Journal of the History of Ideas*, 42 (1981), 53–71.

Stewart, M. A., 'The Scottish Enlightenment', in S. Brown (ed.), *British Philosophy and the Age of Enlightenment* (London, 2001), 274–308.

Turnbull, H. W., *Bicentenary of the Death of Colin Maclaurin (1698–1746), Mathematician and Philosopher, Professor of Mathematics in Marischal College, Aberdeen (1717–1725)* (Aberdeen, 1951).

Vailati, E., 'Clarke's Extended Soul', *Journal of the History of Philosophy*, 28 (1990), 213–28.

Wolterstorff, N., *Thomas Reid and the Story of Epistemology* (Cambridge, 2001).

Yolton, J., *A Locke Dictionary* (Oxford, 1993).

Natural History

Anderson, L., *Charles Bonnet and the Order of the Known* (London, 1982).

Bibliothèque Publique et Universitaire, *Catalogue de la Correspondance de Charles Bonnet: Conservée à la Bibliothèque de Genève* (Geneva, 1993).

Bonnet, Charles, *Science against the Unbelievers: The Correspondence of Bonnet and Needham, 1760–1780*, ed. R. G. Mazzolini and S. A. Roe (Oxford, 1986).

Bowler, P. J., *Evolution: The History of an Idea* (London, 1989).

Cain, J. A., 'John Ray on "Accidents" ', *Archives of Natural History*, 23 (1996), 343–68.

—— 'Thomas Sydenham, John Ray, and Some Contemporaries Species', *Archives of Natural History*, 26 (1999), 85–100.

Cooper, N. (ed.), *John Ray and His Successors: The Clergyman as Biologist* (Braintree, 1999).

Cuvier, G., and Thouars, A. D., 'Notice of Ray, by Cuvier and Aubert Dupetit Thouars (from the *Biographie Universelle*), translated by G. Busk', in E. Lankester (ed.), *Memorials of John Ray* (London, 1846).

Davies, G. L., *The Earth in Decay* (London, 1969).

Derham, W., 'Select Remains and Life of Ray', in *Memorials of John Ray*, ed. Edwin Ray Lankester (London, 1846).

Eddy, M. D., 'Geology, Mineralogy and Time in John Walker's University of Edinburgh Natural History Lectures', *History of Science*, 39 (2001), 95–119.

—— 'Scottish Chemistry, Classification and the Early Mineralogical Career of the "Ingenious" Rev. Dr. John Walker', *British Journal for the History of Science*, 35 (2002), 382–422.

—— 'Scottish Chemistry, Classification and the Late Mineralogical Career of the "Ingenious" Professor John Walker (1779–1803)', *British Journal for the History of Science*, 37 (2004), 373–99.

Foucault, M., *The Order of Things: An Archaeology of the Human Sciences* (New York, 1971/1994).

Gilson, E., *From Aristotle to Darwin and Back Again: A Journey in Final Causality, Species, and Evolution*, trans. John Lyon (Notre Dame, Ind., 1984).

Ginger, J., *The Notable Man: The Life and Times of Oliver Goldsmith* (London, 1977).

Glass, B., 'Heredity and Variation in the Eighteenth-Century Concept of Species', in B. Glass *et al.* (eds.), *Forerunners of Darwin, 1745–1859* (Baltimore, 1959), 144–72.

Huxley, T. H., 'Evolution in Biology', *Darwiniana* (London, 1893).

Kafker, F. A., 'William Smellie's Edition of the *Encyclopaedia Britannica*', in Frank A. Kafker (ed.), *Notable Encyclopedias of the Late 18th Century: Eleven Successors of the Encyclopédie* (Oxford, 1994), 145–254.

Kerr, R., *Memoirs of the Life, Writings, and Correspondence of William Smellie* (Edinburgh, 1811).

Keynes, G., *John Ray: A Bibliography* (Amsterdam, 1976).

King-Hele, Desmond, *Erasmus Darwin: A Life of Unequalled Achievement* (London, 1999).

Knight, D. M., *Natural Science Books in English, 1600–1900* (London, 1972)

—— *Ordering the World* (London, 1981).

Knoefel, P. K., 'The Astronomical and Meteorological Observatory of the Florentine Royal Museum of Physics and Natural History', *Physis: Rivista Internazionale di Storia della Scienza*, 24 (1982), 399–422.

Lyon, J., and Sloan, Philip R., *From Natural History to the History of Nature: Readings from Buffon and His Critics* (London, 1981).

Pilkington, A. E., ' "Nature" as Ethical Norm in the Enlightenment', in L. J. Jordanova (ed.), *Languages of Nature: Critical Essays on Science and Literature* (London, 1986).

Porter, R., 'Creation and Credence', in B. Barnes and S. Shapin (eds.), *Natural Order: Historical Studies of Scientific Culture* (Beverly Hills, Calif., 1979).

Prior, J., *The Life of Goldsmith, M. B. from a Variety of Original Sources* (2 vols. London, 1837).

Raven, C., *John Ray Naturalist: His Life and Works* (Cambridge, 1942).

Reill, P. H., *Vitalizing Nature in the Enlightenment* (London, 2005).

Roger, J., 'The Living World', in G. S. Rousseau and R. Porter (eds.), *The Ferment of Knowledge* (Cambridge, 1980).

Rolfe, W. D., 'Breaking the Great Chain of Being', in W. F. Bynum and R. Porter (eds.), *William Hunter and the Eighteenth-Century Medical World* (Cambridge, 1985).

Rossi, P., *The Dark Abyss of Time: The History of the Earth & the History of Nations from Hooke to Vico*, trans. Lydia G. Cochrane (Chicago, 1984).

Rousseau, G. S. (ed.), *Goldsmith: The Critical Heritage* (London, 1974).

Schiebinger, L., 'Gender and Natural History', in N. Jardine, J. Secord, and E. Spary (eds.), *Cultures of Natural History* (Cambridge, 1996).

Sells, A. L., *Oliver Goldsmith: His Life and Works* (London, 1974).

Sloan, P. R., 'John Locke, John Ray, and the Problem of the Natural System', *Journal of the History of Biology*, 5 (1972), 1–53.

Smith, C. U. M., and Arnott, Robert (eds.), *The Genius of Erasmus Darwin* (Aldershot, 2005).

Stearn, W. T., 'John Wilkins, John Ray and Carl Linnaeus', *Notes and Records of the Royal Society of London*, 40 (1985–6), 101–23.

—— *Botanical Latin: History, Grammar, Syntax, Terminology and Vocabulary* (Portland, Oreg., 2000).

Uglow, J., *The Lunar Men* (London, 2002).

Wiley, B., *The 18th Century Background: Studies on the Idea of Nature in the Thought of the Period* (London, 1980).

Wood, P., 'The Science of Man', in N. Jardine *et al.* (eds.), *Cultures of Natural History* (Cambridge, 1996).

Medicine

Adelmann, H. B., *Marcello Malpighi and the Evolution of Embryology* (Ithaca, NY, 1966).

Beth, E. W., 'Nieuwentyjt's Significance for the Philosophy of Science', *Synthese*, 9 (1953–5), 447–53.

Cuttingham, A., 'Sydenham versus Newton: The Edinburgh Fever Dispute of the 1690s between Andrew Brown and Archibald Pitcairne', *Medical History Supplement*, 1 (1981), 71–98.

Duchesneau, F., 'Vitalism in Late Eighteenth-Century Physiology: The Cases of Barthez, Blumenbach and John Hunter', in W. F. Bynum and R. Porter (eds.), *William Hunter and the Eighteenth-Century Medical World* (Cambridge, 1985), 259–95.

Eddy, M. D., 'Set in Stone: The Medical Language of Mineralogy in Scotland', in D. M. Knight and M. D. Eddy (eds.), *Science and Beliefs: From Natural Philosophy to Natural Science, 1700–1900* (Aldershot, 2005).

Guerrini, A., 'James Keill, George Cheyne, and Newtonian Philosophy', *Journal for the History of Biology*, 18 (1985), 247–66.

Holmes, F. L., *Eighteenth-Century Chemistry as an Investigative Enterprise* (Berkeley, 1989).

Houston, R. A., 'Madness and Gender in the Long Eighteenth Century', *Social History*, 27 (2002), 309–26.

Jacyna, L. S., 'Immanence or Transcendence: Theories of Life and Organization in Britain, 1790–1835', *Isis*, 74 (1983), 311–29.

—— 'Images of John Hunter in the Nineteenth Century', *History of Science*, 21 (1983), 85–108.

King, R., 'John Hunter and the Natural History of Human Teeth: Dentistry, Digestion, and the Living Principle', *Journal of the History of Medicine and Allied Sciences*, 49 (1994), 504–20.

Knight, D. M., 'The Vital Flame', *Ambix*, 32 (1976), 5–15.

Knoefel, P. K., 'Felice Fontana on Generation', *Perspectives in Biology and Medicine*, 23 (1979), 70–82.

—— 'A Barometer and a Thermometer of Felice Fontana', *Annali dell'Istituto e Museo di Storia della Scienza di Firenze*, 5 (1980), 55–67.

—— *Felice Fontana, 1730–1805: An Annotated Bibliography* (Trento: 1980).

—— 'Felice Fontana on Poisons', *Clio Medica*, 15 (1981), 35–65.

—— 'Felice Fontana on Animal Chemistry', *Atti della Accademia roveretana degli Agiati di scienze, lettere e arti*, 33 (1983/4), 159–62.

—— *Felice Fontana: Life and Works* (Trento, 1984).

Maehle, A. H., *Drugs on Trial: Experimental Pharmacology and Therapeutic Innovation in the Eighteenth Century* (Amsterdam, 1999).

Malpighi, M., *The Correspondence of Marcello Malpighi*, ed. H. B. Adelmann (5 vols., Ithaca, NY, 1975).

Needham, N. J. T. M., *A History of Embryology* (Cambridge, 1934).

Porter, R., 'Love, Sex and Madness in Eighteenth-Century England', *Social Research*, 53 (1986), 211–42.

—— *The Greatest Benefit to Mankind: A Medical History of Humanity from Antiquity to the Present* (London, 1999), chapter 10.

—— *Madness: A Brief History* (Oxford, 2002).

—— and Hall, L., *The Facts of Life: The Creation of Sexual Knowledge in Britain, 1650–1950* (New Haven, 1995).

Roe, S. A., *Matter, Life, and Generation: Eighteenth-Century Embryology and the Haller–Wolff Debate* (Cambridge, 1981).

Tröhler, U., *'To Improve the Evidence of Medicine': The 18th Century British Origins of a Critical Approach* (Edinburgh, 2000).

Wilson, P., 'An Enlightenment Science? Surgery and the Royal Society', in R. Porter (ed.), *Medicine in the Enlightenment* (Amsterdam, 1995).

Paley and the Nineteenth Century

Bowler, P. J., 'Malthus, Darwin, and the Concept of Struggle', *Journal of the History of Ideas*, 37 (1976), 631–50.

Cantor, G., and Shuttleworth, S. (eds.), *Science Serialized: Representations of the Sciences in Nineteenth-Century Periodicals* (Cambridge, Mass., 2004).

—— Dawson, G., Gooday, G., Noakes, R., Shuttleworth, S., and

Topham, J. R. (eds.), *Science in the Nineteenth-Century Periodical: Reading the Magazine of Nature* (Cambridge, 2004).

Desmond, A., *The Politics of Evolution: Morphology, Medicine, and Reform in Radical London* (Chicago, 1992).

Fyfe, A., 'The Reception of William Paley's "Natural Theology" in the University of Cambridge', *British Journal for the History of Science*, 30 (1997), 321–35.

—— 'Publishing and the Classics: Paley's *Natural Theology* and the Nineteenth-Century Scientific Canon', *Studies in the History and Philosophy of Science*, 33 (2002), 433–55.

—— *Science and Salvation: Evangelical Popular Science Publishing in Victorian Britain* (Chicago, 2004).

Gould, S. J., 'Nonmoral Nature', *Natural History*, 91 (1982), 19–26.

Heavner, E., 'Malthus and the Secularization of Political Ideology', *History of Political Thought*, 17 (1996), 408–30.

Livingstone, D. N., *Darwin's Forgotten Defenders: The Encounter Between Evangelical Theology and Evolutionary Thought* (Grand Rapids, Mich., 1987).

—— 'Replacing Darwinianism and Christianity', in D. C. Lindberg and R. Numbers (eds.), *When Science and Christianity Meet* (Chicago, 2003), 183–202.

McIver, T. *Anti-Evolution: A Reader's Guide to Writings before and after Darwin* (Baltimore, 1992).

Ospovat, D., 'Darwin after Malthus', *Journal of the History of Biology*, 12 (1979), 211–30.

—— *The Development of Darwin's Theory: Natural History, Natural Theology and Natural Selection, 1838–1859* (Cambridge, 1995).

Radick, G., 'Is the Theory of Natural Selection Independent from its History?', in J. Hodge and G. Radick (eds.), *The Cambridge Companion to Darwin* (Cambridge, 2003).

Schwartz, J. S., 'Charles Darwin's Debt to Malthus and Edward Blyth', *Journal of the History of Biology*, 7 (1974), 301–18.

Topham, J. R., 'Beyond the "Common Context": The Production and Reading of the Bridgewater Treatises', *Isis*, 89 (1998), 233–62.

—— 'Science, Natural Theology, and Evangelicalism in the Early Nineteenth Century: Thomas Chalmers and the Evidence Controversy', in D. N. Livingstone, D. G. Hart, and M. A. Knoll (eds.), *Evangelicals and Science in Historical Perspective* (Oxford, 1999), 142–74.

—— 'A View from the Industrial Age', *Isis*, 95 (2004), 431–42.

Viner, J., *The Role of Providence in the Social Order* (Philadelphia, 1972).

Von Sydow, M., 'Charles Darwin: A Christian Undermining Christianity?', in D. M. Knight and M. D. Eddy (eds.), *Science and Beliefs: From Natural Philosophy to Natural Science* (Aldershot, 2005), 141–56.

EXPLANATORY NOTES

3 *Shute Barrington*: (1734–1826), politician, priest, and natural philosopher. Barrington came from an influential family (his brother was the well-known naturalist Daines Barrington, 1727/8–1800) and was made Bishop of Durham in 1791. Paley lived near Sunderland but still had close ties with Durham's Cathedral and law courts and this is how he was able to foster his relationship with Barrington. For a helpful view of clergy in the Church of England during this period, see Michael Hinton, *The Anglican Parochial Clergy* (London, 1994).

make up my works into a system: though *Natural Theology* was the last of Paley's books, he wanted it to be read first. He then wanted the remaining three to be read in the following order: *A View of the Evidences of Christianity* (1794), *Horae Paulinae* (1790), and *Principles of Moral and Political Philosophy* (1785). Paley held that all of the books all added up to one system of philosophy.

4 *Bishop Wearmouth*: a town located at the mouth of the River Wear on the north-east coast of England. Paley's parsonage was located here. Though it was much nearer to Newcastle and Durham, his parish was under the jurisdiction of the Bishop of Carlisle.

7 *watch*: clocks and watches were used frequently in early modern natural theology arguments. Though Paley's analogy is probably the most eloquent, other versions appeared in the works of other British thinkers like Robert Boyle (1627–91), William Derham (1657–1735), and Joseph Priestley (1733–1804).

8 *mechanism*: *Natural Theology* was published in an era when the word 'mechanism' was applied to many processes and/or objects, four prominent examples being: (1) the complex inner workings of cathedral clocks or the newly invented chronometer, (2) the human body (particularly the shape of the body and the movement of joints), (3) Newtonian conceptions of planetary rotation around the sun, and (4) the hydraulic machines like steam engines and water-driven mills of the early Industrial Revolution. The many definitions of this word allowed Paley to use it in a variety of different ways throughout his argument.

9 *a principle of order*: the early modern period was fascinated with conceptions of natural 'order'. Whereas natural philosophy appealed to Newton's laws of motion, natural history and medicine developed large classification systems that were either based on the three kingdoms of nature (mineral, vegetable, and animal) or on diseases found in the human body.

11 *a stream of water ground corn*: many of the examples that Paley uses in his argument are taken directly from the new machines and processes of the first Industrial Revolution. This example, which refers to the stream of

water that turns the wheel used to grind grain, is comparing two mechanical objects: a watch and a watermill.

12 *There cannot be design without a designer*: a clear formulation of Paley's intent to use teleological argumentation in addition to the cosmological reasoning mentioned a few pages earlier. This means that *Natural Theology* is a hybrid between two different methods used to address the existence of God.

14 *metaphysics*: at the end of the eighteenth century, the word 'metaphysics' was often used to refer to how one acquired and cognized knowledge. It was strongly linked to the theory of ideas propounded by John Locke (1632–1704), and it covered topics that are addressed today by the philosophy of mind.

15 *atheism*: generally a term ascribed to a person who did not accept the Ten Commandments of the Old Testament, and sometimes to one who did not believe in the Resurrection.

16 *comparing a single thing with a single thing: an eye, for example, with a telescope*: using the eye to illustrate divine design had been practised since the ancient Greeks and comparing an eye to a telescope was commonplace in the anatomy textbooks of the seventeenth and eighteenth centuries. Throughout *Natural Theology*, Paley uses analogies to infer that many of his statements or premises about the human body or natural world are valid. This sort of reasoning was used in many of the books and journals that were published in Paley's lifetime.

such laws being fixed: Paley is referring to the laws of optics (the study of light) and not 'laws' in a larger causal sense. Like his use of mechanism, Paley often uses the word 'law' ambiguously and it is sometimes unclear whether he is referring to a specific physical law (gravitation, for example) or the general notion of natural or moral law.

17 *an automatic statue*: automatons were statues that moved because they contained mechanical works inside them that were more complicated versions of the gears used to animate the hands of a clock. By Paley's time, automatons had become quite complex, ranging from a chess-playing Turk to silver swans with moving heads and feet. Automatons were well known to aristocratic patrons (both in Britain and Continental Europe), who often collected them or went to see them in travelling shows.

experience and observation demonstrate: during the last part of the eighteenth century in Britain there was a revival of empiricism and a strong reaction against theoretical speculation. One of the main causes for this situation was that the British political and intellectual hierarchy felt that the French Revolution (1789) had been caused indirectly by French materialistic speculation that sought to eliminate the moral guidance of the religious principles found in either the Bible or organized religion. This led to a revival in works that claimed that they followed the empirical tradition of Francis Bacon (1561–1626) and Locke, two of England's most readily identifiable philosophers. Paley, an Archdeacon in the Church of England, was therefore keen to emphasize that his argument was based on personal experience and observation.

17 *camera obscura*: a precursor of the modern film projector. It was a small, ventilated box that contained a hole for a painted glass slide in one of its sides. When a candle was placed in the box (after the lights had been dimmed), light shot out through the slide and projected an image on the wall.

18 *the adaptation of the organ*: here, 'adaptation' is being used to describe the process by which muscles move the different parts of the eye, namely the iris (which controls the intake of light) and the lens (which effects focus and magnification). The 'sagacious optician' was John Dollard.

19 *furlongs*: one-eighth of an English mile, exactly 220 yards.

20 *refraction of light ... had long formed a subject of enquiry and conjecture*: the movement of light and its effect upon the eye (optics) was not a new subject. It had been discussed by the ancient Greeks and subsequently by Arabic and Western scholars throughout the Middle Ages. During the Renaissance it was applied both to natural philosophy and the arts in works like Leon Battista Alberti's *On Painting* (published as *De Pictura* in 1435). By the time of Isaac Newton (1642–1727), optics was a standard subject in mathematics and natural philosophy.

21 *Harrison's contrivance*: John Harrison (1693–1776) was a watchmaker (horologist) who spent the better part of his career designing a small watch that could keep accurate time on the long voyages taken by the British Navy. In 1765, after many trials, he produced a 'chronometer' that lost only one second per day. This became the prototype for the watches used by the Navy during the last decades of the eighteenth century.

23 *leagues*: a league is roughly three miles.

Sturmius: Johann Christophorus Sturm, *Mathesis Enucleata: Or, The Elements of the Mathematicks* (1700). Sturm (1635–1703) was Professor of Philosophy and Mathematics in the University of Altorf.

coatimondi: a species of Brazilian raccoon.

Mem. Acad. Paris, 1701: *The Memoirs of the Natural History of Animals* (London, 1701) is a translation of *Mémoires pour Servir à l'Histoire Naturelle des Animaux*. The French version was edited by the Académie des Sciences in 1671. A reprint appeared in 1687 and the first English edition came out in 1701. Paley also cites a 1687 edition.

Heister: Lorenz Heister, *A compendium of anatomy. Containing a short but perfect view of all the parts of humane bodies* (London, 1721). Heister (1683–1758) was a German surgeon and anatomist who had studied under Frederik Ruysch (1638–1731).

24 *scientific*: the word 'science' in the early modern period was defined somewhat differently than in the nineteenth century and later. In the Middle Ages the Latin word *scientium* was used to describe the

systematic ordering or acquisition of knowledge. When Paley wrote *Natural Theology*, 'science' still retained this definition and it was applied not only to topics like physics and chemistry, but also to law and philosophy.

nictitating membrane: a membrane that can be drawn over the eye for protection (like an eyelid).

lachrymal humor: tears or mucus produced to protect the eye.

25 *Phil. Trans. 1796*: Paley kept his argument up to date by referencing recent editions of the *Philosophical Transactions of the Royal Society of London*. Throughout *Natural Theology*, he cites papers from the 1796 edition, so he most probably had a bound copy on hand in his vicarage. He also cites scientific and medical publications from the past two hundred years and this was common practice at the time (especially for anatomy and mathematics).

26 *cassowary*: a flightless bird with a striking blue (sometimes purple) head and black feathers that lives in the tropics.

27 *God . . . and to work his ends within those limits*: throughout his career, Paley was accused of being a Deist – a charge that he vehemently rejected. Deism held that the universe was guided by natural laws that had been set in motion by God in a distant point in the past. In this line of reasoning, divine intervention (or revelation) was not really needed and divinity could be inferred from examples taken from the natural world in a process quite similar to that used by natural theology.

28 *The ear*: like the eye, the ear was a common example cited in natural theology arguments during the Enlightenment.

29 *stapes . . . membrana tympani*: the latter is the eardrum and the former is one of the bones in the inner ear.

to propagate the impulse in a direction towards the brain: since most gross anatomical features of the body had been catalogued by 1600, it was known that the nerves played a role in carrying sensation through the body. A widely held view (famously articulated by Locke) was that there was some sort of 'pulse' that carried the sensation to the brain. This was supported by eighteenth-century experiments in which the legs of dead frogs were made to move by the application of an electric current. However, many of these theories were attached to philosophical and religious commitments on the nature of matter itself and this is why Paley does not explore the topic in detail.

pneumatic principles: pneumatics was an early modern branch of chemistry that concentrated on the composition and qualities of air (and sometimes combustion).

30 *Mr Everard Home*: [Sir] Everard Home (1756–1832), a surgeon, trained by the influential Scottish anatomist William Hunter (1728–93). Home published several texts used by London surgeons, two being *A Dissertation on the Properties of Pus* (1788) and *Practical Observations on the*

Treatment of Ulcers on the Legs (1797). Additionally, he wrote more than a hundred articles in the *Philosophical Transactions*.

32 *Can any distinction ... between the producing watch, and the producing plant?*: earlier, Paley concentrated mainly on the eye and the ear, two parts of the body that could be described mechanistically so that they may be compared with the construction of a watch. In this chapter, he pushes the comparison further by asking the reader to accept his analogy between a watch and a *plant*. Later in the work he assumes that his reader has accepted the watch/plant analogy. This method is used throughout the book for a wide variety of analogies.

33 *From plants we may proceed to oviparous animals; from seeds to eggs*: oviparous animals lay eggs (ducks and chickens, for example). By comparing a seed to an egg, Paley is extending the watch metaphor from plants to animals.

34 *animals which bring forth their young alive*: the watch metaphor has reached its final destination: humans.

35 *all the organized parts of the works of nature*: Paley extends his watch metaphor to any object or phenomenon in the world that is 'organized', that is, composed in a manner that might suggest divine order.

occasional irregularities: from Antiquity onwards one of the reoccurring counterexamples used to challenge teleological arguments was the issue of pain, perceived disorder, or the existence of evil (theodicy). Several decades before Paley published *Natural Theology*, the Scottish philosopher David Hume summarized the challenge in part V of his *Dialogues Concerning Natural Religion* (1779): 'Many worlds have been botched and bungled, throughout eternity, ere this system was struck out: Much labour lost; many fruitless trials made.' Indeed, Paley's comments on 'irregularities' have often been interpreted as a response to Hume's writings. However, Hume's thoughts on this matter, though known to most philosophers, did not become influential until later in the nineteenth century (particularly with T. H. Huxley's biography, published in 1878) and there was no pressing need for Paley to respond directly to Hume *per se*. Paley's comments on instances where the watch may 'go wrong' need to be considered in light of much older teleological challenges advanced and/or criticized in the Bible (particularly in the books of Job, Jeremiah and Ecclesiastes), or in the writings of classical thinkers like Democritus, Epicurus, or Cicero.

36 *When the argument respects his attributes*: the attributes of God, i.e. the qualities that can be attributed to God based upon the texts of the Bible and then subsequent reflection upon the natural world, formed the backbone of Paley's own perception of how God (creator) relates to nature (creation). Though Paley does not state this predisposition up front, the importance of the attributes of God become much more apparent later in *Natural Theology*, especially in Chapters XXII to XXVI.

lymphatic system: a network of connected glands (nodes) and vessels (capillaries) that stretch throughout the human body and contain lymph. In Paley's time, lymph was understood as to be 'a fine fluid, separated in the body from the mass of blood, and contained in peculiar vessels'. See 'Lymph', in *Encyclopaedia Britannica* (1771).

38 *I desire no greater certainty in reasoning . . . the natural world*: Paley has already demonstrated on p. 35 that he is willing to use probability to support his argument. Here it must be noted that Paley's definition of chance is stated just below 'the operation of causes without design'.

39 *which the lapse of infinite ages has brought into existence*: since the ancient Greeks there had been many different opinions about the age of the earth, the solar system, or even matter. In Paley's day one of the main theorists on this topic was Georges-Louis Leclerc, Comte de Buffon (1707–88), a French aristocrat who popularized his ideas in two books: *Les Époques de la nature* (1788) and *Histoire naturelle* (revised in multiple editions from 1749 until he died in 1788; Paley does not indicate which edition he used). Buffon suggested that the planets were originally hot masses of matter that a comet had blasted off from the sun. As the earth had cooled, Buffon further hypothesized, microscopic organisms had spontaneously formed and then changed into plants and animals. The one thing that was needed for this theory to be plausible was a long span of time, a concept that is now called 'historical time'. In his published work, Buffon argued that the world had to be at least 70,000 years old. Though this figure seems quite conservative when compared to later Darwinian theories, the empirical mindset of late-eighteenth-century Britain (especially in Scotland) often allowed natural philosophers to dismiss Buffon's ideas because they were inconsistent with laboratory-based chemistry and mineralogy.

40 *stocking-mills*: knitting machines, sometimes called stocking-looms. See note to p. 52.

fish-skin: a form of sandpaper; usually a type of sharkskin called chagrin (or shagreen).

43 *Euclid's Elements or Simpson's Conic Sections*: Euclid, *Elements of the Conic Sections*, trans. A. Marshall (Edinburgh, 1775). Euclid's *Elements* was a common geometry text at this time. Robert Simpson (1687–1768) was Professor of Mathematics at the University of Glasgow.

45 *hydraulic machine*: a water pump.

46 *an argument separately supplied . . . cumulative in the fullest sense of the term*: this is one of the clearest formulations of Paley's method of argumentation. He states that *Natural Theology* is not a work of logic, rather, it is synthetic, inductive, and selective. Such a method was not uncommon and allowed him to draw from techniques used in classical rhetoric.

47 *the distinction, here proposed . . . of animals and vegetables*: such a distinction implicitly allows Paley to focus on anatomy (primarily that of

animals, but also, to a lesser extent, on plants), a well-established subject matter used in natural theology. Likewise, he concentrates more on morphology (parts of the body) and less on physiology (bodily processes like circulation and respiration); this is most probably because his knowledge of experimental chemistry was limited and because he did not want to introduce the chemical theology proposed by Robert Boyle's *Some Physico-Theological Considerations about the Possibility of the Resurrection* (London, 1675) and Joseph Priestley's *Disquisitions Relating to Matter and Spirit* (London, 1777).

47 *fluid, gaseous, elastic, electrical*: qualities of matter traditionally associated with chemistry during the seventeenth and eighteenth centuries.

48 *automaton on the Strand*: the Strand in London was a street full of map, book and instrument shops. Such a crowded area was an excellent place to set up an automaton display.

magnetic effluvium: in early modern chemistry, magnetism, like air and electricity, was often described as being a 'fluid'.

nervous agency: for Locke's remarks on the nervous impulse, see note to p. 29.

49 *voluntary motion, of irritability, of the principle of life, of sensation, of animal heat*: theories used in philosophy and medicine that sought to explain the cause of animation and thought in humans and other life forms. *Voluntary motion* was of central importance to the interaction of the body and mind and as such it was related to the larger dualist tradition that stretched back through Descartes, Neoplatonism, St Augustine, and Plato's theory of forms. *Sensation* was a term closely associated with John Locke's argument that the human mind was a blank slate at birth and that it filled up with ideas as life went along. These ideas originated from data communicated to the brain from the five sense organs (see note to pp. 14 and 29). The 'principle of life', 'irritability', and the 'animal heat' were theories that were often employed by materialists, i.e. those who sought to explain life and/or thought purely in terms of material composition. Though the concept of 'irritability' had been raised as early as Galen's *De motu musculorum* (second century AD), it was brought to centre stage during the Enlightenment by the German anatomist Albrecht von Haller in his *De partibus corporis humani sensilibus et irritabilibus* (1752). He argued the contraction of muscle fibres (irritability) could be used to explain physiological processes. Haller's irritability theory was taken in many directions and some even proposed that it could explain the act of thinking itself (thus eliminating the category of the 'mind' or even the 'soul'). For the most part, however, irritability was linked to theories relevant to curing and preventing disease, as can be seen in Robert Whytt's *Observations on the Sensibility and Irritability of the Parts of Men and other Animals* (1755). *The principle of life*, simply stated, was the 'principle' or 'vital force' that gave life to the material form of the human body. Questions on this topic were ancient in origin;

notably Aristotle's *De anima* (mid-fourth century BC) set forth a highly empirical analysis that inspired or enraged scholars for the next two millennia. In response to the materialistic connotations of Haller's irritability thesis, a mid-eighteenth-century movement called 'vitalism' emerged that sought to explain the animating 'force' or 'spark' in living matter. One of the better known vitalists was Dr John Hunter. Throughout his writings, he argued that the 'vital spark' of life was superadded to matter (much like Isaac Newton's belief about gravity and matter) and was responsible for creating *animal heat*, that is, the heat produced by a living body. However, Hunter's writings were taken in many directions and by the 1790s some scholars used his work to argue that the vital spark was inherent in matter itself. A conservative backlash followed, particularly in the wake of John Thelwall's *An Essay toward a Definition of Animal Vitality . . . in which Several of the Opinions of the Celebrated John Hunter are Examined and Controverted* (London: 1793). Overall, all the theories about the mind and body mentioned above were hotly disputed by Paley's contemporaries and he wisely sidestepped the debate by concentrating on anatomical mechanics.

menstrua: fluids.

50 *than a caustic alkali or mineral acid, than red precipitate or aqua fortis itself*: pre-Lavoisierian chemistry divided chemical substances into six categories: salts, earths, metals, inflammables, water, and airs. Salts were subdivided into alkalis and acids. The chemical substances mentioned by Paley above were highly corrosive (hence the word 'caustic'). The modern equivalent of aqua fortis is nitric acid (HNO_3).

51 *No chymical election, no chymical analysis*: for most of the eighteenth century medical and industrial chemistry used principle-based 'chymistry' (see previous note). It was believed that there were two basic attractions that held substances together: chemical affinity and mechanical affinity. The former was the stronger and more permanent; the latter was weaker and more easily disrupted. Étienne-François Geoffroy (1672–1731) and later chemists like Torbern Bergman (1735–84) proposed tables that tried to explain the strength of these affinities and these arrangements were in turn used in experiments on human and animal physiology. Paley returns to the topic of 'chymical' and mechanical 'unions' in Chapter XIX.

emulgent artery: an artery (*vas efferens*) leading away from the *glomeruli* (sing. *glomerulus*) of the kidneys. *Glomeruli* are bunches of blood vessels that filter urine out of the blood.

papillæ: nipples, or nipple-like projections (sing. 'papilla').

52 *We see the blood carried by a pipe, conduit, or duct, to the gland*: comparing blood vessels to objects like pipes, conduits, or ducts drew from a much larger tradition, stretching back to Antiquity, in which anatomists transferred the names of human-made objects to internal and external features of the human body. Such a practice formed an excellent pool of

examples from which natural theologians like Paley could select analogies of divine design in human and animal bodies.

52 *stocking-loom, a corn-mill, a carding-machine, or a threshing-machine*: examples of mechanical processes that, in Paley's day, were powered by the turning of a water wheel. Stocking-looms consisted of a large frame strung with thread through which wooden (or iron) arms moved in and out to create a piece of cloth (textile); corn-mills ground corn into flour; carding-machines separated wool, cotton, or other fibres from their natural state so that they could be spun into thread; threshing-machines separated grains or seeds from the pod, husk, and chaff of the original plant.

rovings: also 'roves'. A rove is a 'sliver of any fibrous material (esp. cotton or wool) drawn out and very slightly twisted' (*OED*).

mill-wright: a person who designs, operates, or repairs a mill or its machinery.

54 *the example being capable of explanation without plates or figures, or technical language*: although Paley did not think that *Natural Theology* needed any visual examples, many illustrations were added to the book throughout the nineteenth century. This trend was set off by an 1826 edition, with illustrations by the surgeon James Paxton (1786–1860), who went on to make a career out of anatomical illustration, principally in his *An Introduction to the Study of Human Anatomy* (1831).

the quadrant: an instrument used in astronomy or surveying that was one-fourth of a circle, or 90 degrees.

tenon and mortice: terms used by joiners, carpenters, or stonemasons to describe the parts of a stone or piece of wood that fit together to make a frame, wall, or foundation. A tenon is a notch at the end of one piece of wood or stone that fits into a mortice (a cavity) of another.

56 *medullary substance*: the spinal cord.

59 *compages*: a system made out of several different types of parts.

luxation: dislocation.

chine of a hare: Paley was a keen outdoorsman and he most probably witnessed anatomical dissections of local game in person.

60 *tribe*: during the mid-eighteenth century, Carl Linnaeus (1707–78) proposed a system of mineral, vegetable and animal classification based on class, order, genus, and species in the many editions of his *Systema Naturæ*. Paley's use of the term 'tribe' in this instance is synonymous with a Linnaean 'order'. However, later in the book he uses the term to denote a smaller group of organisms in a fashion that overlaps with Linnaeus' concept of 'species'. This inconsistent usage of terms was not abnormal, partly because there were local traditions based on pre-Linnaean nomenclatural terms that stemmed from earlier writers like John Ray (see note to p. 181) and William Derham.

fusee: 'A conical pulley or wheel, esp. the wheel of a watch or clock upon which the chain is wound and by which the power of the mainspring is equalized' (*OED*).

Der. Phys. Theol.: Derham's *Physico-Theology, or, A Demonstration of the Being and Attributes of God from his Works of Creation* (London, 1713). William Derham was a clergyman who wrote *Artificial Clock-Maker* (1696), *Astro-Theology* (1714), and *Christo-Theology* (1729) and edited the works of Robert Hooke (1635–1703) and John Ray (see note to p. 181).

61 *Keill*: James Keill (1673–1719), a physician who had benefited from a medical education in Edinburgh, Paris, Leiden, Oxford, and Cambridge. In 1699 he purchased a MD from Aberdeen, but his reputation was more enhanced by the honorary MD he received from Cambridge in 1705. The source here is Keill's *Anatomy of the Human Body* (1698). Later in *Natural Theology*, Paley cites Keill's third edition (1708).

Keill and William Cheselden (see note to p. 64), Paley's two major anatomical sources, were both 'iatromechanists', believing that the body could be reduced to Newtonian mechanics. Also included in this group of physicians were Archibald Pitcairne (1652–1713), David Gregory (1659–1708), and George Cheyne (1671–1743). For them, the key for Newtonian interpretations of the human anatomy and physiology was to realize that 'forces' of attraction were not inherent to the body (a material object). Matter, whether it was a planet or a blood cell, was sustained by God. Keill's *Account of Animal Secretion* (1708) argued for attractive forces in the body based on Newton's theory of attraction in matter. This reference to unpinpointed 'attractive forces' served well for Keill's perception of natural theology and would later contribute to physiological inquiry into the mechanically elusive forces of vitalism, sensibility and irritability.

Schelhammer: Gunther Christoph Schelhammer (1649–1716), a German physician who wrote on a wide range of medical topics, including anatomy, botany, and chemistry. Schelhammer wrote so many works that it is hard to know which one Paley is citing.

patella, or knee-pan: knee-cap.

62 *ossification*: a process in which soft tissue hardens into a bone.

os hyoides: the bone to which the tongue is anchored (also called the hyoid bone, *os hyoideum*, or the lingual bone).

64 *ginglymus*: this is not a type of bone, rather it is the general name given to a hinge joint (the elbow, for example).

Cheselden: William Cheselden (1688–1752), an entrepreneurial London surgeon known for his popular lectures. In 1718 he moved to St Thomas's Hospital where he delivered four courses a year. Using visual aids, Cheselden's lectures promoted anatomy as an entertaining and enlightening subject. His enthusiasm and methods, however, did ruffle some feathers of the London medical establishment. In 1714 the Barber-Surgeons Company accused him of dissecting the corpses of criminals in his own house. Even so, in the 1720s and 1730s Cheselden's fame increased via his dextrous removals of painful bladder stones. This process, called a lithotomy, usually took twenty minutes, but Cheselden could do it in under five. He charged £500 for the operation and wrote up the procedure in the *Treatise on a High Operation for the Stone* (1723). This work became a standard anatomical treatise for several decades. Additionally, Cheselden had a strong publication record that included the widely used anatomical atlas *Osteographia* (1733) and Paley's choice, *The Anatomy of the Human Body* (1713). The latter is cited here from the 7th edition (1756).

66 *gristle*: a colloquial term for cartilage.

68 *the axis, the nave, and certain balls upon which the nave revolves*: the nave is the hub of a wheel and the axis is the straight piece of wood (or iron) to which a wheel is fixed on each end.

    lamella: a thin layer of bone or tissue.

69 *sartorius*: the longest muscle in the human body, which begins at the outer hip and runs across the front of the thigh and down to the inner tibia; 'sartorius' comes from the Latin *sartor*, a tailor.

70 *brachiæus internus*: a muscle of the upper arm that moves the elbow joint.

    brachiæus externus, and the anconæus: muscles in the upper arm that move the elbow joint.

    the cubit: an ancient unit of measurement, more commonly known for its use in the Old Testament. Traditionally, it was held to be the length between the tip of the middle finger and the elbow—roughly 17 to 22 inches (43 to 56 centimetres).

71 *deglutition*: the act of swallowing.

    the number of muscles, not fewer than four hundred and forty-six in the human body: as fn. 1 indicates, this figure is taken from the third edition of James Keill's *The Anatomy of the Humane Body Abridg'd: A Short and Full View of All Parts of the Body. Together with Their Several Uses, Drawn from Their Compositions and Structures* (1708).

72 *It appears to be a fixed law . . . towards its centre*: here, Paley links an act of individual, organic design (teleology) with the idea of an overarching law.

73 *says this writer*: Bernard Nieuwentyjt; spelt Nieuentyt in text below. The quotation is from his *A Religious Philosopher*. Nieuwentyjt (1654–1718) was a Dutch physician who was educated at the universities of Leiden and Utrecht. In addition to publishing books, he practised medicine in Westgraftdijk and Purmerend. Unlike many of his contemporaries, who wrote in Latin, he wrote most of his works in Dutch because he claimed that he wanted them to be useful to his own countrymen. His writings usually addressed medicine, mathematics, or physico-theology. He was particularly influenced by the writings of John Keill (1671–1721) and

Christian Wolff (1679–1754) and his best-known book in England was *A Religious Philosopher, or the Right Use of Contemplating the Works of the Creator*, which was translated from the original 1714 Dutch version (*Het Recht Gebruik*) into English in 1718 by John Chamberlayne (1668/9–1723). The book was very popular and went through several Dutch and English editions. It was read by leading natural philosophers, including William Derham, whose *Physico-Theology* drew heavily from it. Nieuwentyjt, like John Ray and Derham, held that the natural world was ordered and overseen by God and by the end of the eighteenth century his work was a well-known entry in the British natural theology canon.

74 *nine hundred and ninety-nine persons out of a thousand*: the justification of a claim based on an appeal to a prevalent view held by society has larger links to Paley's commitment to utilitarianism, a school of thought that held that the needs of the many outweigh the needs of the few. His commitment to utilitarianism is most strongly evinced in his *Principles of Moral and Political Philosophy*, a text that was required reading at both Oxford and Cambridge at the turn of the nineteenth century. *Natural Theology* did not receive official status on the Cambridge university curriculum, but individual colleges did recommend it to divinity students.

75 *dividing the pneumatic part from the mechanical*: here, 'pneumatic' refers to respiration, a process that involved air, which meant that it was relevant to pneumatic chemistry. Paley most probably mentioned pneumatics because the therapeutic value of gases was quite fashionable around 1800. For instance, Dr Thomas Beddoes (1760–1808) set up the Medical Pneumatic Institution in Bristol and performed, with the help of his assistant Humphry Davy, a set of highly publicized experiments on laughing gas (nitrous oxide).

77 *the Leipsic Transactions*: started in 1682, *Acta Eruditorum Lipsiensium* was one of Europe's oldest academic journals. Paley does not state which volume he was using. In English works it was often cited as the *Leipsic Transactions* (because it was printed in Leipzig, Saxony). Its pages contained the names of many acclaimed European authors. Nieuwentyjt published a series of articles in the journal between 1694 and 1700 that challenged Gottfried Leibniz's (1646–1716) new calculus.

steelyard: a small, portable balance that consisted of two arms, one shorter than the other, with a weight that slid along the long arm.

lever: an ancient, simple mechanical device that consisted of a bar balanced over a central, fixed point (a block of wood, iron ball, etc.). It was used to move or raise heavy objects.

Phil. Trans. part i. 1800: Everard Home, 'The Croonian Lecture. On the structure and uses of the membrana tympani of the ear', *Philosophical Transactions*, 90 (1800), 1–21.

78 *husbandman*: someone who owned or operated a farm, bred animals, or attended to agricultural crops. Synonyms included 'farmer' or 'granger'.

78 *his scythe, his rake, or his flail*: tools associated with manual agricultural labour. Scythes, which consisted of a curved blade attached to a shaft, were used to cut down (reap) grain or grass. Rakes were shaped like their modern equivalent and they were employed primarily to cultivate the earth or remove debris from a field. A flail was used to beat grain out of plants (a process called threshing) and it consisted of a central staff to which a small piece of wood, stone or metal was attached via a piece of leather or a chain.

digastric: a muscle involved in lowering the lower jaw and in elevating the hyoid bone (see note to p. 62).

80 *appetency . . . through an incalculable series of generations*: later (Chapter XXIII), Paley defines appetency to be a quality inherent to matter that allows life-forms to arise spontaneously and then develop into more complex organisms over a series of generations. Here, and elsewhere, Paley seems to indicate that he believes that spontaneous generation (a point in time) and time-driven morphological variation (a process) were one and the same. This conflation would not have been accepted by contemporary theorists, several of whom are cited by Paley throughout the book. Though Paley does not explain why he chose to use the word 'appetency', it is quite clear that he uses it as a synonym for Erasmus Darwin's (see note to p. 189) definition of a filament. This being the case, it is worth noting that Paley introduces his opposition to the concept well before he even mentions Darwin. In general, most naturalists in Enlightenment Britain (and other parts of Europe) were concerned primarily with the empirical classification of the natural world and they were amused, but not convinced, by the theories proposed by writers like Darwin and other earlier thinkers who used the word 'seed' or 'germ' in a similar fashion.

renitency: resistance to physical forces.

Bishop Wilkins: John Wilkins (1614–72), a founding fellow of the Royal Society who served as its secretary from 1663 to 1668. He held several prestigious appointments at the universities of Oxford and Cambridge and in the Church of England and was eventually made the Bishop of Chester in 1668. He wrote several books, including *A Discourse Concerning a New Planet* (1640), *Mercury, or The Secret and Swift Messenger* (1641), *Mathematical Magick* (1648), and *An Essay towards a Real Character and a Philosophical Language* (1668). Though he wrote on a wide variety of topics, he was keenly interested in the philosophy of language and its impact upon methods of empirical description (classification in particular).

Galen: (*c.* 130–200), a Greek physician from Asia Minor (modern Turkey). Early in his career he lived in Rome, where he cared for injured gladiators. He was interested in the organization of the human body and he used the horrific injuries of his patients to improve his anatomical knowledge and to propose a medical method based on bodily humours and Aristotle's elements. Though he wrote many treatises, his anatomical and therapeutic views were preserved for posterity in *On the Natural Faculties* and this work remained authoritative all the way through the early modern period.

I have sometimes wondered . . . in a watch or a mill: after giving many analogies that compared mechanical objects to anatomical examples, Paley finally makes an explicit connection between morphology and his two favourite examples of mechanism: the watch and the mill.

81 *are nothing short perhaps of logical proofs*: the operative word in this statement is 'perhaps'. In order to establish his argument, many of Paley's premises depend upon the use of analogy and, to a lesser extent, metaphors.

an able anatomist: Niels Stensen (1638–86); Steno was the Latinized form of his name. He studied medicine in Copenhagen and Leiden and eventually became physician to Ferdinand II and then to Cosimo II in Florence (which led him to convert to Catholicism). In 1672 he became professor of anatomy in Copenhagen and then spent the remainder of his life as a bishop (in Titiopolis, Hanover, and Hamburg). He is one of the few Catholic scholars quoted by Paley. 'Blas. Anat, Animal.', cited in the footnote, is untraced.

'Imperfecta . . . omnem superant admirationem': 'this imperfect description of the muscles is no less monotonous for the reader than their preparation was enjoyable for those observing them. This is because the most elegant creations of engineering, frequently evident in these things, are but obscurely articulated in words. When exposed to the eye, these designs of mechanical works of art surpass every admiration by means of their body's symmetry, their vivid colour, and the sense of proportion in their implantation and division.'

82 *engine*: from the fifteenth to the nineteenth century, 'A mechanical contrivance, machine implement, tool' or, in a collective sense, 'apparatus or machinery' (*OED*).

that of the water pipes in a city: early modern anatomy thrived on the use of colloquial metaphors and similes. This practice was common in textbooks as well as physico-theological accounts of the human body. As evinced in Paley's main anatomical sources, James Keill and William Cheselden, examples were often drawn from everyday experience. For instance, Keill's *Anatomy* compares the colon to a large 'bag-pipe' and the entrance of the auricles into the heart to 'little Ears'; at one point he states that 'The figure of the [lung] Lobes together resembles a Cows foot being a little concave betwixt the Lobes', and when describing the cerebellum's foldings he suggests that they 'resemble the Segments of Circles, or edges of Plates laid on another'. Cheselden's *Anatomy* states that the Omentum is 'somewhat like a net-work . . . and resembles an apron tucked up', the orifice connecting the fallopian tubes to the uterus is 'about the size of a hog's bristle', and that the cornea is 'concavo-convex, like glasses of that kind'. Both anatomists also had pet similes and metaphors that recur throughout their books. Keill repeatedly uses such words as 'pear', a 'goose-quill', an 'egg', 'the white of an egg', 'pipes', 'canals', 'trunks', and 'branches'. Similarly, Cheselden continuously refers to bird 'quills', 'nuts', 'eggs', 'berries', 'tubes', 'branches', and 'trunks'.

84 *For our purpose . . . the heart acts*: Paley once again sidesteps the vociferous debates (see note to p. 49) regarding the material or immaterial nature of the human mind and the origin of life itself.

85 *Dr Hunter's account of the dissection of a whale*: John Hunter, 'Observations on the Structure and Economy of Whales', *Philosophical Transactions*, 77 (1787), 350–71. See notes to pp. 49 and 94.

it is found . . . its impure part: by 1802, the 'pure' part of air (also called 'vital air' or 'respirable air') had been renamed 'oxygen' by most laboratory-based chemists.

86 *a separate and supplementary artery*: the pulmonary artery.

brought back by a large vein: the pulmonary vein.

87 *nor can any short and popular account do this*: Paley was keenly aware of his audience. Books at this time were expensive and this meant that his initial readers came from the aristocracy, gentry, and rising professional classes. Compared to an anatomy textbook, his book was therefore 'a popular account'. However, as the nineteenth century progressed and publication prices fell, literacy rates rose across Europe and this allowed *Natural Theology* to be read by a much wider section of society.

Hamburgher: untraced.

90 *I say nothing; because it is chymistry, and I am endeavouring to display mechanism*: throughout *Natural Theology*, Paley oscillates on the teleological value of chemical examples. Though he claims chemical ignorance on p. 43, this does not stop him from discussing digestion throughout the rest of the book.

93 *alimentary system*: the digestive system.

Abbé Spallanzani: Lazzaro Spallanzani (1729–99), an Italian priest and polymath whose research remains foundational in the fields of physiology and natural history to this day. Spallanzani studied and taught mostly in Lombardy. Some of his most significant achievements included the discrediting of spontaneous generation, the refutation of animalcules, the establishment of *arteriovenous anastomoses* in warm-blooded animals and coining the concept of gastric juice. Like Paley, Spallanzani disagreed with many aspects of Buffon's writings—especially Buffon's spontaneous generation claims. Because the English-speaking world of the eighteenth century had its own natural philosophers who wrote about these areas, and because of the lack of English translations of Spallanzani's works, his ideas did not initially enjoy popular recognition in England. However, he was known to the halls of the Royal Society and his work was translated as *Dissertations Relative to the Natural History of Animals and Vegetables. Translated from the Italian of the Abbé Spallanzani . . . To which are added two letters from Mr Bonnet to the author. And (to each volume of this translation) an appendix, the first containing a paper written by Mr Hunter, FRS and the experiments of Dr Stevens on digestion; the second a translation of a memoir of Mr Demours . . .* (1784). Paley cites Hunter, Spallanzani and Stevens in his section on digestion and this suggests that he was using the entire compendium. The essays in *Dissertations* address Spallanzani's experiments on birds and other animals (and, at times, himself) that took place during the 1770s. Building on research conducted in the 1670s by the Tuscan physician and poet Francesco Redi (1626–97), Spallanzani not only offered convincing evidence against spontaneous generation, but also went on to give an informed account of fowl gizzards that shed much light on the digestive powers of gastric juice. Paley used Spallanzani's research to demonstrate how digestive systems were uniquely tailored to fit the habitat and diet of individual animal species.

not a simple diluent, but a real solvent: a 'diluent' was a generic name given to a substance that diluted another solid or liquid—the adding of common water to seawater, for example. In the chemistry of Paley's day, this was a mechanical process, i.e. it did not involve a chemical change. A 'solvent' was much more powerful. It was a substance that chemically altered most of the matter with which it came into contact. The two most powerful solvents *c.* 1800 were acids and alkalis (see note to p. 50).

By experiments out of the body: Paley is referring to *in vitro* experimentation, i.e. research on the products or parts of an organism conducted in the artificial conditions of a laboratory setting. In Britain, the leading site for this type of experimentation was the University of Edinburgh. Medically orientated *in vitro* experiments were the central focus of laboratory-based chemistry well into the nineteenth century. This engendered a wide variety of biochemically related books that addressed topics ranging from drunkenness to natural theology. Two good examples are Thomas Trotter, *An Essay, Medical Philosophical, and Chemical on Drunkenness and its Effects on the Human Body* (1804) and William Prout, *Chemistry, Meteorology and the Function of Digestion Considered with Reference to Natural Theology* (1834).

putrefaction . . . fermentative process . . . digestion of heat: for chemically inclined physicians during the eighteenth century, there were two possible chemical causes used to explain digestion: heat and acids. The physical decomposition of food that resulted from either of these causes was called fermentation (although the term was also applied to other types of experiments on plants and bodily secretions). The final state of fermentation was putrefaction: 'When a body is in a putrefying state, it is easy to discover, by the vapours which rise from it, by the opacity which invades it.' See 'Chemistry', *Encyclopaedia Britannica* (1771).

94 *trituration*: the process by which a solid, dry substance is ground down into a powder.

Dr Stevens: Edward W. Stevens (1718–84) was born in St Croix, Virgin Islands. The work cited here is *Dissertatio inauguralis de alimentorum concoctione* (London, 1777). Reputedly the half-brother of the American Alexander Hamilton, Stevens took an AB from King's College, New York, and then an MD from the University of Edinburgh in 1777. He was admitted to the Royal Medical Society (RMS) of Edinburgh one year before he received his degree. While his fellow colonists were fighting the British in the American Revolution, Stevens served as the RMS president in 1779 and in 1780. To graduate from Edinburgh's Medical School at this time, students were required to write a final dissertation. Stevens took the task seriously and performed a set of experiments that allowed him to isolate human 'gastric juice'. In the following decades Lazzaro Spallanzani used this study when performing his digestion experiments.

Dr Hunter: as mentioned in the note to p. 49, John Hunter was part of a larger vitalist movement in the late eighteenth century. One of the many Scotsmen who migrated to the potentially lucrative medical world of London, Hunter trained with his brother William and with William Cheselden. He then served in the military and returned to London to set up practice as a surgeon. Among his many students were Edward Jenner (1749–1823) and John Abernethy (1764–1831). In 1767 he was elected a fellow of the Royal Society and his reputation as a lecturer and surgeon eventually secured him the post of Surgeon General in 1790. Often seen as an important figure in the history of medicine, Hunter actually wrote more about natural history. In some of his works, Hunter departed from the Baconian practice of straightforward description and began to write about the 'vital force' that animated a living organism. His *Lectures on the Principles and Practice of Surgery* (given in 1785, but published posthumously in 1833) stated: 'Every individual particle of the animal matter, then, is possessed of life, and the least imaginable part which we can separate is as much alive as the whole.' Additionally, Hunter's treatise *The Natural History of Human Teeth* (1771) shows a clear connection between digestion and his perception of a 'living principle'. He considered one of the distinguishing internal features of an animal to be the act of digestion as performed in the stomach.

95 *ductus hepaticus*: the hepatic duct.

Phil. Transac. vol. lxii: John Hunter, 'On the digestion of the stomach after death', *Philosophical Transactions*, 62 (1772), 447–54.

Malpighius: Marcello Malpighi (1628–94), a physician who taught medicine at the universities of Bologna, Pisa, and Messina. In his final years, he was the personal physician to Pope Innocent XII. Though he wrote about many topics in anatomy, botany, and chemistry, he was chiefly remembered in the eighteenth century for his research on embryos and body tissue. In particular, he used observations taken via his microscope (a new instrument at the time) to argue that capillaries were integral to blood circulation. In England his research on this topic was published as *Anatome Plantarum Idea* (1675). He published widely (mostly in Latin and Italian), including letters in the *Philosophical Transactions*.

96 *parotid gland*: lymphatic glands (*lymphoglandulae parotideae*) are located at the back of the jaw and are involved in draining the eyes, nose, and face.

buccinator muscle: a muscle located under the cheek.

98 *Bonnet*: Charles Bonnet (1720–93), a Genevan who originally studied law but went on to establish himself as a natural historian and a Christian apologist. Within evolutionary history, Bonnet is often singled out as a forerunner of Darwin, primarily for two reasons. First, his *Traité d'insectologie* (1745) gave one of the first published accounts of parthenogenesis, i.e. how female insects (aphids) could fertilize themselves without the help of a male. Secondly, he held that the growth of organisms was overseen by a microscopic 'germ', which for Bonnet was a miniature structure passed on at fertilization. His preliminary thoughts on this topic were presented in *Considérations sur les corps organisés* (1762) and then more fully in *Contemplation de la nature* (1764–5), a book that was translated into English as *The Contemplation of Nature* in 1766. In recognition of his work, he was made a corresponding member (but not a full member) of the Academy of Sciences in Paris and a fellow of the Royal Society of London. Additionally, he maintained correspondence with many well-known natural philosophers, including René-Antoine Ferchault de Réaumur (1683–1757) and Albrecht von Haller (1708–77). Bonnet was keenly interested in fusing his research and theories with his faith in God. When he was young he read the teleologically underpinned *Spectacle de la nature* (1732), a work of natural theology written by Abbé Noël-Antoine la Pluche (1688–1761), and translated as *Nature Displayed*. Drawing from this book and his own research, Bonnet held that the universe possessed uniformity and this led him to use philosophical ideas similar to those promoted by ancient Neoplatonist authors like Plotinus. In particular, he suggested that organisms in the chain of being were slowly changing, based on a predetermined pattern being passed on from generation to generation via germs. This strain of thought was developed not only in his *Contemplation* but also in *La Palingénésie philosophique* (1769). Though many British intellectuals read French at the end of the eighteenth century, it is quite likely that Paley's polite audience would have known Bonnet via his theologically focused writings. For instance, in 1785 the Methodist leader John Wesley published a version of Bonnet's *Contemplation* in his *A Compendium of Natural Philosophy Being a Survey of the Wisdom of God in the Creation* (1763). Paley was most probably familiar with Bonnet's work because two of his letters were included in a widely read 1784 compendium of Spallanzani's works (see note to p. 93).

100 *its lymphatics, exhalants, absorbents; its excretions and integuments*: all these were substances that were of direct relevance to monitoring one's health. Building on new understanding gained in laboratory chemistry and ancient beliefs in the four humours of the body, a new therapeutic method, sometimes called Neo-Humoralism by modern historians, was developed at the end of the eighteenth century. It sought to balance bodily fluids (blood, lymph, and excretions) with the hardness of bodily tissue (integuments and absorbent blood vessels in the body).

102 *it is found where it conduces to beauty or utility*: the history of natural theology from ancient times is interwoven with aesthetic notions of beauty or perfection. John Brooke and Geoffrey Cantor address this in *Reconstructing Nature: The Engagement of Science and Religion* (Oxford, 1998); see esp. chapter 7, 'From Aesthetics to Theology'.

103 *hydrostatical*: hydrostatics was the study of liquids (and sometimes gases) and the pressures that they produced. Two well-known books on this topic during the eighteenth century were Robert Boyle's *New Experiments Physico-Mechanicall, Touching the Spring of Air and Its Effects* (1660) and the Revd Stephen Hales's *Vegetable Staticks: Or, An Account of Some Statical Experiments on the Sap in Vegetables* (1727).

105 *the umbilical vein, which, after birth, degenerates into a ligament*: the umbilical vein is located in the umbilical chord and it transports blood from the placenta to the foetus. After birth it becomes a ligament that helps to keep the liver secure in the abdomen. For more on embryology, see the following note and notes to pp. 95 and 138.

The bladder is tied to the navel by the urachus transformed into a ligament: the urachus is a vessel that transports urine out of the foetus into the cloaca. After birth it becomes a ligament (*ligamentum umbilicale medianum*) that helps keep the urinary bladder in place.

omentum: a fold in the peritoneum (the membrane that lines the interior abdominal cavity), connecting the stomach to other organs.

107 *cellular or adipose membrane*: fatty tissue that lies under the skin.

108 *There are parts also of animals ornamental*: colourful or shapely characters of animals had long had a place in natural history. In the early modern period they were noted for classification purposes, or simply as objects of wonder. This can be seen in the 'peacock' and 'ruffe' entries of Francis Willughby's *The Ornithology of Francis Willughby of Middleton in the County of Warwick . . .* (1678). By the end of the nineteenth century ornamental characters of animals were strongly associated with natural selection, a point investigated by Charles Darwin in chapters 13 and 14 of *The Descent of Man* (1871).

irides: plural of 'iris' (also irises); the coloured part of the eye that surrounds the pupil.

corolla: collective name given to all of the petals of a flower.

the ascent of the sap: a popular topic within the early modern period was the movement of sap (or plant 'juices').

109 *when not vitiated by habits forced upon it*: there were several definitions of 'habit' during the eighteenth century, especially in natural history. Paley defined the word simply as the actions learned by an organism (*a posteriori*). Throughout *Natural Theology*, he criticizes the theories that held that habits led to the development of completely new body parts (as opposed to the modification of a previously existing one). A good example of this occurs in Chapter XXIII, where he disagrees with Oliver Goldsmith's use of the word 'adapted' to describe the pelican's bill. See notes to pp. 126 and 160.

the produce of numerous and complicated actions . . . and of the mind upon its sensations: here, Paley appeals to John Locke's epistemology, which held that the building blocks of thought were 'ideas'. The mind created a primary idea based upon input from the five senses. It then combined these into more complex secondary ideas. For example, if the mind had two primary ideas like 'pink' and 'elephant' (taken in by the eye), it could combine them into a secondary idea of a 'pink elephant'.

111 *May it not be said to be with great attention, that nature hath balanced the body upon its pivots?*: early modern perceptions of 'nature' varied and Paley's use of the term throughout *Natural Theology* reflects this context. To understand Paley's argument, it is crucial to explain the nuances represented in his use of the word. As can be seen by *OED*'s entries for 'nature', the word implied a variety of different meanings. During the early decades of the eighteenth century, the term 'nature' was used to refer to the immutable and designed natural world created by God and ordered by mechanical laws that God had set in motion. God, not 'nature', was the active force behind matter. As Paley points out in Chapter XXIII, 'Divine Nature' was 'a Being, infinite, as well in essence, as in power'; it was active and observable, and it closely resembled an attribute of God. Paley oscillated between this sense of 'nature' and another that emerged during the second half of the century, arising from the permissive exegetical and doctrinal climate of Hanoverian England. The application of Newtonian physics demonstrated that the universe ran rather well on its own and really did not need frequent twiggings from God. In addition, a growing influx of specimens from the Americas suggested to some naturalists that certain animal and plant species might be mutable. Thus, 'nature' became an active force of its own—but a force still fully subordinate to God. Paley's also employs this notion of 'nature', especially when referring to the fixed observable world around him, as he states, that which was 'sometimes called nature, sometimes called a principle' (Chapter XXIII).

112 *periosteum*: a fibrous membrane that surrounds a bone and acts as an attachment point for muscles and tendons.

114 *evidence, which the most completely excludes every other hypothesis*: up to this chapter, Paley's style has been politely suggestive. However, from this point forward, his tone becomes more forceful and he presents inductive conclusions from prior chapters as factual statements.

Arkwright's mill: constructed by Sir Richard Arkwright (1732–92) in 1771 and located in England's Peak District, this mill was the first to use a water wheel to power a cotton-spinning machine. It laid the foundation for industrial textiles and the large steam-powered factories of the nineteenth century.

115 *tribe*: see note to p. 60.

117 *glutinous*: in Enlightenment chemistry, the word 'gluten' was used to describe the sticky sap distilled from plants or the substance that allowed soft matter to cement into rocks.

118 *which is easy to see with the microscope*: it was common for eighteenth-century naturalists (who had the funds) to use a microscope to look at mineral, plant, and animal specimens. This practice had been popularized the previous century by Robert Hooke's *Micrographia: Or Some Physiological Descriptions of Minute Bodies Made by Magnifying Glasses with Observations and Inquiries Thereupon* (1665).

119 *rictus*: the opening of a jaw.

121 *similitude*: a similarity between two objects or processes.

By what habit . . . the substance of which it is composed?: Paley is pointing out a widely held criticism of the theories proposed by thinkers like Buffon and Erasmus Darwin (see note to p. 189). At the time there was no empirically viable mechanism that could explain morphological change (see notes to pp. 39 and 80). As evidence was gathered from the fossil record during the early nineteenth century, a wider evidential base was created for theories that hypothesized large periods of time to explain morphological change.

cotton-card: a spiky-toothed tool used for separating wool into strands.

exuviæ: the skin that is shed by insect larvae.

122 *valvulæ conniventes*: the circular folds on the interior membrane of the jejunum (part of the small intestine).

123 *cæteris paribus*: with all facts or phenomena remaining the same; a Latin term used in philosophy arguments.

124 *specific gravity*: a ratio used for measuring density, especially for objects that are the same size but have different weights.

a gravid uterus: a uterus carrying live young or eggs, the anatomy of which was a very popular topic in eighteenth-century medicine. See William Hunter's *The Anatomy of the Human Gravid Uterus, Exhibited in Figures* (1774).

cletch or covey: a flock of birds.

cubitus: the hinge joint between the upper and lower forelimb of most mammals and reptiles. In humans it is also called the elbow.

126 *the keel . . . out-riggers . . . oars*: parts of a boat, well suited for extending Paley's use of mechanical metaphors.

Goldsmith's Hist. of An. Nat.: Oliver Goldsmith (1728–74), after earning a

BA from Trinity College, Dublin, attended medical school in Edinburgh and then began a life as a writer. In 1769 he was contracted to write *An History of the Earth and Animated Nature*. It took five years to complete and was published two months after his death in April 1774. Eight volumes long, *Animated Nature* is a well-written summary of eighteenth-century natural history. It drew from classical writers like Aristotle, Lucretius, Diodorus Siculus, and Pliny the Elder, as well as contemporary authorities like John Ray, Henri Louis Duhamel du Monceau (1700–82), Carl Linnaeus, Georges-Louis Leclerc, Comte de Buffon and René-Antoine Ferchault de Réaumur. The book is also cited as *Nat. Hist.* Goldsmith's writing skill and his former medical training made *Animated Nature* easy to read.

Like Paley, Goldsmith was fond of discussing natural theology. As he stated in his introduction to Richard Brookes's *A New and Accurate System of Natural History* (1763), he felt that the improvement of natural knowledge was conducive to the improvement of religion and piety, 'it was thought expedient to make this work as cheap as possible, that it might fall within the compass of every studious person, and that all might be acquainted with the great and wonderful works of nature, see the dependence of creature upon creature, and of all upon the Creator'. This approach contributed to the work's success. An anonymous review of the book published in the *Critical Review* in 1774 favourably compared it to la Pluche's *Nature Displayed*. Like Goldsmith's novels, *Animated Nature* continued to be popular well into the 1840s, so much so that James Prior's 1837 biography of Goldsmith asserted that it was a larger and equally popular version of Gilbert White's *Natural History of Selborne* (1789) and that it was attractive to readers who would normally have been 'repelled' by more complicated works. A quick glance through *Natural Theology*'s footnotes reveals that Goldsmith was one of Paley's main natural history sources. Whether writing about the movement of pectoral and dorsal fins, muscles of the opossum pouch, the upper chap of a parrot or sparrow fledglings, Paley cites Goldsmith with the same confidence as he does with medical authorities like Cheselden or James Keill.

127 *had this part worn away by treading upon hard ground*: here Paley addresses the possible degeneration of a morphological part, a point that would be considered in much more detail during the nineteenth century.

130 *conatus*: in classical Latin, a noun meaning 'an effort, exertion, struggle, endeavour' or 'an impulse, inclination, tendency' (Liddell and Scott). Paley follows the Latin definitions by using the word to connote an internal impulse or tendency. As so defined, a conatus was not necessarily bad. However, in this instance Paley is referring to a 'blind conatus', something which his teleological view would not permit. *OED* defines conatus as 'A force, impulse, or tendency simulating a human effort; a nisus', and cites this passage from Paley; however, suggesting that Paley's definition of 'conatus' included 'a human

effort' is quite misleading, as this sentence clearly refers to birds, not humans.

131 *diving machine*: a diving bell, or a bell-shaped iron casing, which lowered trapped air down into the sea, thereby allowing a diver to breathe underwater. The chemical understanding of this process (especially in relation to pressure) is addressed in the 'Pneumatics' entry in the *Encyclopaedia Britannica* (1771).

three great kingdoms: natural history at this time was most often divided into three kingdoms: mineral, vegetable, and animal.

132 *os pubis*: the pubic bone, located on the front of the pelvis.

134 *taken from Buffon*: even though Paley has discussed several ideas addressed by Buffon earlier in the book, this is the first time that he mentions him by name. See note to p. 39.

the babyrouessa, or Indian hog: as Paley was writing *Natural Theology*, the British East India Company's power was being (grudgingly) handed over to the Crown and Parliament. Throughout *Natural Theology*, Paley refers to exotic plants and animals located throughout the British Empire. This was common practice in late-Hanoverian natural history books.

136 *viviparous animals*: animals that give birth to live offspring.

137 *lacteal system*: the anatomical vessels used by the body to produce milk.

138 *foetal thorax*: the thorax (the part of the body that lies between the neck and the diaphragm) of a foetus. Embryology, which included the comparative morphology of foetuses, was a feature of eighteenth-century vitalist and materialist theories (see notes to pp. 49, 94).

foramen ovale and ductus arteriosus: the heart consists of four chambers: the right and left atrium (pl. atria) and the right and left ventricles. The *foramen ovale* is an opening in the septum between the two atria of the heart; it normally closes after birth. The *ductus arteriosus*, also known as the *ductus botalli*, is another anatomical feature found in a foetus. More specifically, though foetuses have lungs, they do not begin to function until after birth. The *ductus arteriosus* is a duct between the arch of the aorta and the pulmonary trunk that allows the blood to bypass a foetus's unfunctioning lungs. The duct normally closes after birth (when the baby first begins to breathe) and then turns into a ligament. Paley addresses foetal anatomy earlier in the book, especially morphological parts that transform into different structures after birth: see p. 105.

140 *relation*: John Locke's section on 'Relation' in *An Essay Concerning Human Understanding* (1689) gives the following definition: '[Relation] be not contained in the real existence of Things, but something extraneous, and superinduced: yet the *Ideas* which relative Words stand for, are often clearer, and more, distinct, than those Substances to which they do belong.' Paley's notion of the word was more concrete and *OED*'s historical definition 3.a for 'relation' more closely matches Paley's use of it:

'That feature or attribute of things which is involved in considering them in comparison or contrast with each other; the particular way in which one thing is thought of in connexion with another; any connexion, correspondence, or association, which can be conceived as naturally existing between things.' Paley often employs a 'relation' to link examples that remain beyond the plausibility of analogical comparison or inductive inference. For this reason, his use of the word should not be confused with the later nineteenth-century definition: 'A constituent of a proposition or propositional function that connects two terms (a dyadic relation) or more (triadic, n-adic, etc.)' (*OED*, 3.d).

141 *stum*: partially fermented wine, beer, or cider. In Paley's day the chemistry of fermentation was often used to understand digestion.

142 *gallinaceous*: large terrestrial birds like pheasants, turkeys, and grouse.

craw or crop: a sac-like part of a bird's gullet in which partially digested food is stored so that it can be regurgitated. Craw was an older word that had been an anatomical term in English anatomy texts since at least the seventeenth century.

hopper: a container used for storing or funnelling grain.

experiments . . . with perforated balls: these were conducted by Lazzaro Spallanzani and detailed in his *Dissertations*; see note to p. 93.

143 *graminivorous*: grass-eating.

144 *The sexes are manifestly made for each other*: apart from being a reference to animal procreation, this sentence could also be applied to plants because the most popular botanical classification system at the time (that of Linnaeus; see note to p. 170) was based on the characters of sexual organs. The dual relevance of sexual traits in this context often allowed authors that preceded Paley to use the word rather loosely, or in some cases, quite lewdly.

145 *palmated*: web-like.

146 *corking pin*: a large pin used to attach a woman's headdress (or sometimes a veil) to a cork mould.

147 *proboscis*: a trunk.

course of generations . . . (which is the general hypothesis . . . for the forms of animated nature): here Paley is most likely referring to the filament theory of Erasmus Darwin (see note to p. 189). More remotely, he may be countering the early thoughts of Jean-Baptiste-Pierre-Antoine de Monet, Chevalier de Lamarck (1744–1829), as expressed in his *Système des animaux sans vertèbres* (1801). Lamarck suggested that if a body trait changed in an organism's lifetime, it could be passed on to its offspring. His views on this, however, were more clearly voiced after Paley died, in *Philosophie zoologique, ou Exposition des considérations relatives à l'histoire naturelle des animaux* (1809).

148 *numerous tribes*: since so many different types of new life forms had been encountered throughout the early modern period, there was a prevalent

belief that more were waiting to be discovered. Such an idea worked against the notion of extinction because many thought that species which had 'disappeared' in Europe had most probably migrated to another part of the world.

149 *When this lattice work was first observed*: Paley is most probably referring to Robert Hooke's *Micrographia*, section 34.

   Adams tells us: George Adams (junior), *Essays on the Microscope . . . A General History of Insects, Their Transformations, Peculiar Habits and oeconomy . . .* (1787). Adams (1750–95) was a popular maker of scientific instruments on London's Fleet Street.

150 *nictitation*: winking; opening and closing the eyes rapidly.

151 *accession*: growing larger by the slow addition of a hardening substance. From the mid-eighteenth century, this process had been closely studied by *in vitro* experiments on bladder stones conducted in Edinburgh.

   mutation: a change or alteration, as connoted by the Latin masculine noun *mutatus*.

152 *h. e.*: hic est; (Latin) this is.

153 *herbivorous*: plant-eating.

   Spal. Diss.: see note to p. 93.

154 *annuli*: the plural noun from which the adjective 'annular', defined earlier in the paragraph, derives.

155 *strict relation to the elements by which they are surrounded*: early modern natural history was quite interested in the natural environment of a plant or an animal. For instance, Linnaeus's widely consulted *Systema Naturae* often contained a 'Habitat' entry that detailed the geographical location and other environmental factors relevant to the morphology of a given species. See chapters 16 ('Habitat') and 17 ('Geographical Names') in W. T. Stearn, *Botanical Latin: History, Grammar, Syntax, Terminology and Vocabulary* (Portland, Ore., 2000).

157 *The earth in its nature is very different from the sea, and the sea from the earth*: a good example of an antimetabole, a figure that repeats the same word or idea, but in a different order. The use of antimetaboles and other rhetorical figures like ploches and polyptotons occurred quite frequently in both natural theology and the natural sciences from the early modern period through the nineteenth century.

160 *An* INSTINCT *is a propensity, prior to experience, and independent of instruc-tion*: i.e. an *a priori* state. Paley also believed that instinct should not be equated or linked with the conception of sensation, a notion that had been taken in materialistic directions by the disciples and interpreters of Albrecht von Haller. Later in this chapter, he states: 'I am not ignorant of a theory, which resolves instinct into sensation; which asserts, that what appears to have been a view and relation to the future, is the result only of the present disposition of the animal's body, and of pleasure or pain experienced *at the time*.' The distinction '*at the time*' indicates that Paley

held the sensation theory to be an *a posteriori* argument and, therefore, inconsistent with his *a priori* view of instinct. It is also important to note that Paley does not equate instinct with habit. Unlike the *a priori* nature of instinct, a habit was learned, and it was consequently *a posteriori*. Since instinct was created to coincide with the fixed morphological parts of a designed organism, there was no need for change.

This position is consonant with the general commitments on instinct during the eighteenth century, especially in moral philosophy. To give an example for his definition of instinct, Paley notes how birds seem to know inherently how to incubate eggs. Although this was a common trope for early modern design arguments, Paley got his information specifically from an essay on instinct written by Joseph Addison (see note to p. 162) in *The Spectator* on 18 July 1710. Like Paley, Addison argues that instinct is inherently found in animals and humans. It is not learned. At the end of the essay, Addison concludes that instinct is mysterious and 'cannot be accounted for by any properties in matter'. He likens it to 'the principle of gravitation in bodies, which is not to be explained by any known qualities inherent in the bodies themselves' and states that it is 'an immediate impression from the first mover, the Divine energy acting in the creatures'. Not only does this explanation dovetail with Paley's stance on instinct, it also harmonizes with his view that matter contains no inherent organizing principle.

161 *cicatrix*: in Latin, a scar, a mark of incision in a plant, or the seam of a patch in a shoe. Paley is using the word to describe an egg's newly formed blood vessels ('red streaks') that divide (incise) 'the white from the yolk'—an action that resembles a needle running through a shoe patch to make a seam.

162 *says Addison*: Joseph Addison, *The Spectator*, no. 120 (Wednesday, 18 July 1711). Addison (1672–1719) was one of the best-known men of letters in England during the early modern period. Paley was most influenced by the essays that Addison wrote for the gentleman's serial entitled *The Spectator*. Addison saw himself as the spokesperson for the English gentleman and as a champion of the design argument. Like Paley, he was keenly interested in natural history and believed in the providential design of animals. Indeed, in this *Spectator* article he wrote that 'the arguments for Providence drawn from the natural history of animals' were 'demonstrative'. Paley probably first had a hard look at Addison's work when he was a lecturer at Cambridge. In his comments on design, Addison borrowed many of his arguments and examples from previous thinkers like John Ray and William Derham. He also believed there were no inherent organizing principles to be found in matter. Although he only quotes twice from Addison in *Natural Theology*, Paley cites him in his other works, especially in his *Evidences*.

163 *not by chance*: see note to p. 38.

    of which memory she shews no signs whatsoever: memory played a very important role in perceptions of personal identity during the eighteenth century. In Britain, theories of memory were strongly affected by the idealism of John Locke's *An Essay Concerning Human Understanding*. As the century progressed, there were increasing debates as to whether animals had memories, and, if they did, whether or not they had a 'mind'. The heart of these debates centred on whether the evidence used to discuss memory in humans could be imputed to mammals (or even insects). Most of Paley's contemporaries felt that such a move was not warranted. However, some, like David Hume, disagreed. For example, his 'Of the reason of Animals' in book i, part iii, section xvi, of his *A Treatise of Human Nature* (1739) suggests that animals do indeed possess minds. For natural theology, memory became a key concern for authors like Joseph Priestley who, in *Disquisitions Relating to Matter and Spirit*, sought to empirically explain the resurrection. Paley's reference to a butterfly's memory here shows that he was not willing to admit that it had retained any memory of its former life as a caterpillar. Rather, the butterfly obtained knowledge of its previous state only by *analogically* comparing its offspring to what it must have been in the past.

164 *parturition*: the act of giving birth.

165 *an abstract anxiety for the general preservation of the species ... a solicitude lest the butterfly race should cease from the creation*: in early modern natural history it was readily observed that animals took specific measures to preserve their young; a state of affairs that was generally interpreted as evidence of a divinely implanted instinct that led to the continuation of the species. This interest in the 'preservation' of species was still generally accepted in Britain when Charles Darwin attended Edinburgh and Cambridge during the 1820s, and over the next thirty years he would go on to reinterpret the concept within a more agnostic framework, more interested in the mechanism that explained how the preservation of certain species (and extinction of others) took place over a long period of time. In Chapter 4 of *On the Origin of Species by means of Natural Selection, or the Preservation of Favoured Races in the Struggle for Life* (1859), he concluded: 'This preservation of favourable variations and the rejection of injurious variations, I call Natural Selection.'

166 *Nor does parental affection ... if such a thing were intelligible*: Paley is referring to what he later calls a 'law of autonomous generation', a notion, promoted by Erasmus Darwin (see note to p. 189) and others, that suggested that organisms could become more complex over a period of time.

   I am not ignorant of a theory, which resolves instinct into sensation: see note to p. 49.

167 *sugescent*: Paley uses this word to denote 'pertaining to or adapted for sucking', though *OED* says this is a misuse (it cites only this passage from Paley from the period).

   appetencies: see note to p. 80.

168 *A gardener lighting up his stoves*: greenhouses in Enlightenment Britain often contained a wide variety of tropical plants imported from its colonies. Furnaces or stoves were therefore required to keep the plants from dying of exposure.

when a male and female sparrow come together . . . perpetuating their species: this account of animal procreation draws from wider Enlightenment medical theories that argued sexual activity drained the body of its energy and that it was therefore wiser to limit one's sexual encounters.

prospection: the act of looking forward to something or making provisions for the future.

says Derham: see note to p. 60.

169 *I recognise an invisible hand*: this modified metaphor of an 'invisible hand' is taken from Adam Smith's *An Inquiry into the Nature and Causes of the Wealth of Nations* (1776), which, in chapter 2, states that 'By directing that industry in such a manner as its produce may be of greatest value, he [the worker] intends only his own gain, and he is in this, as in many other cases, led by an invisible hand to promote an end which was no part of his intention'. This metaphor fits within Smith's larger utilitarian approach to economics and moral philosophy.

Harvey: William Harvey (1578–1657) studied medicine in Cambridge and Padua. Today he is remembered as being the physician-extraordinary to James I and as the first person to propose a blood circulation theory that most resembles the modern understanding of the process. His ideas on circulation were first published in Frankfurt as *Exercitatio Anatomica de Motu Cordis et Sanguinis in Animalibus* (1628). At first the work was heavily criticized as being unempirical but it began to be accepted near the end of the century as new evidence was presented by physicians. However, Paley's citation is not from *De Motu*, but from Harvey's *Anatomical exercitations, concerning the generation of living creatures* (London, 1653; originally published as, *Exercitationes de generatione animalium quibus accedunt quaedam de partu, de membranis ac humoribus uteri & de conceptione*, London, 1651; the Latin edition was republished in 1662, 1666, and 1680; William Hunter (1728–93) also republished the English edition in 1737). The book addressed the anatomy and physiology of a hen's egg (including incubation) and was aimed at unseating several Aristotelian notions of reproduction, spontaneous generation, and materialism.

170 *We are not writing a system of natural history*: early modern naturalists were quite keen to classify nature. The largest categories were the three kingdoms of nature: mineral, vegetable and animal. During the late seventeenth century, a wide number of arrangement systems were used within different linguistic and national contexts. Within Britain the works of John Ray became a standard reference point for determining the species of a plant, particularly his *Methodus Plantarum Nova* (1682). Ray's system classified plants based upon several parts of the organism,

including the morphology of seeds. Though helpful, this approach did provide much room for confusion because naturalists often could not agree on which parts should be singled out as representative for a given species. In 1735 a new classification scheme was offered by Linnaeus in his *Systema Naturae*. The work, based on his medical dissertation, suggested that all plants should be classified based on their reproductive organs. Over the next three decades, the subsequent editions of the book proposed that plants should be classified based on a descending scale: kingdom, class, order, genus, and species. Once a plant was fitted into these categories, its scientific name was determined by the Latin words used to describe its genus and species. This new naming system came to be known as 'binomial nomenclature' and by the 1750s Linnaeus had extended his classification system to animals and minerals (both being classified based on their external characters). The sheer practicality of the system made it very popular and it became a common reference point for most naturalists seeking to propose a classification system. However, though Linnaeus's taxonomy was helpful, there was widespread disagreement on which classification characters should be privileged over others. Throughout *Natural Theology*, Paley astutely exploits this situation by not promoting a specific classification system when he offers specific plants and animals as examples for his design argument. In doing this, he was able to draw examples freely from any natural history source, even those based on incompatible classification systems.

170 *for that minuteness we can, in some measure, follow with glasses*: the microscope; see note to p. 118.

We want a ground of analogy: see note to p. 16.

171 *I need not add, that . . . . they deposit their eggs in the hole*: this instance of contrivance, and that of the bee's sting in the following paragraph, are clear examples of suffering induced by design. Paley's approach to this topic was based on the utilitarian views evinced in his earlier works on political and moral philosophy. He envisioned an organism's contentment as an average between the occurrences of pain (suffering) and pleasure (happiness) in a person's lifetime. He held that the scales were usually tipped toward the latter and that made up for any inconvenience caused by the former. This was a common calculus promoted by many utilitarians, including Jeremy Bentham (1748–1832) in his *An Introduction to the Principles of Morals and Legislation* (1789). A clear use of this pain–pleasure calculus is exhibited throughout *Natural Theology*, but can clearly be seen on p. 241.

peristaltic motion: waves produced by muscles that push objects through a tubular bodily cavity.

spy-glass: telescope.

172 *chymistry*: see note to p. 51.

telum imbelle: (Latin) a useless weapon, unfit for war.

173 *monopetalous*: one unit formed from several petals (a flower).

grubs: wormlike insect larvae. One of the most influential books on this topic in Paley's day was Jan Swammerdam's *The Book of Nature, or The History of Insects*, ed. Herman Boerhaave and trans. T. Floyd (1758). Though Swammerdam had done much of his work at the end of the seventeenth century, the work was not published until the 1730s when it was edited by the Leiden medical professor Herman Boerhaave as *Biblia naturae*.

177 *panorpa*: scorpion flies.

St Pierre: Jacques Bernardin Henri de Saint Pierre (1737–1814), an engineer by training, who travelled widely through Europe and to Africa in his early adulthood; this inspired him to write *Voyage à l'Île de France* (1773). This two-volume work established him as a man of letters and as a defender of religious ideas. In 1784 he published the three-volume *Études de la nature*, a work of natural theology aimed at rebutting the materialist views of nature espoused by the *encyclopédistes*. It sold well and he expanded it into four volumes in 1788. Following the success of *Études*, he published two novels that promoted teleological views of nature: *Paul et Virginie* (1789) and *La Chaumière indienne* (1791). Both of these books were quickly translated into English and they went on to be best-sellers in both France and England. In *Natural Theology*, Paley cites St Pierre's *Études*, a work that was translated by Henry Hunter into English as *Studies of Nature* in 1796. Like his novels, St Pierre's *Studies* sold very well. In fact, there was a printing of it every year between 1796 and 1802.

phosphoric: able to produce light.

178 *tapers*: long wax-covered wicks used to light oil lamps or candles.

sky-rockets: firecrackers launched into the sky.

Adams: see note to p. 149.

179 *The will of the animal could not determine the quality of the excretion*: see note on Erasmus Darwin, p. 189.

180 *chymically speaking . . . a kind of exsiccation, like the drying of clay into bricks*: the process by which cement and bricks hardened was a central concern of chemistry during the mid- to late eighteenth century. One of the most cited publications on this topic was Joseph Black's *Experiments upon Magnesia Alba, Quick-Lime, and Other Alcaline Substances* (1754). Since most laboratory chemistry at the time used aqueous solutions (water, acids, and alkalis) to test the composition of an object, Paley's statement about the congelation of animal excretions under water is in line with contemporary chemistry. Aqueous-based chemical experiments also formed the intellectual backdrop for theories of the earth concerning the impact or probability of the biblical deluge or a worldwide inundation in general.

181 *Cicero*: (106–43 BC), a Roman writer who wrote prolifically on rhetoric and philosophy. It is not clear which of Cicero's works Paley is citing. Like many eighteenth-century writers, Paley was profoundly influenced

by Cicero's comments on inductive arguments and figures of speech. Additionally, next to the Bible, Cicero's *De Natura Deorum* (*On the Nature of the Gods*) was probably the most cited work of natural theology from the classical time period onwards.

181 *Ray*: John Ray (1627–1705), a nonconformist who graduated from Trinity College, Cambridge, in 1648. He later refused to sign the act of uniformity in 1662 and this cost him his Cambridge fellowship. With the collaboration and patronage of Francis Willughby (1635–72), Ray was able to collect specimens and write about natural history. In 1660 Ray published *Catalogus plantarum circa Cantabrigiam nascentium*, an attempt to classify all of the plants in Cambridgeshire. From the 1660s to the 1690s his publishing record was prodigious. He wrote several books, including *Historia Generalis Plantarum* (1686), and many articles in the *Philosophical Transactions*. In 1690 he published *Synopsis*, a book held in high esteem for the next hundred years. Like Linnaeus and, later, Georges Cuvier (1769–1832), and even Paley, Ray believed that species are fixed, but he also believed that limited transmutation was possible. In the early 1690s Ray published two works of natural theology that would later influence Paley: *The Wisdom of God Manifested in the Works of the Creation* (1691) and *Three Physico-Theological Discourses* (1693). These two books effectively laid the foundation for the physico-theological genre that would thrive in England for the next century. The *Discourses* are important for this tradition because they demonstrate that Ray did not subscribe to literal one-week creation as described in Genesis. *The Wisdom of God* was immensely popular, going though numerous editions during the eighteenth century.

182 *nymphæ*: (singular: nympha, or nymph) insect larvae.

hydrocanthari: the larva of a burrowing water beetle. The modern term for the species is *hydrocanthus* and it is found in the *Noteridae* family.

Derham: see note to p. 60.

183 *fecundation*: fertilization.

Phil. Trans. part iii. 1796: Corrêa de Serra, 'On the fructification of the submersed Algæ', *Philosophical Transactions*, 86 (1796), 494–502.

185 *farina . . . pistil . . . stamina . . . antheræ*: parts of a flower's reproductive system. These characters were used to classify plants in the Linnaean system (see notes to pp. 60, 170).

stone fruits: fruits that contain a stone or pit.

186 *senna*: medicinal plant used as a laxative.

187 *not a single species, perhaps, has been lost since the creation*: see note to p. 148.

pericarpium: the ripened section(s) of a plant's ovary (the modern term is pericarp).

plantule: the embryo of a plant formed after germination.

germ: during the early modern period, this word was used to denote the earliest stages of an organism. It was used both for plants and animals and was modified by several authors in various theories of heredity. Bonnet discusses it in his *La Palingénésie philosophique* and Buffon his *Histoire naturelle*.

188 *pippins*: seeds of fleshy fruit (like apples).

a close analogy between the seeds of plants, and the eggs of animals: a comparison commonly made in early modern embryological studies.

189 *adapted the objects?*: Paley takes care not to attack Erasmus Darwin's filament theory directly. He simply 'removes it a little further back' by concentrating on *adaptation*, Darwin's principle of change. He is so confident that his reader realizes the answer to the question he poses here that he simply lets it stand as it is. Additionally, Darwin's theory was not necessarily meant to unseat the existence of or need for God. In fact, Charles Darwin would later defend his grandfather's belief in God in *The Life of Erasmus Darwin* (London, 1879).

Darwin's Phytologia: after surreptitiously countering the biological ideas of Erasmus Darwin (1731–1802) in earlier sections of the book (see notes to pp. 80, 121, 147, 166), Paley finally cites Darwin here for the first time. A succssful physician, Erasmus Darwin (the grandfather of Charles Darwin) was the author of natural philosophical poems that included footnotes and lengthy appendices to explain the scientific terms that he used. In addition to poems, Darwin wrote about digitalis's effects on dropsy and was instrumental in founding Birmingham Lunar Society, the Lichfield Botanical Society, and the Derby Philosophical Society. He promoted the idea that the morphology of organisms could change over a series of generations if given enough time to do so. His theory revolved around the concept of a *filament* (which was guided by a force that Paley called an 'appetency'), a piece of matter, similar to the eighteenth-century conception of a seed or germ, that guided the process along. Many early modern philosophers, including Paley, thought that this approach was too theoretical and could not be substantiated with convincing evidence. Additionally, some of Paley's contemporaries rejected Darwin's theory for political reasons. Darwin was a self-avowed liberal, supporting the ideological aims of both the American and French revolutions. However, his books sold well. This was not necessarily because the reading public accepted his theory; rather, it was because Darwin's books supplied interesting commonplaces (points of discussion) for polite conversation. Darwin avoided being associated with atheism by making reference to 'the Great First Cause'. Paley could therefore afford to cite Darwin because he was a popular author whose 'filaments' could be interpreted within the larger remit of British natural theology. The book cited here is *Phytologia: or, The philosophy of agriculture and gardening; with the theory of draining morasses, and with an improved construction of the drill plough* (London, 1800).

191 *Dr Darwin's Botanic Garden*: Erasmus Darwin, *The Botanic Garden, Part I, Containing the Economy of Vegetation, a Poem with Philosophical Notes* (London, 1791).

With.: William Withering (1741–99) held a medical doctorate from the University of Edinburgh and, like Erasmus Darwin, was a member of Birmingham's Lunar Society. He was also a fellow of the Royal Society of London, the Linnaean Society, and the Royal Academy of Sciences of Lisbon. He wrote widely read books on chemistry and botany, his most influential works being *Experiments and Observations on the Terra Ponderosa* (1784), *An Account of the Foxglove and Some of its Medical Uses, &c.* (1785) and a translation of Torbern O. Bergman's *Outlines of Mineralogy* (1783). These works generally appealed to a medical or industrial audience who had been trained in chemistry. His botanical works, on the other hand, enjoyed a much wider readership because they were relevant to pharmacy and to gardening. As one of the strongest promoters of the Linnaean system in Britain, Withering was firmly committed to natural theology's appeal to an ordered natural world. The book cited here is *A Systematic Arrangement of British Plants* (London, 1787, 1792), the first edition of which was published as *The Botanical Arrangement of All the Vegetables Naturally Growing in Great Britain* (London, 1776). The references from William Withering that Paley actually chooses to identify (e.g. he sometimes fails to include an edition and page number) come from the second and third editions.

193 *dionæa muscipula*: a Venus flytrap.

Smellie's Phil. of Nat. Hist.: *Philosophy of Natural History*, by William Smellie (1740–95), a prominent Edinburgh printer, editor, and natural historian. As an adolescent, Smellie attended medical and arts lectures at the University of Edinburgh, but did not take a degree because he could not afford to pay the matriculation fees. His literary career began in the 1760s when he edited the gentlemanly *Scots Magazine* and this experience allowed him to establish one of the most influential printing presses in the city from the 1760s to the 1780s. Concurrent with his printing activities, he was the editor (and compiler) of the first edition of the *Encyclopaedia Britannica* (1771) and he translated Buffon's *Histoire Naturelle* as *Natural History, General and Particular* in nine volumes (1780–5). During his lifetime he was also the editor of at least six different magazines or journals. Near the end of his career he published his own ideas in *Philosophy of Natural History* (vol. i, 1790, vol. ii, 1799). After he died the work went on to become an influential book in nineteenth-century America, especially at Harvard University, where it was a set text for Thomas Nuttall's lectures. Though he was a member of the Edinburgh Philosophical Society, he made an unsuccessful bid for the chair of natural history at the University of Edinburgh during the late 1770s. After this incident, he became increasingly marginalized by the university's scientific community. His explication of Buffon's cosmological accounts in *Natural History* and his interest in the history of botanical and animal sexuality in *Philosophy of Natural History* served to isolate him even further.

194 *the elements*: Paley is referring to Aristotle's elements (air, fire, water, and earth), not those of Lavoisier's oxygen theory. From his previous comments on acids, alkalis, and digestion, it is clear that Paley had at least a basic understanding of eighteenth-century chemistry (see note to p. 50). In this chapter he uses Aristotle's elements more as literary categories, under which he includes general points of information relevant to chemistry and associated topics (like meteorology).

Addison: see note to p. 162. This quotation has not been traced.

lately discovered: in the September issue of the 1800 edition of the *Philosophical Transactions*, Alessandro Volta published an article about a 'voltaic pile', essentially the first electric battery. In May of the same year, two British chemists, William Nicholson (1753–1815) and Anthony Carlisle (1768–1840), channelled the pile's electric current into water and found that the shock created two gases: hydrogen and oxygen.

Εξ ὑδατος τα παντα: all things emanate from water.

an atmosphere's investing our globe: meteorology was very popular during the early modern period. Rain tables were published in nearly every gentleman's magazine and in specialist journals that touched on medicine, natural history, or natural philosophy.

195 *power of evaporating fluids . . . action upon the sea*: evaporation was a perennial interest of natural philosophers from ancient times forward. A staple author quoted on this topic was Edmund Halley (1656–1742), who used evaporation and ocean salinity levels to determine the age of the earth.

elective power in the air: eighteenth-century chemistry promoted the concept of 'elective affinity', the idea that substances were attracted together in varying degrees (see note to p. 51).

cullender: a strainer (colander).

respiration: see note to p. 75.

196 *oleaginous*: greasy or oily.

197 *fluxing ores*: metals that melt into a liquid with application of heat.

Of LIGHT, (whether we regard it as of the same substance with fire, or as a different substance): the nature of light in the early modern period was by no means settled. Some argued that it was a substance (a particle) and others that it was an immaterial attribute of matter.

a force sufficient to shatter to atoms the hardest bodies: an 'atom' in the late eighteenth century was simply the smallest physical unit into which a piece of matter could be reduced. For many, it was a hypothetical concept that could not be substantiated by empirical evidence. The modern perception of an 'atom' did not gain full acceptance until the end of the nineteenth century.

198 *spissitude*: thickness.

that variety of colours: the fact that white light can be split in different colours when shone through a prism—an experiment popularized by Isaac Newton.

199 *Astronomy . . . is not the best medium through which to prove the agency of an intelligent Creator*: this disavowal of the usefulness of astronomy for natural theology is somewhat misleading. Though Paley only devotes a chapter to planetary astronomy (which was also called sidereal astronomy), the underlying material and metaphysical assumptions of Newtonian natural theology greatly influenced his thoughts on the attributes of God (see note to p. 213).

orreries, planetaria: an orrery was a model of the solar system, consisting of metallic representations of planets (usually smaller than the human hand), each mounted on a small bar that connected to a central mechanical wheel that made them move around in circular motion meant to copy the rotation of the planets around the sun. The name was taken from the 4th Earl of Orrery, who commissioned artisans to make such devices at the beginning of the eighteenth century. Sometimes the words 'orrery' and 'planetaria' were used interchangeably.

the Lord Bishop of Elphin: John Law (1745–1810) was made Bishop of Elphin (in the Anglican Church of Ireland, not the Roman Catholic Church) in 1795. He and Paley had met at Cambridge and they remained lifelong friends. John was the oldest son of Edmund Law (1703–87), Bishop of Carlisle from 1768 to 1787. Paley's contact with John Law spanned several decades and, even after the publication of *Natural Theology*, Paley was still entertaining suggestions from him. Because of this, it is not surprising to learn that they corresponded about the scientific proofs that could be used to support a natural theological argument. In 1797 Law wrote a letter to Paley that stated: 'In your chapter on divine contrivance, you must have an article on the solar system, which no one can describe more forcibly or eloquently.' Law even suggested a few authors, including John Ray, Bernard Nieuwentyjt, Colin Maclaurin (see note to p. 211), William Derham, Richard Bentley (1662–1742), and Buffon.

The letters of John Law and Paley were part of an elaborate correspondence network set up primarily by Bishop Edmund Law, but which also included John Douglas (1721–1807) and Thomas Percy (1729–1811). In his Cambridge days, Edmund Law was master of Peterhouse. Along with several others, he had introduced suggestions for liberal educational reforms at Cambridge and for political reform in London during the 1760s. He failed to implement his plans and he subsequently used his political connections to secure the bishopric of Carlisle. Paley recognized Edmund Law's patronage when he dedicated his *Principles of Moral and Political Philosophy* to him. Paley and John Law were part of Edmund's attempted reform while they were tutors at

Cambridge and they became part of his correspondence network after they left. Some of their letters are printed in Edward Paley (ed.), *An Account of the Life and Writings of William Paley* (London, 1825).

Rev. J. Brinkley: John Brinkley (1763–1835), the Astronomer Royal for Ireland. Paley's correspondence with him was mediated by his friend John Law (see note above). Whereas Brinkley and Law were both keen mathematicians, Paley's mathematical studies waned after he left Cambridge. Both Brinkley and Law corresponded with each other and with fellows of the Royal Society on mathematics and natural philosophy. Brinkley was eventually elected to the society in 1803 and elected as Ireland's first Astronomer Royal in 1792. He had graduated from Caius College, Cambridge, in 1788. He then went to the Royal Observatory in Greenwich as an assistant to Nevil Maskelyne, and was ordained to the Church of England. This eventually allowed him to become the Bishop of Cloyne in 1826. While Brinkley was helping Paley with *Natural Theology*, he was also busy writing up his astronomy lectures into what would later become his influential *Elements of Plane Astronomy* (1808), a book that became a popular academic astronomy text in the following decades. In light of the complicated mathematical formulae being used to determine planetary physics at the end of the eighteenth century, Paley needed someone like Brinkley (and John Law) who could cut to the chase and explain Newton's theories to him.

200 *devoid of any inert matter either fluid or solid*: one of the key assumptions in Isaac Newton's planetary astronomy was that space was a vacuum.

sensitive natures, by which other planets are inhabited: the existence of life on other planets was debated during the seventeenth century and would become a popular topic again in the nineteenth.

intellectual agency in three of its principal operations: as with his teleological examples in the previous chapters, Paley attributes intellectual agency to the laws of physics and, in so doing, highlights the fact that the astronomy chapter is a cosmological argument. In this case, he ascribes three qualities to God (choosing, regulating, and determining) that are closely related to the divine attribute of omniscience. Throughout the rest of the chapter, Paley employs examples from astronomy and physics to illustrate various attributes of God, a practice that had a long history in Christian theology (see note to p. 213).

201 *at least, seven*: up to the 1780s there were six known planets: Mercury, Venus, Earth, Mars, Jupiter, and Saturn. In 1781 Sir William Herschel (1738–1822) discovered a new planet and named it *Georgium Sidus* (George's Star, in honour of George III); however, it eventually came to be known as Uranus.

those who reject an intelligent Creator: e.g. Buffon, who argued that the earth had formed from a piece of matter that a comet had blasted off the sun (see Buffon's *Natural History*, chapter I, article I, 'Proofs of the Theory of the Earth; Of the Formation of Planets' (translated by William

Smellie, 1780–5), volume i (1781), 59–96). Paley's treatment of Buffon here, and his refutation of Buffon later in the chapter, once prompted his friend John Law to write: 'As to Buffon's silly hypothesis, there cannot be a better refutation of it than your own. None but Frenchmen (and those very weak ones) ever give the least credit to it.'

203 *Calculations were made a few years ago of the mean density of the earth*: Henry Cavendish, 'Experiments to determine the density of the earth', *Philosophical Transactions*, 88 (1798) 469–506.

204 *a body continues in the state in which it is, whether of motion or rest*: a succinct rephrasing of Newton's first law of physics.

an attraction which varies reciprocally as the square of the distance: a succinct rephrasing of Newton's law of gravity.

205 *Cotes*: Roger Cotes (1682–1716), a personal friend of Newton who helped to publish the 1713 second edition of *Philosophiae Naturalis Principia Mathematica*. At the beginning of the eighteenth century, many of Newton's mathematical and astronomical propositions were yet to be proved. Over the first half of the century mathematicians from Britain and abroad worked to demonstrate the soundness and the utility of Newtonian physics, calculus, and geometry. Cotes was among these scholars. So were Colin Maclaurin (see note to p. 211), Richard Bentley, Samuel Clarke (1675–1729), and, interesting enough, Buffon, who translated Newton's *The Method of Fluxions and Infinite Series* into French in 1740. Cotes was a brilliant mathematician and the scientific world was sorrowful when he died at the age of 33. It is said that upon hearing of the death of Cotes, Newton exclaimed: 'If Cotes had lived, we might have known something.' After graduating from Trinity College, Cambridge, Cotes had served his Alma Mater as the first Plumian Professor of Astronomy and Natural Philosophy. In this capacity, he promoted Newton's scientific method and made several original contributions to the field of mathematics. This being the case, Paley and other gentlemen studying mathematics at Cambridge would have probably been required to at least be familiar with Cotes's works. It is also highly probable that Paley's friend John Brinkley recommended Cotes to him. While Paley was writing *Natural Theology*, Brinkley was busy proving what was known as 'Cotes's theorem', a fluxion formula postulated in Cotes's 1722 posthumous work *Harmonia mensurarum*. In 1797 Brinkley proved this theorem in his essay 'A General Demonstration of the Property of the Circle Discovered by Mr Cotes Deduced from the Circle Only' (*Transactions of the Royal Irish Academy*, 7 (1797), 151–9). Because of this state of intellectual affairs, it is highly probable that Brinkley either explicitly or implicitly motivated Paley to mention Cotes in *Natural Theology*'s astronomy chapter. On a theological note, Cotes promoted the Newtonian view that the intricacy of the world pointed to a Deity. In his preface to the second edition of Newton's *Principia*, Cotes wrote: 'Newton's distinguished work will be the safest protection against the attacks of atheists, and nowhere more surely than this quiver can one

draw forth missiles against the band of godless men.' Such a conviction made him an excellent mathematical reference for *Natural Theology*.

a primordial property of matter: ever since Newton had proposed the concept of gravity, there had been many debates as to whether it was an inherent (primordial) or superimposed quality of matter. Paley, like Newton, held that it was superimposed by God; a move that allowed them both to argue for God's direct role in physical laws via a continued intervention upon matter.

206 *æthereal fluid*: from ancient times to the Renaissance the movement of the planets was often explained by using a geocentric model, i.e. the notion that the earth was at the centre of the universe and that the planets and stars rotated around it. This changed in the sixteenth century via what is often called the 'scientific revolution', when geocentrism was replaced by heliocentrism, the idea that the planets rotate around the sun. One of the problems with this new approach was that it was quite hard to explain the physical forces (or lack thereof) that allowed the planets to move through space. In the 1640s the French philosopher René Descartes (1596–1650) proposed that the planets were surrounded by 'aether' (ether), an invisible, unquantifiable substance that spun around the sun (much like a whirlpool) and pulled the planets along with it. These ideas were most clearly laid out in 1644 when he published *Principia Philosophiae* (*Principles of Philosophy*). When Newton published his *Principia* in 1687 he eliminated the need for Descartes's ether by offering gravity and physical laws to explain the orbits of the planets. Instead, Newton's ether acted only at a short range in the attraction and repulsion of (hypothetical) microscopic particles. Since Newton was unclear on this topic, and since Descartes's theory continued to be used well into the eighteenth century (especially in France), it is not surprising to find Paley addressing the properties of an ethereal fluid. After the publication of *Natural Theology*, ether would continue to be used in physics until the end of the nineteenth century.

perturbing forces: building on the work of Johann Kepler (1571–1630), Newton had proposed that the path of the planets around the sun was elliptical. However, based on astronomical observations (with telescopes) and mathematical calculations, it became clear to Newton's followers that the planets 'wobbled' a bit out of the elliptical path at various stages of their orbit. This wobbling was called a 'perturbance' and over the next century mathematicians would work on equations to show that these deviations were caused by the gravitational pull of the other planets.

208 *quiescent*: inactive or at rest.

eccentricity: deviation from circularity.

ecliptic: 'in astronomy . . . that path or way among the fixed stars, that the earth appears to describe to an eye placed in the sun'; *Encyclopaedia Britannica* (1771).

210 *Buffon*: see note to p. 201.

211 *Georgium sidus*: now called Uranus; see note to p. 201.

But what is the fact?: the idea of a 'fact' in late-eighteenth-century Britain was often contrasted with that of a 'theory'—the former being empirically verifiable and the latter being mere speculation.

Maclaurin's Account of Newton's Phil.: Colin Maclaurin (1698–1746) was a gifted mathematician from Scotland. Educated at the University of Glasgow, he was among a group of mathematicians who explained and popularized Newtonian philosophy via textbooks. Likewise, Maclaurin (and another member of this group, Roger Cotes) advanced the study of synthetic geometry in England. Maclaurin was elected a fellow of the Royal Society in 1719. Like his French contemporary Pierre Louis Moreau de Maupertuis (1698–1759), he created a Newtonian mathematical framework for proving that the earth was an oblate spheroid—a shape mentioned several times by Paley. The most enduring of his academic contributions was his posthumous *An Account of Sir Isaac Newton's Philosophical Discoveries* (London, 1748). As a mathematics student at Christ's College, Cambridge, a mere twenty years later, Paley would have at least been familiar with Maclaurin's exposition of Newton's mathematics and philosophy. Maclaurin, like Paley, was an avid promoter of natural theology. The last chapter of *Discoveries* (dictated by Maclaurin from his deathbed) contains a mini natural theological exposition that favourably mentions Aristotle and Plotinus, labels Descartes's cosmology as disgusting to the 'sober and wise part of mankind', dismisses Spinoza's and Leibniz's absolute necessity, applauds the theological expediency of Newton's perception of causation, and attempts to rectify theological misinterpretations of Newton's philosophy. To the latter goal, Maclaurin sought to wrestle the concept of gravity from the hands of the materialists by asserting that 'Its action is proportional to the quantity of solid matter in bodies, and not to their surfaces, as is usual in mechanical causes: this power, therefore, seems to surpass mere mechanism'. Lest anyone be theologically confused, Maclaurin goes on to state: 'But, whatever we say of this power [gravity], it could not possibly have produced, at the beginning, the regular situation of the orbs and the present disposition of things ... The same powers, therefore, which at present govern the material universe ... are very different from those which were necessary to have produced it from nothing.' Unlike his ambiguous position on vitalism, Paley's comments on matter throughout *Natural Theology* agree with Maclaurin on this point (particularly in Chapter V).

213 *Amongst other things it proves the personality of the Deity*: throughout *Natural Theology*, Paley uses the notion of a 'proof' rather loosely. On the whole, he does not use the term in the logical sense of the word, rather he uses it either as a synonym for a teleological analogy (or example) or to connote a conclusion made from inductive assent. Following from this

usage, Paley's use of the verb 'to prove' here should be read as 'to suggest' or 'to imply', but not in the airtight sense of the verb as it is used in logic. See note on 'relation', p. 140.

Now that which can contrive, which can design, must be a person: within the history of Christian theology the nature of God, physical or spiritual, has been defined via various controversies and doctrines. Central to these was the question of what sort of evidence could be used to define the divine. Throughout the first 400 years of the Church, many councils were held on this topic and the answer was the same: the evidence was to be taken from the Bible, i.e. the truth of God that had been given to humans via divine revelation in Old and New Testaments. From these documents, theologians forged doctrines on key topics like salvation, the Trinity, and creation from which they attempted to formulate a description of God. To do this, they decided to approach God as a personality and they selected qualities like wisdom, omniscience, and immutability (just to name a few) in an effort to create a working definition. These personal qualities became known as the 'attributes of God' (or the 'divine attributes'). Once the attributes were established, evidence from the natural world was then used to supplement the descriptions. For an Anglican like Paley, two guides would have been the Nicene Creed (or 'The Creed of Athanasius') and the Thirty-Nine Articles of the Church of England.

From ancient times up to the early modern period, natural philosophy was used by theologians to provide secondary empirical examples that helped to illustrate the divine attributes. Since these descriptions of God were made by humans (who were perceived to be fallible), they were most often seen as approximations, to be guided by the personality of God as revealed in the Bible. In 1705 Samuel Clarke synthesized this tradition with Newtonianism in *Demonstration of the Being and Attributes of God*. Though this book had a profound influence on eighteenth-century perceptions of the attributes of God, it was often not cited directly by authors because of Clarke's association with Arianism (the ancient theological belief that held that though Christ was divine, his essence was different from that of God). Likewise, though Paley does not cite Clarke, it is quite evident that he is propounding arguments similar to those advanced in *Demonstration*, a work which used Newtonian mechanics to supplement the following attributes of God: eternality, independent being, necessary existence, incomprehensible essence, infinitude, omnipresence, intelligence, free agency, unity, all-powerfulness, wisdom, goodness, justness, and truthfulness. On the whole, Paley's assumptions about the attributes of God were guided by the Bible, the Thirty-Nine Articles, and Newtonian natural theology. A historical perspective on the development of Paley's views on the divine attributes, especially goodness, can be found in the sermons he wrote during the 1780s and 1790s, namely: 'The being of God demonstrated in the works of creation', 'Unity of God', 'The ills of life do not contradict the goodness of God', and 'The goodness of God proved from the light of nature and revelation'. These can all be found in *The Works of William Paley D. D.*, ed. Edmund Paley (London, 1825).

213 *sensorium . . . may comprehend the universe*: the notion that the universe is the *sensorium* of God is taken from Isaac Newton. In Query 31 of the fourth edition of *Opticks* (1730) he states: 'Such a wonderful Uniformity in the Planetary System must be allowed the Effect of Choice. And so must the Uniformity in the Bodies of Animals . . . be the effect of nothing else than the Wisdom and Skill of a powerful ever-living Agent, who being in all Places, is more able by his Will to move the Bodies within his boundless uniform Sensorium, and thereby to form and reform the Parts of our own Bodies.' This idea was promoted and modified by many of Newton's eighteenth-century followers, including Samuel Clarke.

No man hath seen God at any time: John 1: 18.

Priestley's Letters to a Philosophical Unbeliever: Joseph Priestley was a natural philosopher and a Unitarian theologian. During the 1770s he travelled with Lord Shelburne (William Petty, 1737–1805) on the Continent and his conversations with French philosophers on this trip inspired him to write *Letters to a Philosophical Unbeliever* (London, 1774), a tract that criticized David Hume's scepticism of natural religion as presented in *A Treatise of Human Nature*. In addition to *Letters* (which went through multiple printings), he wrote several works that addressed natural theology, including *Disquisitions*, a work that gave a Newtonian explication of the resurrection.

214 *There may be senses suited to the perception of the powers, properties, and substance of spirits*: Paley is suggesting that angels may have a higher sense of perception than humans.

215 *ab extra*: (Latin) from without.

the common sense of mankind: this notion of common-sense reasoning was taken from Thomas Reid (1710–96) who in chapter 7, section 4, of *An Inquiry into the Human Mind, on the Principles of Common Sense* (1764) stated that sensory experiences are 'a part of that furniture which Nature hath given to the human understanding. They are the inspiration of the Almighty, no less than our notions or simple apprehensions . . . They make up what is called the common sense of mankind; and, what is manifestly contrary to any of those first principles, is what we call absurd.' Reid was part of a larger group of philosophers in Scotland who collectively criticized Humean scepticism.

216 *the effects of volcanos or inundations*: in Paley's day it was commonly believed that large-scale changes to the earth's surface were caused by quick catastrophes like floods or volcanic eruptions. The notion of long periods of time (similar to those developed during the nineteenth century) was often met with scepticism. (See also note to p. 180.)

the misapplication of the term 'law': see note to p. 16.

218 *second causes*: for Paley and most of his contemporaries, the first cause

was God. Second causes were forces or laws of nature (gravitation, electricity, etc.) through which God could also manifest power. See note to p. 9.

elective attractions: see note to p. 195.

219 *trains of mechanical dispositions, fixed beforehand by an intelligent appointment*: although Paley does not develop his thoughts on these 'mechanical dispositions', the idea that morphological change could be guided along by a divine power via inherent organic laws proved to be quite resilient during the nineteenth century, both before and after Darwin's *On the Origin of Species*.

221 *Otaheite*: Tahiti.

222 *the old systems of atheism and the new agree*: Paley is most probably referring to older ideas of spontaneous generation that had been proposed as early as Aristotle and his followers under the names of *generatio aequivica, generatio primara, archegenesis, autogenesis*, and *archebiosis*.

the antiquated system of atoms: the conception of atomism, the belief that all matter is made up of tiny indivisible particles, goes as far back as the ancient Greek philosopher Leucippus of Miletus (480–420 BC) and it was then developed by Democritus (b. 460 BC), Epicurus (342–270 BC) and Lucretius (95–51 BC). See note to p. 197.

223 *internal molds*: Buffon suggested that an internal mould guided the formation of particles into an embryo.

vermes: worms or worm-like creatures.

224 *essential forms*: Plato argued that all objects originate from an idealized metaphysical 'forms' that serve as the prototypes for all objects in the sublunar world. His most well-known expositions of the forms occur in the *Republic*, especially in the divided line example and the allegory of the cave (*Rep.*, books 6 and 7).

appetencies: see note to p. 80.

perhaps in a hundred millions of years: see notes to pp. 39, 165, 189.

225 *the changes in Ovid's Metamorphoses*: the poem by Publius Ovidius Naso (43 BC–*c.* AD 17) retells the stories of several classical myths in which the protagonists undergo a series of physical or spiritual changes.

mammæ: teats or nipples.

inusitation: disuse.

nec curtorum, per multa sæcula, Judæorum propagini deest præputium: although the foreskin has been removed from the offspring of the Jews for many centuries, it remains unshortened.

Mr. Everard Home: see note to p. 30.

226 *grumous*: semisolid or coagulated.

active habit . . . passive habits: see notes to pp. 80, 109.

227 *unauthenticated by testimony*: in chapter 1 of his *Evidences of Christianity*, Paley avers that 'testimony' is satisfactory evidence given by a reliable witness. His interest in this topic fell within the larger realm of late-eighteenth-century debates over the types of evidence that could be used to support the occurrence of miracles. The touchstones for these arguments were the writings of David Hume, especially section 10 ('On Miracles') in his *An Enquiry Concerning Human Understanding* (1746).

229 *conatus*: see note to p. 130.

230 *The attributes of such a Being*: see note to p. 213.

the revelations which we acknowledge: the Bible.

231 *omnipotence, omniscience, omnipresence*: all-powerfulness, all-knowingness, present everywhere. See note to p. 213, especially the list of divine attributes named by Samuel Clarke and description of God in the Thirty-Nine Articles.

a foreknowledge of their action upon one another, and of their changes: such omniscience explains why Paley thought 'trains of mechanical dispositions' (see note to p. 219) could be guided by God.

233 *Bishop Wilkins's Principles of Nat. Rel.*: John Wilkins, *Of All Principles and Duties of Natural Religion* (London, 1675). Wilkins (1614–72) was both a bishop and one of the founders of the Royal Society. His *Principles* went through numerous editions in the seventeenth and eighteenth centuries.

234 *the same element of light does*: see notes to pp. 20, 197.

235 *cetaceous tribe*: marine mammals.

237 *The proof of the divine goodness*: from a historical perspective, the divine attributes sometimes have been divided into those that are negative (simplicity, infinity, immutability) and those that are positive (unity, truth, goodness, beauty, omnipotence, omnipresence, intellect, and will).

the design of the contrivance is beneficial: Paley's linking of the divine attribute of goodness to beneficial (and benevolent) design provided the springboard for many criticisms in the nineteenth century, especially after the publication of Darwin's *On the Origin of Species*.

238 *Nor is the design abortive. It is a happy world after all*: for faulty design, see note to p. 35. Paley's definition of 'happy' here is based strongly upon his utilitarian view that the sufferings of human existence are on the whole outweighed by times of contentment (see note to p. 213). Additionally, the happiness afforded to humans by God, for Paley, carried a moral obligation. This is stated in book II, chapter V of his *Principles of Moral and Political Philosophy*: 'Since God hath called forth his consummate wisdom to contrive and provide for our happiness, and the world appears to have been constituted with this design at first; so long as this constitution is upholden by him, we must in reason suppose the same design to continue. . . . We conclude, therefore, that God wills and wishes the

happiness of his creatures. And this conclusion being once established, we are at liberty to go on with the rule built upon it, namely, "will of God, concerning any action, by the light of nature, is to inquire into the tendency of that action to promote or diminish the general happiness." '

Paley's thoughts on happiness were influenced by the larger remit of utilitarianism in the late eighteenth century. As Jeremy Bentham, one of the leading supporters of the movement, stated in chapter 1 of his *An Introduction to the Principles of Morals and Legislation*, the very definition of 'utility' is based upon the idea of 'happiness'. By utility is meant that 'property in any object, whereby it tends to produce benefit, advantage, pleasure, good, or happiness, (all this in the present case comes to the same thing) or (what comes again to the same thing) to prevent the happening of mischief, pain, evil, or unhappiness to the party whose interest is considered: if that party be the community in general, then the happiness of the community: if a particular individual, then the happiness of that individual.' Even though Bentham's work was often seen as a more liberal response to Paley's *Principles*, Bentham, like Paley (see note to p. 251), linked his notion of happiness to divine benevolence (see Chapter 1 in his *Morals*). In the years after the publication of *Natural Theology*, Bentham's notion of utility had a profound influence on political philosophy and, consequently, Paley's appeal to happiness remained popular well into the nineteenth century.

240 *It is well described by Rousseau*: Jean-Jacques Rousseau (1712–78), French philosopher who wrote books and novels that addressed numerous social topics, including education and political philosophy. One of the key assumptions of his work was that human beings are inherently good and that this quality is corrupted by society; children are born innocent and it is the evils of society that corrupt them. Though his ideas were widely circulated, his bombastic personality and the anti-establishment tenor of his ideas often led him to flee from one town to another. Based on these travels and other experiences he wrote an account of his life in *Les Confessions* (published posthumously in 1782). In book I, he recounts his adolescent days in the town of Bossey. Though his time there was less than enjoyable, he felt that such inhospitable experiences had been tempered by time. He stated, '[N]ow that I have passed the age of maturity and am descending toward old age, I sense that it is these memories which, while others fade, grow brighter, and are imprinted on my memory with the clarity of detail that grows every day more charming and potent. It is as though, already sensing life slipping away, I were trying to catch hold of it again at its beginnings.' When considering this quotation in the light of Paley's life and thought, it is worth noting that he wrote *Natural Theology* at the end of his life and at a time when he was in poor health. Though it was published in 1802, he had originally intended to finish it during the mid-1790s, but fits of illness prevented him from doing so.

Father's Instructions, by Dr. Percival of Manchester: Thomas Percival, *A Father's Instructions to His Children: Consisting of Tales, Fables, and*

Reflections; Designed to Promote the Love of Virtue, etc. (London, 1775). Percival (1740–1804) was a Unitarian physician who lived in Warrington. He published many books on medicine, including *Essays medical and experimental* (London, 1772) and *Medical jurisprudence; or a code of ethics and institutes, adapted to the professions of physic and surgery* (Manchester, 1704). His *Instructions* went through numerous augmentations and editions during the last three decades of the eighteenth century.

241 *is a case of millions*: as in previous chapters, Paley is using his utilitarian calculus of happiness. See note to p. 171.

prepollency: predominance.

242 *in my Moral Philosophy*: book II, chapter V, 'The Divine Benevolence'.

244 *Abbé Fontana*: Felice Gaspar Ferdinand Fontana (1730–1805), a physician who, though born in Pomarolo, lived the better part of his life in Florence. He wrote on many topics, including chemistry (especially gases), mineralogy, zoology, and physics. He is remembered today for the wax museum that he assembled and maintained in Florence. He wrote a book on poisons, *Treatise on the Venom of the Viper* (London, 1787). Though he cites Fontana for his work on poisons, Paley had probably first encountered him via his treatise *De' moti dell' iride* (1765), which addressed the muscles of the iris. Because of the similarity of the symptoms caused by snake bites and other illnesses (particularly fevers), the venom of snakes was quite a popular chemical topic from the 1750s to around 1800. Fontana addressed this issue during the 1760s with research on several poisonous snakes from the Americas.

245 *says Adanson*: Michel Adanson, *A voyage to Senegal, the island of Goree, and the river Gambia* (London, 1759). The book was originally published as *Histoire naturelle du Senegal* (1757). Adanson (1727–1806) was a French natural historian who wrote on botany and zoology.

Let them enjoy their existence: let them have their country: Paley's comments on the 'wickedness of man' in this section stand in contrast to environmental writers (Lynn White for example) who have argued that the roots of Western environmental abuses lie in Judaeo–Christian objectifications of nature.

248 *says Pallas*: Peter Simon Pallas, *Travels through the Southern provinces of the Russian Empire, in the years 1793 and 1794* (London, 1802). This book was first published as *Reise durch verschiedene Provinzen des russischen Reiches* (1771–6). It was translated into French, *Voyages de m. P.S. Pallas: en differentes provinces de l'empire de Russie, et dans l'Asie septentrionale* (1788), and then into English. Pallas (1741–1811) was a German naturalist who was a professor in the St Petersburg Academy of Sciences. Under the patronage of Tsarina Catherine II of Russia, he travelled to Siberia and collected natural history specimens.

249 *All superabundance supposes destruction, or must destroy itself*: the relationship between natural resources and human populations was treated by

Paley in his *Principles of Moral and Political Philosophy*. His thinking on this subject was further augmented when he read Thomas Malthus's comments on scarcity in *An Essay on the Principle of Population* (1798) (esp. Ch. 1), a work that would also have a strong impact on Charles Darwin.

the loss of certain species: see notes to p. 148.

250 *the Deity has added pleasure*: a restatement of the utilitarian calculus used to factor a balance of pleasure and pain; see note to p. 171.

251 *the pure benevolence of the Creator*: Paley firmly links the attribute of God's goodness (see notes to pp. 171, 237) to a utilitarian calculus of benevolence and happiness (see note to p. 238).

Dr Balguy's treatise upon the Divine Benevolence: Thomas Balguy, *Divine Benevolence Asserted; and Vindicated from the Objections of Ancient and Modern Sceptics* (London, 1781). Balguy (1716–95) was an Anglican clergyman who wrote several works that defended the authority of the established Church of England against its detractors. When David Hume's *Dialogues Concerning Natural Religion* was published (posthumously) in 1779, Balguy decided to write a book against it. The result was *Divine Benevolence*. It addressed natural religion and drew several of its arguments from Cicero's *De Natura Deorum* and Joseph Butler's *The Analogy of Religion* (1736).

253 *Persons in fevers, and I believe, in most maniacal cases*: the nature of madness in the eighteenth century was heavily influenced by Locke's philosophy of mind and theories of bodily fibres and fluids.

independently of habit: see note to p. 109.

254 *simple and original perception*: Paley is referring to the philosophy of mind, as outlined in Locke's *An Essay Concerning Human Understanding*. Locke argued that the senses produce primary ideas in the mind and these laid the foundation for more complex secondary ideas. See notes to pp. 29, 49, 163.

I have been a great follower of fishing myself: Paley was quite fond of fishing and it is likely that he dissected the fish that he caught.

quantum in rebus inane: how much vanity exists in things!

fens: a fen is a flat, swampy piece of land, particularly in Cambridgeshire and Lincolnshire.

255 *ascribe to the Deity the character of benevolence*: see note to p. 238.

Of the ORIGIN OF EVIL no universal solution has been discovered: see note to p. 35. In early Christianity, the notion of good and 'evil' played a central role in the debates on the nature of God. Throughout the history of Christian thought, two recurring classifications of evil have been employed. (1) *Physical evil*, brought about by natural events over which humans have no immediate control; Paley incorporates natural and physical evils, or 'external evils', under this category (see p. 260). (2) *Moral evil* engendered by the immoral actions of humans to each other or to

God; Paley seems to incorporate 'civil evils', or the 'evils of civil life' under this category (see p. 261). The two distinctions of evil are inferred from the Bible and were elaborated by early Church Fathers, two of the most influential being Irenaeus of Lyons (second century) and Augustine of Hippo (354–430). In the Middle Ages numerous authors addressed the problem of evil (theodicy), including Anselm of Canterbury (d. 1109) and Meister Johann Eckhart (d. 1329). The Reformation saw refined versions of evil offered by both Roman Catholic and Protestant thinkers. During the early modern period, Continental writers, especially Leibniz and Descartes, offered rationalist explanations, while in England the existence of evil was treated throughout the eighteenth century, notably in Samuel Clarke's *Demonstration of the Being and Attributes of God* (and in his correspondence with Leibniz), William Derham's *Physico-Theology* (1712), William King's *Essay on the Origin of Evil* (1732), and Joseph Priestley's *Doctrine of Philosophical Necessity* (1782). In the following sections, as in other parts of *Natural Theology*, Paley draws from Derham and, most likely, Clarke.

256 *The doctrine of imperfections*: Paley's comments in this paragraph draw from wider eighteenth-century debates that addressed the relationship between the goodness of God, the order of nature, and the moral responsibility of sentient beings. See chapter 6 in Arthur O. Lovejoy, *The Great Chain of Being* (Cambridge, Mass., 1936). The order of the scale of nature, however, was by no means blindly accepted within the different religious and philosophical communities (Protestant or other) in Britain.

258 *the stone or gout*: bladder stones ('the stone') and inflammation of the joints (gout) were two painful illnesses during Paley's lifetime.

dispensary: an institution where medicine was distributed and where patients were sometimes kept for treatment. There were numerous dispensaries in Paley's environs and he probably is referring to either the Newcastle Dispensary or Sunderland Dispensary (Infirmary).

food . . . exercise . . . sleep . . . atmosphere: all frequently prescribed as medical cures during the early modern period.

259 *despumation*: the process of removing impurities from the body.

Sydenham: Thomas Sydenham (*c.*1624–89), an Oxford-educated statesman and physician who promoted the use of medical chemistry over classical humoral theories of illness. His most recognized work was *Methodus curandi febres, propriis observationibus superstructa* (1666), which went through several revisions, but it is not clear which work was used by Paley. Though Paley confines his discussion to Sydenham's work, there were contemporary theories of wellness and disease based upon balancing the body's 'fluids' and 'solids'—many of which were promoted by physicians trained by William Cullen at the University of Edinburgh.

260 *instincts*: see note to p. 160.

physical or natural evils: see note to p. 255.

the late Mr Tucker: Abraham Tucker (1705–74), a political philosopher educated at Merton College, Oxford. He wrote numerous anonymous articles and two books that influenced Paley: *Freewill, Foreknowledge and Fate* (1763) and *The Light of Nature Pursued* (1768). In the preface to *Principles of Moral and Political Philosophy*, Paley praised Tucker for 'more original thinking and observation upon the several subjects that he hath taken in hand, than in any other, not to say, than all others put together'.

261   *Civil evils, or the evils of civil life*: see note to p. 255.

  a late treatise upon population: Malthus, *An Essay on the Principle of Population*.

263   *et in maximâ quâque fortunâ minimum licere*: even in great fortune, minimal fortune is possible.

264   *There must always therefore be the difference between rich and poor*: Paley was quite interested in the plight of the poor. It was often alleged (particularly by his friend John Law) that Paley was never made a bishop because George III felt that *Principles of Moral and Political Philosophy* intimated that the lands of the aristocrats should be redistributed to the rest of the population. This passage therefore needs to be considered in light of this context. Throughout his works, Paley sometimes used the word 'poor' as a synonym for the working classes. As he stated in his *Reasons for Contentment: Addressed to the Labouring Part of the British Public* (1792): 'I do not now use the terms poor and rich; because that man is to be accounted poor, of whatever rank he be, and suffers the pains of poverty, whose expenses exceed his resources; and no man is, properly speaking, poor but he. But I, at present, consider the advantages of those laborious conditions of life which comprise the great proportion of every human community.' The issue for Paley was the advantages, the virtue, of the work that was being done; not the amount of money in a person's pocket. Though eloquent, this position was challenged, particularly in a series of published letters that Thomas Holt White (anonymously) edited in 1796.

265   *Mr Hume in his posthumous dialogues*: Though Paley has indirectly addressed several of Hume's arguments throughout *Natural Theology* (see notes to pp. 35, 163, 213, 215, 227), this is the first time that he mentions his name. Paley is specifically referring to *Dialogues Concerning Natural Religion*.

  vis inertiæ: force of inertia.

  the appearance of chance: see note to p. 38.

267   *Operumque laborem . . . trahebat*: work and labour were divided in equal parts, or a lot was drawn to decide (Virgil, *Aeneid*, i. 507–8).

268   *ex hypothesi*: by hypothesis.

270   *more will be said under the next article*: this does not appear in the published book.

doctrine of divine Providence: providence, simply summarized, is the notion that there is a superhuman being who oversees the world and who beneficently guides creatures towards a final purpose. The Greeks, Romans, and Jews believed in divine providence and the idea subsequently played a central role in the history of Christian thought. Near the end of the eighteenth century, providence was secularized and incorporated into political economy, i.e. the practice by which the state passes laws to regulate the economy for the good of the entire population.

273 *but still he is inferior to his slave*: Paley supported slavery as the punishment for a crime; however, he was in principle opposed to the African slave trade. See book III, part II, chapter III of *Principles of Moral and Political Philosophy*.

lights of revelation: the Bible.

274 *In dividing the talents*: the parable of the talents (Matthew 25: 14–30), in which a master gave his servants an unequal amount of money but expected them to do the most with what they had been allotted.

279 *expatiate*: give a detailed account.

The works of nature want only to be contemplated: the use of nature as an object of spiritual meditation has a long history in Christian thought and was a key tenet of early modern natural history.

281 *They who have taken up the opinion*: most likely referring to Joseph Priestley's materialist account of the resurrection in *Disquisitions Relating to Matter and Spirit* (London, 1777).

282 *carries identity and individuality*: see note to p. 163. Priestley's *Disquisitions* argued that a person's identity would be remembered by God at the resurrection.

libellula: a dragonfly.

283 *See Search's Light of Nature*: see note to p. 260. Tucker originally published his book using the pseudonym of 'Edward Search'.